Animal Reproduction Handbook

Animal Reproduction Handbook

Editor: Charlotte Johnson

MURPHY & MOORE
www.murphy-moorepublishing.com

Murphy & Moore Publishing,
1 Rockefeller Plaza,
New York City, NY 10020, USA

Visit us on the World Wide Web at:
www.murphy-moorepublishing.com

ISBN: 978-1-63987-051-6 (Hardback)

Cataloging-in-Publication Data

Animal reproduction handbook / edited by Charlotte Johnson.
p. cm.
Includes bibliographical references and index.
ISBN 978-1-63987-051-6
1. Reproduction. 2. Animals. 3. Animal breeding. 4. Veterinary endocrinology. I. Johnson, Charlotte.
QP251 .A55 2022
612.6--dc23

Table of Contents

Preface

Over the recent decade, advancements and applications have progressed exponentially. This has led to the increased interest in this field and projects are being conducted to enhance knowledge. The main objective of this book is to present some of the critical challenges and provide insights into possible solutions. This book will answer the varied questions that arise in the field and also provide an increased scope for furthering studies.

Reproduction is the process of producing offspring of the same kind. Reproduction in animals can be either sexual or asexual reproduction. Sexual reproduction is the process wherein the male and female gametes fuse together to form a new individual. This process involves fertilization which can be internal or external. Internal fertilization takes place inside the body of the female whereas external fertilization takes place outside the body of the female. Asexual reproduction is the process in which a new individual is formed by the involvement of a single parent. The offspring produced are genetically and morphologically identical to the parent. Asexual reproduction occurs in unicellular organisms. The cells are divided by mitotic division and no fertilization takes place. It includes binary fission, budding, fragmentation, regeneration, parthenogenesis. This book elucidates the concepts and innovative models around prospective developments in relation to animal reproduction. Different approaches, evaluations, methodologies, and advanced studies on this topic have been included in it. This book is a vital tool for all researching or studying animal reproduction as it gives incredible insights into emerging trends and concepts.

I hope that this book, with its visionary approach, will be a valuable addition and will promote interest among readers. Each of the authors has provided their extraordinary competence in their specific fields by providing different perspectives as they come from diverse nations and regions. I thank them for their contributions.

Editor

1

Germ Cell Determinant Transmission, Segregation, and Function in the Zebrafish Embryo

Celeste Eno and Francisco Pelegri

Abstract

Animals specify primordial germ cells (PGCs) in two alternate modes: preformation and epigenesis. Epigenesis relies on signal transduction from the surrounding tissues to in-struct a group of cells to acquire PGC identity. Preformation, thought to be the more de-rived PGC specification mode, is instead based on the maternal inheritance of germ cell-determining factors. We use the zebrafish as a model system, in which PGCs are specified through maternal inheritance of germ plasm, to study this process in vertebrates. In ze-brafish, maternally inherited germ plasm ribonucleoparticles (RNPs) have co-opted the cytoskeletal machinery to reach progressive levels of multimerization, resulting in the formation of four large masses of aggregated germ plasm RNPs. At later stages, germ plasm masses continue to use components of the cell division machinery, such as the spindles, centrosomes, and/or subcellular organelles to segregate asymmetrically during cell division and subsequently induce germ cell fate. This chapter discusses the current knowledge of germ cell specification focusing on the zebrafish as a model system. We al-so provide a comparative analysis of the mechanism for germ plasm RNP segregation in zebrafish versus other known vertebrate systems of germ cell preformation, such as in amphibian and avian models.

Keywords: Germ cells, germ plasm, zebrafish, cell division, cytoskeleton, RNP segregation

1. Introduction

1.1. Fundamentals of germ cell specification: Epigenesis vs. preformation

One of the most fundamental early cell fate decisions in animal embryos is the specification of primordial germ cells (PGCs) from the somatic tissue. PGCs are the precursors to gametes and thus hold the information to recreate the species in each generation. Consequently, individuals from every population must employ a robust mechanism for PGC induction. Specification occurs by one of two mechanistic modes: *epigenesis* or *preformation* (Figure 1). Epigenesis refers to a mechanism by which certain cells receive PGC-inducing signals from surrounding tissues. Epigenesis is found in mammals, as well as species in many other clades. Preformation describes a cell autonomous specification of germ cells from maternally inherited germ plasm deposits that specify the PGCs. This mechanism of preformation is used in a large number of organisms distributed in many animal lineages, including major model systems for biological research, such as the fruit fly *Drosophila melanogaster* and the nematode *Caenorhabditis elegans*, amphibians, such as *Xenopus laevis* and *X. tropicalis*, and the chick.

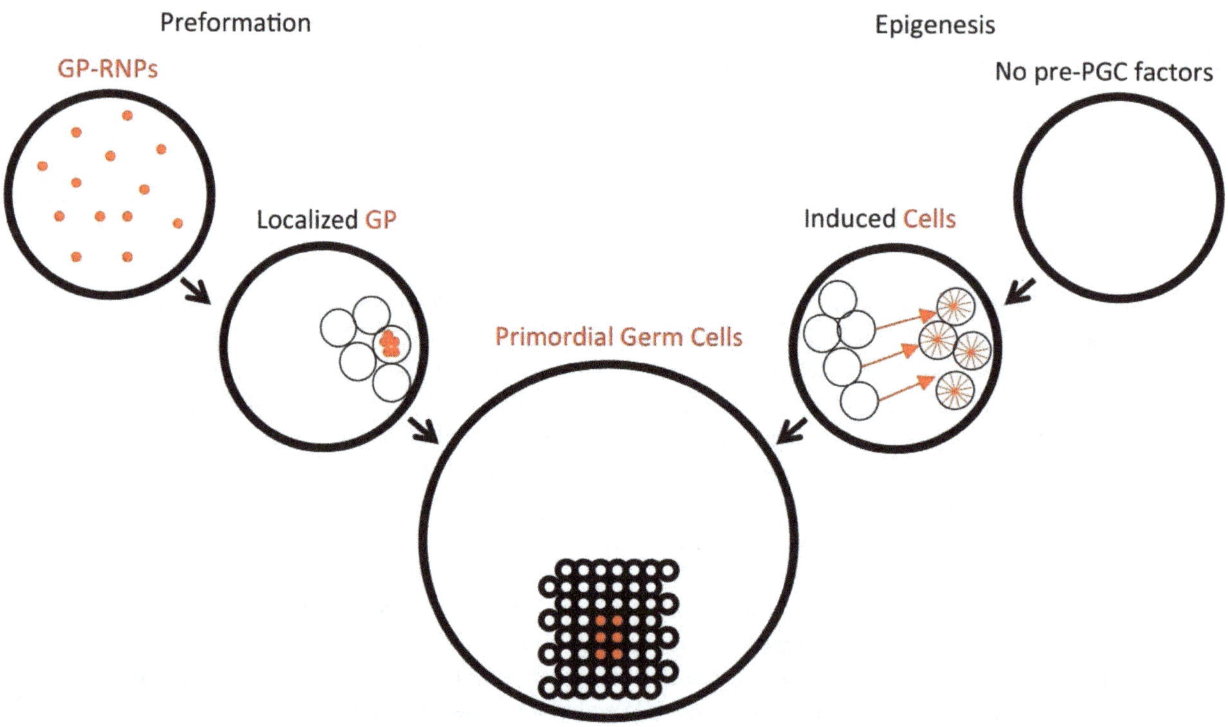

Figure 1. Mechanisms of primordial germ cell specification. Species employing preformation deposit germ plasm in oocytes. Throughout development, germ plasm localizes into select cells that eventually become the germ cells. In species employing epigenesis, cells receive inductive signals from neighboring tissue to become the primordial germ cells.

When comparing epigenesis and preformation as PGC-determining mechanisms, it was originally pointed out that epigenetic germ cell determination is an exception, and most animals use germ plasm [1, 2]. Subsequently, the analysis of distribution of these mechanisms within the evolutionary tree has led to the hypothesis that epigenesis is an ancestral PGC-determining mechanism and that preformation has arisen multiple times from this basal mode. Many clades employ both mechanisms, for example, within amphibians axolotls (urodeles) [3] use epigenesis and frogs (anurans) employ preformation [4]. Similarly, species in the group reptilia, such as turtles, use epigenesis [5], and species of the related group aves (birds) use preformation [6].

When protein-coding sequences of vertebrate species using epigenesis are compared with sister taxa employing preformation, genes of the latter evolve more rapidly. For example, when comparing protein-coding sequences from species in the amphibian and actinopterygian (ray-finned fishes) taxons, in all cases, species using preformation grouped further away from mammals than those using epigenesis. Evans et al. [7] found that no biological factor, including genome size, longevity, and generation time, correlates with sequence rate changes as optimally as the mode of PGC induction. This has led to the suggestion that the use of germ plasm relieves constraints on somatic development by dissociating PGC development from somatic development during early embryogenesis. This is because in animals using preformation mechanisms, changes in gene networks involved in somatic tissue development can occur without causing deleterious effects on the germ line, which would cause sterility and would therefore be selected against. This developmental flexibility allows for a faster rate of evolution of developmental programs. For example, ancestral gene regulatory networks (GRNs) for pluripotency and mesoderm specification found in axolotls are conserved through the evolution of mammals, whereas in frogs, pluripotency GRNs have been lost and the number of key regulators (i.e. multiple copies of Nodal and Mix) have increased in the mesoderm specification GRN [2]. Since axolotls specify PGCs through epigenesis within the mesoderm, the GRN is under constraints preventing variation—constraints that are not present in frogs. However, frogs carry preformed germ plasm, whose direct role in PGC induction allows the embryo to undergo drastic changes to the mesoderm specification GRN, such as Nodal expansion [8]. Consistent with this hypothesis, the use of germ plasm coincides with increased morphological variation and enhanced speciation within a clade [7].

Although these two PGC-determining mechanisms differ, many of the factors involved in PGC development are highly conserved regardless of the mechanism employed to differentiate PGCs. For example, the gene product of *vasa* is found in germ cells of a wide range of organisms from planaria [9] to humans [10]. Importantly, both mechanisms, epigenesis and preformation, in addition to promoting germ cell fate in PGCs, must also block somatic fate in this cell type [11].

1.1. Epigenesis, a mechanism by which cells receive signals from their surroundings

Most animal clades include lineages that use epigenesis to specify PGCs. Within the amphibian clade, axolotl (salamander) PGC specification occurs in primitive ectoderm (animal cap) cells in response to ventral mesoderm-inducing signals [12], including fibroblast growth factor (FGF) and BMP4 [13]. PGC precursors arise in the ventral marginal zone and migrate over the blastopore during gastrulation and by tailbud stage are detected in the posterior of the dorsal–lateral mesoderm expressing PGC-specific genes, such as *dazl* [3] and *vasa* [14].

All mammals specify PGCs through inductive signaling, and this mechanism is most well studied in mouse. In these organisms, extraembryonic tissue signals the most proximal region of the postimplantation epiblast to become the PGCs, whereas cells that do not receive the signals give rise to somatic tissue [15]. The mouse PGC precursors are detected just prior to primitive streak formation, and by the end of primitive streak formation, a population of around 40 PGCs are generated. Key instructive signals are the bone morphogenic proteins

(BMP2, BMP4, and BMP8B), which act through SMAD1 and SMAD5 [16, 17,18]. The population of PGCs is heavily reliant on BMP dosage as loss of BMP4 reduces their induction [18]. Some RNAs present in the germ plasm in organisms with a preformative mechanism, such as *nanos*, *dead-end* (*dnd*), and *dazl*, also have roles in germ cell development in the mouse [19]. For example, mouse Dazl protein has a complex role in PGC development by targeting mRNAs for translational inhibition, thus preventing somatic differentiation, pluripotency, and apoptotic death [20]. In addition, mouse Dazl also has a role at later stages of germ cell development, during spermatogonia differentiation [21]. This latter role is likely conserved in humans, having led to the original identification of this gene (DAZ or "deleted in azoospermia") as being affected in 13% of human male infertility cases [22, 23].

1.2. Preformation: Germ plasm deposits determine the germ cell fate

Preformation describes a mechanism whereby differentiation as germ cells is decided by the acquisition of maternally inherited determinants following fertilization [24]. The common term to describe such determinants is germ plasm, which is considered as a specialized cytoplasm enriched for factors that function in PGC determination. Germ plasm contains RNAs (both coding and noncoding) and proteins, which in many organisms appear to assemble together in punctate structures known as ribonucleoparticles (RNPs). In many organisms, germ plasm closely associates with the cytoskeleton and/or mitochondria. Proposed functions of germ plasm include the translational regulation of germ plasm RNAs, the establishment of a partially repressive chromatin in the germ line and the prevention of activation of somatic development by repressing mRNA transcription (reviewed in [25, 26]).

In *Drosophila*, arguably one of the most well-characterized species that use germ plasm as a means of PGC determination, the progenitors of the germ line are four to five pole cells found at the posterior pole of the embryo before blastoderm formation [27]. Pole cells acquire PGC identity through inheritance of the germ plasm (reviewed in [28]) called pole plasm in this organism [29, 30]. Polar granules appear at the posterior tip during mid-oogenesis, coincident with the time when vitellogenesis occurs. Toward the end of oogenesis, the polar granules associate with other germ plasm-associated structures (i.e. mitochondria) [31]. Germ plasm components are released from cortical actin at the posterior of the oocyte, and are transported via dynein motors on astral microtubules to centrosomes associated with nuclei in posterior regions of the early embryo [32]. These posterior nuclei subsequently separate from the syncytial mass through cell cleavage to form the PGC precursors or pole cells (reviewed in [28]). Embryological manipulations confirm that pole plasm is a true germ cell determinant as it can recreate pole cells when transplanted ectopically, for example, in the anterior region of the embryo. Interestingly, pole cell number relies on the dosage of a single RNA, *oskar* [33], which is required for proper induction of germ cells. However, the function of *oskar* in PGC determination is restricted to dipterans [34], so that other factors must take its role in other organisms.

P-granules are another type of RNA-rich cytoplasm structures involved in the specification of the *C. elegans* germ line. These granules are scattered evenly throughout the cytoplasm before and just after fertilization [35]. At a time coincident with pronuclear fusion, under the influence of cell polarity factors (PAR factors) and implemented by microfilament-dependent cytoplasmic streaming [36, 37], P-granules translocate to and accumulate at the posterior of the early embryo. As a result of asymmetric positioning of the spindle during cell divisions, each division generates daughters of unequal size, with the smaller daughter inheriting most of the P-granules asymmetrically until the generation of the P_4 cell, after four cell divisions. However, the minority of P-granules remaining at the anterior end or in non-P_4 cells is degraded [38, 39]. Interestingly, P-granule localization during these early divisions (which generate P_2, P_3, and P_4 cells) relies on attachment to the nuclear periphery coupled to centrosome rotation, which leads to asymmetric placement of the P-granules with respect to the nucleus, specifically at the ventral side of the P cell [38]. Although "local gathering mechanisms" that each of the P cells use to deliver P-granules to the proper location appear to vary, asymmetric cell division and directional P-granule segregation occur through the first four cell cycles, generating P-granule-containing cell P_4. This cell type, whose nucleus is surrounded by large, coalesced P-granules, is the first cell whose descendants are restricted to the germ line [40, 41] and whose ablation results in sterility [42]. The P_4 cell divides symmetrically during embryogenesis resulting in Z2 and Z3, which after the first larval stage continue to divide symmetrically to produce about 1000 germ cells in the adult gonad [43]. P-granules continue to associate with the nuclear envelope of germ cells throughout embryonic and larval divisions and only detach from this structure during gamete maturation. However, the significance of this dynamic association remains poorly understood.

Although research on PGC specification is sparse for colonial ascidians, studies of a single species, *Botryllus schlosseri*, describe aggregation and ventral localization of maternally inherited germ plasm RNAs, in particular *vasa* mRNA [44]. The solitary ascidian (*Ciona intestinalis*) also uses a specialized cytoplasm (myoplasm) containing maternally inherited determinants of germ cells [45]. The first pre-PGCs form at the posterior end of the early embryo [46] and express conserved germ line gene, such as *vasa* [47]. Interestingly, when the PGCs are ablated, a new collection of *vasa*-expressing cells appears in the juvenile after metamorphosis. Because these cells did not inherit myoplasm, this suggests an alternative, compensatory mechanism for PGC specification in this organism, possibly similar to an inductive mode of PGC specification [48], which may be occurring as a safe–fail mechanism in this organism. Other well-studied systems that use germ plasm to specify PGCs, now within vertebrate lineages, include zebrafish, which is the focus of this chapter, *Xenopus* and chicken (see below).

2. Germ cell determinant segregation in zebrafish

2.1. Maternal inheritance of germ plasm RNPs

The fish model, *Danio rerio*, employs the preformative model of germ cell specification wherein the mother deposits factors important for PGC specification into the oocytes, which are transmitted to the embryo. RNPs containing RNAs for all known germ plasm factors in this organism originally localize to the same structure during oogenesis. This is the mitochondrial cloud or Balbiani body (Bb), a large aggregate of organelles found in the oocytes of many species, that translocates from a location near the nucleus to the vegetal pole during oogenesis. Germ plasm RNA and mitochondrial enrichment to the Bb depend on the functional product of *buckyball* (*buc*), a novel vertebrate-specific protein that is the first gene identified as being required for the formation of this structure [49]. The Bb disassembles by stage II oogenesis, when the germ plasm RNAs localize to the vegetal pole via a pathway similar to the messenger transport organizer (METRO) pathway originally described in *Xenopus*. Subsequently, different germ plasm RNAs undergo different patterns of distribution in the oocyte: *vasa* becomes redistributed along the cortex, *nanos* becomes evenly distributed in the oocyte and *deleted in azoospermia* (*dazl*) remains localized to the vegetal pole (reviewed in [50]; see also [51]). These various patterns of redistribution during oogenesis result in the creation of two types of localized germ plasm RNPs, those present in the forming blastodisc at the animal region (animal germ plasm RNPs) and those remaining in the vegetal region (vegetal germ plasm RNPs).

2.1.1. *Animal region germ plasm factors*

Known animal region germ plasm factors include *vasa,* a gene whose RNA or gene product is one of the most widely used markers for germ cells or germ plasm in metazoans [52]. As mentioned above, *vasa*-containing germ plasm RNPs localize along the cortical region of the oocyte at stage II and after egg activation are rapidly transported to the animal pole, where they are found dispersed evenly around the blastodisc. The protein product for this gene is a DEAD-box ATP-dependent helicase involved in RNA metabolism that facilitates interactions between RNA–RNA and RNA–protein [53]. Loss of function mutations in *vasa* lead to reduction in number or functionality of germ cells in almost all organisms [54–58]. In zebrafish, *vasa* morphant embryos (knock down of protein using a morpholino-conjugated oligonucleotide (MO)) do not exhibit defects in the establishment of the germ line [59]. Moreover, injection of *vasa* RNA into one-cell zebrafish embryos does not lead to an increase in PGCs; instead this RNA is degraded, suggesting that endogenous *vasa* RNA levels are carefully controlled by an intrinsic cellular mechanism [60]. These studies suggest that *vasa* is generally required, but is not sufficient, for germ cell development. However, a more rigorous set of tests using a knockdown or ectopic expression initiated during oogenesis, or *vasa/ddx4* maternal mutants, has not yet been reported.

Another animal germ plasm RNA is that of the gene *nanos*. Before egg activation, *nanos* mRNA is not localized; however, shortly after activation *nanos* is found to colocalize with *vasa* RNA. *nanos* RNA and/or gene product are also commonly associated with germ cell development and present in germ plasm. First described in *Drosophila, nanos* encodes a RNA-binding zinc finger protein. In *Drosophila, nanos* is not generally required for PGC formation; however, *Drosophila* PGCs deficient in *nanos* activity have abnormal development, including failure to migrate to the gonad, reduction in egg number, early activation of germ cell genes, deregulated expression of somatic mRNAs, and irregular morphology [61–63]. In zebrafish, knockdown of *nanos1* using antisense morpholinos injected at the one-cell stage proves that it is essential for germ cell development [64]. In these morphants, PGCs are specified, yet these undergo aberrant migration and exhibit reduced number.

Other germline-specific RNAs found at the animal pole include *dnd*, *askopos* (*kop*), and *TDRD7*. *dnd* encodes a RNA-binding protein that blocks negative regulation through mRNA target degradation caused by binding of several microRNAs (miRNAs). Dnd protein achieves this protective function against miRNA-mediated mRNA degradation by binding mRNAs at miRNA target sites contained within these transcripts, thus blocking miRNA binding. This mechanism is used in the zebrafish germ line by PGC factor Dnd1 to protect several mRNAs, including *nanos* and *TDRD7*, from miR-430 repression, which is achieved through Dnd binding to U-rich regions in mRNA targets [65, 66]. Similar to *nanos1*, knockdown knockdown of *dnd* in morphants results in defects in PGC migration exhibits defects in PGC migration and viability [67].

kop mRNA was found to localize to zebrafish germ plasm and encodes a novel nuclear protein [68] that is not yet well studied. Some tudor domain-containing genes are expressed in germ cells, *RNF17*, *TDRD1*, *TDRD6*, and *TDRD7*. The latter (*TDRD7*) also localizes to germ plasm as it aggregates to the forming furrows, and TDRD7 protein colocalizes with Vasa in PGCs [69]. In *Drosophila*, homologous tudor proteins have been proposed to serve as a platform for polar granule assembly [70], a function that is consistent with the presence of TDRD7 protein in the zebrafish germ plasm. Zebrafish *TDRD7* morphants lack normal germ cell granule integrity at late somite stage [69]. Interestingly, TDRD7 is highly expressed in the developing lens, and mutations of TDRD7 in human, mouse, and chick lead to cataracts [71].

For tested animal germ plasm RNAs (*vasa*, *nanos*, and *dnd*), animal RNPs appear to colocalize and exhibit a similar size [72]. These observations suggest that germ plasm RNPs may be composites containing a set of germ plasm factors essential for germ plasm determination.

2.1.2. Vegetal region germ plasm factors

Vegetally localized germ plasm RNPs contain RNAs for the genes *deleted in azoospermia-like* (*dazl*) and *bruno-like* (*brul*). These germ plasm RNPs are initially localized to the vegetal pole of the egg and upon egg activation translocate toward the blastomeres forming at the animal pole [73–75]. Both *dazl* and *brul* encode RNA-binding proteins present in germ plasm across a wide variety of animal lineages. Although Dazl protein is not on its own sufficient to promote translation, this factor promotes RNA polyadenylation to enhance the stability of PGC transcripts, such as those for *TDRD7* and *dazl*. Thus, Dnd1 and Dazl may work additively in this fashion to protect RNAs from miRNA repression [76]. Studies focusing on *dazl* RNA aggregates show that these are restricted to the vegetal cortex in the mature egg and move animally after egg activation, so that 45 minutes post fertilization *dazl* RNP aggregates begin to be observed at the site of germ plasm accumulation at the first cleavage furrow [75, 77]. During animally directed movement, aggregates do not appear to be of uniform size and may translocate using two pathways: an ooplasmic streaming pathway mediated through axial streamers in internal regions of the zygote and a cortical pathway along meridional streamers in more superficial regions [73, 75, 77]. During furrow formation, vegetally derived germ plasm RNPs associate with large germ plasm masses of animal germ plasm RNPs forming at the furrows, in a cortical location near the distal-most end of the furrow and close to the yolk membrane cortex (see below). This location suggests that the animally directed cortical pathway may be the primary pathway used by vegetal germ plasm RNPs fated to become a part of a reconstituted and functional germ plasm mass.

Recent studies have shown that vegetal RNPs appear to be differentially distributed in the vegetal cortex, specifically at a cortical depth that is deeper than that of factors that also localize to the animal pole but are required for axis induction [78]. This differential distribution along cortical depth, with germ plasm RNPs at deeper levels and dorsal factors at more superficial ones, likely facilitates the symmetric animally directed movement of germ plasm RNPs [75, 79] in spite of the asymmetric movement of the bulk of the cortex required for axis specification [80, 81].

2.2. Repackaging germ plasm: Gradual multimerization and recruitment

As mentioned above, after egg activation and prior to the first cell division, animally localized RNPs are present in single particles, which we refer to as singletons, spread throughout the developing blastodisc. Germ plasm RNP singletons aggregate in a wave-like fashion, where the wave of aggregation emanates from the center of the blastodisc outwards, toward the edge of the blastodisc.

Germ plasm RNP multimerization at this stage depends on the interplay between microtubule and microfilament networks during blastomere cell division (Figure 2A). After fertilization in zebrafish, paternally derived centrioles act as a template to reconstruct the centrosome using maternally derived components. This newly formed centrosome nucleates a sperm aster or monoaster prior to initiation of the first cell division cycle. In this structure, plus ends of the astral microtubules interact with germ plasm RNPs at the cortex and help direct germ plasm RNP multimerization in a process of **pre-aggregation** (the prefix of this term indicates that this process occurs prior to furrow formation). The growing tips of astral microtubules thus generate an aggregation front of multimerizing RNPs, initiating near the center of the blastodisc and moving outward. The inner rim of RNP aggregates remains surrounded by a field of RNPs that have not yet been influenced by astral microtubules and therefore remains as singletons until the front arrives [72, 75]. As this process ensues, central regions of the blastodisc become depleted of germ plasm RNPs, generating an outwardly expanding RNP-free zone (Figure 2B). When embryos are treated with the microtubule inhibitor nocodazole, germ plasm RNPs do not undergo multimerization and instead remain dispersed throughout the cortex. This effect is also observed in embryos deficient for the Chromosomal Passenger Complex protein, Birc5b/Survivin, encoded by the gene *motley* [72]. In wild-type embryos, Birc5b protein localizes to the tips of astral microtubules (plus ends) and interacts with germ plasm RNPs. In motley-mutant embryos, mutated Birc5b/Survivin product localizes to germ plasm RNPs but not to microtubule plus ends, demonstrating Birc5b is an important linker between the two and explaining the *motley*-mutant phenotype. The fact that in these mutants germ plasm RNPs do not multimerize indicates that this linkage is necessary for the aggregation process [72].

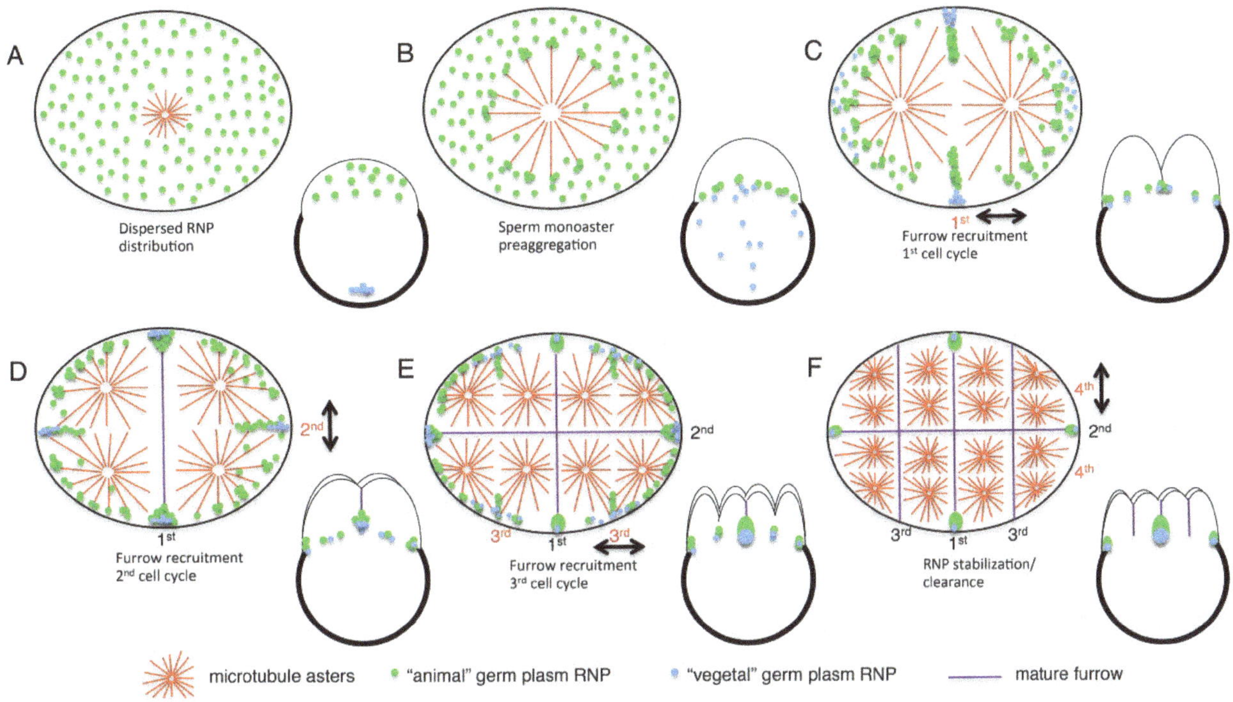

Figure 2. Germ plasm accumulation during the first zebrafish cell cycles. A) In the egg, animal pole (AP) germ plasm RNPs are dispersed at the blastodisc cortex, whereas vegetal pole (VP) germ plasm is localized at the vegetal pole. B) Prior to the initiation of cell cycling, during pronuclear fusion, astral microtubule growth moves AP germ plasm RNPs outward, whereas VP germ plasm begins to travel toward the AP. During the next several cell cycles, aggregated RNPs are collected in the (C) first furrow, (D) second furrow and (E) third furrow. F) Germ plasm RNPs in the first and second furrows are stabilized, whereas all other RNPs are cleared. Figure adapted from [86].

Another maternal factor important for germ plasm RNP recruitment during these early stages is Buc. In addition to a role for Buc to assemble the Bb during oogenesis, this factor is also required for recruitment of germ plasm during early embryogenesis, and its overexpression in the early embryo leads to an increase in PGCs [49]. Recently, a microtubule motor protein, Kif5Ba (kinesin), was found to bind to Buc and mediate its recruitment (thereby recruitment of other germ plasm molecules) to the cleavage furrow. Germ plasm in *kif5Ba* mutants is spread throughout the blastomeres, and Kif5Ba is required for the excess of PGCs that form when Buc is overexpressed [82]. This is reminiscent of the role of kinesins in germ plasm movement in *Xenopus*, where Xklp1 (kinesin-like protein) has also been shown to be required for germ plasm localization, and depletion of Xklp1 leads to arrest of germ plasm aggregation [83]. In *Xenopus*, another kinesin has been shown to have a role in later PGC development: KIF13B functions in migration and polarity of germ cells, as manipulations of xKIF13B result in erroneous PGC migration and reduced numbers [84].

Unlike the first cell cycle, which contains a monoaster involved in pronuclear fusion, blastomere cell cycle divisions contain bipolar microtubule asters nucleated by the pair of centrosomes at spindle poles. When spindle microtubules originating from opposing asters overlap at the spindle midzone, they signal furrow formation along the length of the blastomere. The furrow initiates as a microtubule-free zone that forms at the region of overlap between asters from opposite sides of the furrow. During this process, germ plasm RNP multimers and associated F-actin continue their outward movement, which in the midzone between spindle poles coincides with the microtubule-free zone at the site of furrow formation. This movement results in the accumulation of both F-actin and germ plasm RNPs along the forming furrow, the latter forming a rod-shaped structure [72, 79, 85, 86] composed of individual multimeric groups of RNPs. We refer to this accumulation of germ plasm RNPs at the furrow as the process of **recruitment** (Figure 2C).

During cell division associated with the first several cell cycles, recruitment of germ plasm RNPs to the forming furrow occurs from both sides due to the bipolar nature of the asters. In the furrow, germ plasm RNPs are still connected to the tips microtubules, which form parallel to one another and perpendicular to the furrow forming a structure known as the furrow microtubule array (FMA). Both pre-aggregation and recruitment may facilitate germ plasm RNP multimerization using the same basic mechanism: in both cases, radially expanding microtubule growth (from a monoaster during the first cell cycle and bipolar asters in subsequent cell cycles) facilitates germ plasm RNP multimerization. Because of their intrinsic arrangements, the monoaster does not result in furrow formation and therefore can only contribute to pre-aggregation, whereas asters from bipolar spindles contribute to both continued pre-aggregation and implement recruitment. Thus, germ plasm furrow recruitment employs the normal cell division machinery, in particular astral microtubules, to mediate the aggregation and local gathering of RNP multimers to the forming furrow [86]. This simple mechanism normally couples furrow induction and germ plasm RNP furrow recruitment. Under certain mutant conditions, however, germ plasm RNP furrow recruitment is partially dissociated from furrow formation, as occurs in mutant embryos that fail to initiate a furrow and that nevertheless show *vasa* RNA accumulation at the presumptive furrow site [87, 88].

2.3. Insuring a tight fit: High-order RNP multimerization

2.3.1. Compaction of germ plasm in a modified midbody

As the contractile ring forms and leads to the division of cytokinesis, the FMA tilts distally and moves to the edge of the blastodisc [89, 90]. During this process, the rod-shaped RNP arrangement compacts into tight masses at the edges of the blastodisc, in a process of **distal compaction** (Figure 2D) [85, 91]. Electron microscopy analysis of the distal cleavage furrow shows electron-dense germinal granule-like structures, in which *vasa* RNA is present [92]. Ablation of this area leads to PGC loss later in development [77], providing evidence for these accumulated masses as PGC determinants. Analysis of both *nebel* (maternal-effect mutant) and microtubule-inhibited embryos argues that the dynamic nature of the FMA is needed for proper distal compaction of *vasa* RNA [93]. In addition, non-muscle myosin II function is required in this process, as inhibition of this motor protein leads to FMA stabilization and defects in germ plasm RNP distal compaction [91].

Through distal compaction, animal germ plasm RNPs acquire a distal position at the furrow, and their aggregate is transformed from a rod-like structure to a round and more compact mass, possibly driven by an increase in neighbor-to-neighbor RNP contact and concomitant reduction in germ plasm mass volume. The extent by which this process is driven through cytoskeletal rearrangements, RNP–RNP interactions, or both remains to be determined. The subcellular cues that result in the redistribution of FMA microtubules and germ plasm RNPs to the distal end of the furrow also remain largely unknown, although, as mentioned above, a reduction in myosin activity results in FMA stabilization and a lack of proximodistal reorganization [91]. Analysis of *nebel* mutants [85], which exhibit defects in the distal placement of germ plasm RNP aggregates, may help identify the source of this signal.

The process of aggregation, recruitment, and distal compaction of germ plasm RNPs that occurs in the furrow for the first cell cycle becomes repeated during furrow formation for the second and third cell cycles (Figure 2E,F). The observed pattern of germ plasm RNP recruitment supports a model in which astral microtubules of the spindle apparatus mediate the local gathering of cortical germ plasm RNPs. Since in each subsequent spindle apparatus covers half the cortex as in the previous one, this predicts the accumulation of germ plasm RNPs of about half each subsequent cell cycle. This prediction is indeed observed during the first three cell cycles [86]. Thus, local germ plasm RNP furrow recruitment, coupled to the alternating (by 90 degrees) cleavage orientation pattern, gradually allows the gathering of germ plasm RNPs from the blastodisc cortex into the forming furrows. The adaptation of the cell division machinery for germ plasm RNP recruitment constitutes a simple and effective system to amass inherited single germ plasm RNP aggregates.

A consequence of this mechanism is that germ plasm continues to undergo recruitment to forming furrows for as long as there are germ plasm RNP aggregates in the cortex. This manifests in recruitment of germ plasm RNPs to the third furrow, temporarily generating embryos with eight visible germ plasm masses. However, the four aggregates collected during the third cell cycle do not undergo the subsequent step of ingression and instead appear to become degraded. At the same time, the outward movement of germ plasm RNPs remaining at the cortex that do not become recruited to the furrows, which is also mediated by the cycles of growing astral microtubules, result in the accumulation of these RNPs to the periphery of the blastodisc, where they similarly appear to become degraded. After the first several cycles, only the four larger germ plasm masses remain, corresponding to those recruited during the first and second cell cycles and which encompassed larger regions of the cortex and therefore amassed the largest numbers of germ plasm RNPs. The underlying basis for the selective stabilization of the first four aggregates is not known, but it is possible that these aggregates contain an amount of germ plasm RNPs above a certain threshold that allows their stabilization or further routing into the germ plasm segregation pathway [86].

Studies have shown that when RNA constructs containing a green fluorescent protein (GFP)-coding region coupled to a *nanos1* 3′ untranslated region (GFP-*nanos1* 3′UTR) are injected into the embryo, the GFP degrades quickly in somatic tissues while being stabilized in PGCs. Thus, one can hypothesize that if *nanos* RNA is not packaged properly in one of the four large aggregates, it will be degraded efficiently and the degradation information is in its 3′UTR [64]. These control mechanisms appear to be conserved in *Drosophila,* as non-localized *nanos* RNA also undergoes degradation [94].

2.3.2. Animal meet vegetal RNPs: Generating a full-complement of zebrafish germ plasm

During formation of the first furrows, only animal pole germ plasm RNPs are found in the forming germ plasm masses; however, at the end of the distal compaction phase as the first furrow is completed, vegetally localized RNPs, such as *dazl1* and *bruno-like,* which have translocated to the animal pole, become attached to the compacted mass of animal germ plasm RNPs [75]. At the distal end of the furrow, animal germ plasm RNPs, containing RNAs for *vasa, nanos,* and *dnd,* appear in a single, fully overlapping, RNA localization domain. Localized *dazl* RNA, however, occupies a partially overlapping yet distinct domain, situated in the distal-most region of the germ plasm aggregate. The localization of *dazl* RNA in the distal-most region of the furrow and at the blastomere–yolk cell boundary suggests that *dazl* RNPs that join the germ plasm travel animally along meridional arcs. This idea is consistent with the observation of tracks of *dazl* RNA along meridional arcs of the yolk cell cortex. This germ plasm organization, with animal and vegetal RNPs occupying distinct domains and with vegetal RNAs in a more distal location of the blastodisc, is maintained at least until the 64-cell stage [75]. Further research will be required to determine why germ plasm RNPs use two separate pathways of recruitment, one for animal germ plasm RNPs and one for vegetal ones, as well as distinct domains of localization within the final germ plasm mass. Some precedent exists with regards to germ plasm subcompartmentalization in other systems. In *Drosophila,* noncoding RNAs reside in the polar granules, whereas other RNAs localize to the matrix in which these are embedded [95]. In *Xenopus,* some germ plasm RNAs, such as *Xcat2,* are present in germinal granules corresponding to the germ plasm, others, such as *Deadsouth* and *Xpat,* associate closely with these granules and still others, such as *Xdazl,* localize to the surrounding matrix [96]. The conservation of germ plasm subcompartmentalization across species suggests a functional role for such substructure, a role that remains to be determined.

2.4. Maintaining germ plasm potential during cell division

2.4.1. Ingression into cells and asymmetric segregation

At about the 16-cell stage, the four germ plasm masses, which formed during the first two cell cycles and do not undergo degradation, translocate from their location at the blastomere–yolk cell boundary in each of four corners of the blastodisc into four cells [75, 79, 85, 92], a process that roughly coincides with cellularization of the blastomeres (Figure 3). Although this process of germ plasm ingression has not been yet characterized in the zebrafish, one might hypothesize that it is similar to *Xenopus* germ plasm ingression, which depends on intact microtubules and microfilaments [97].

Once the germ plasm has ingressed into four PGCs, these cells continue to divide and during the cell division process their germ plasm segregates asymmetrically (Figure 3). Although the mechanism by which asymmetric segregation of germ plasm occurs is not completely understood, the germ plasm aggregates form a cup-shape structure that associates with one of the two spindle poles [92, 98], suggesting that segregation might rely on the spindle apparatus as proposed in *Xenopus* germ plasm segregation [92, 99], as well as the asymmetric segregation of a maternally inherited mRNA in mollusc embryos [100]. This pattern of subcellularly

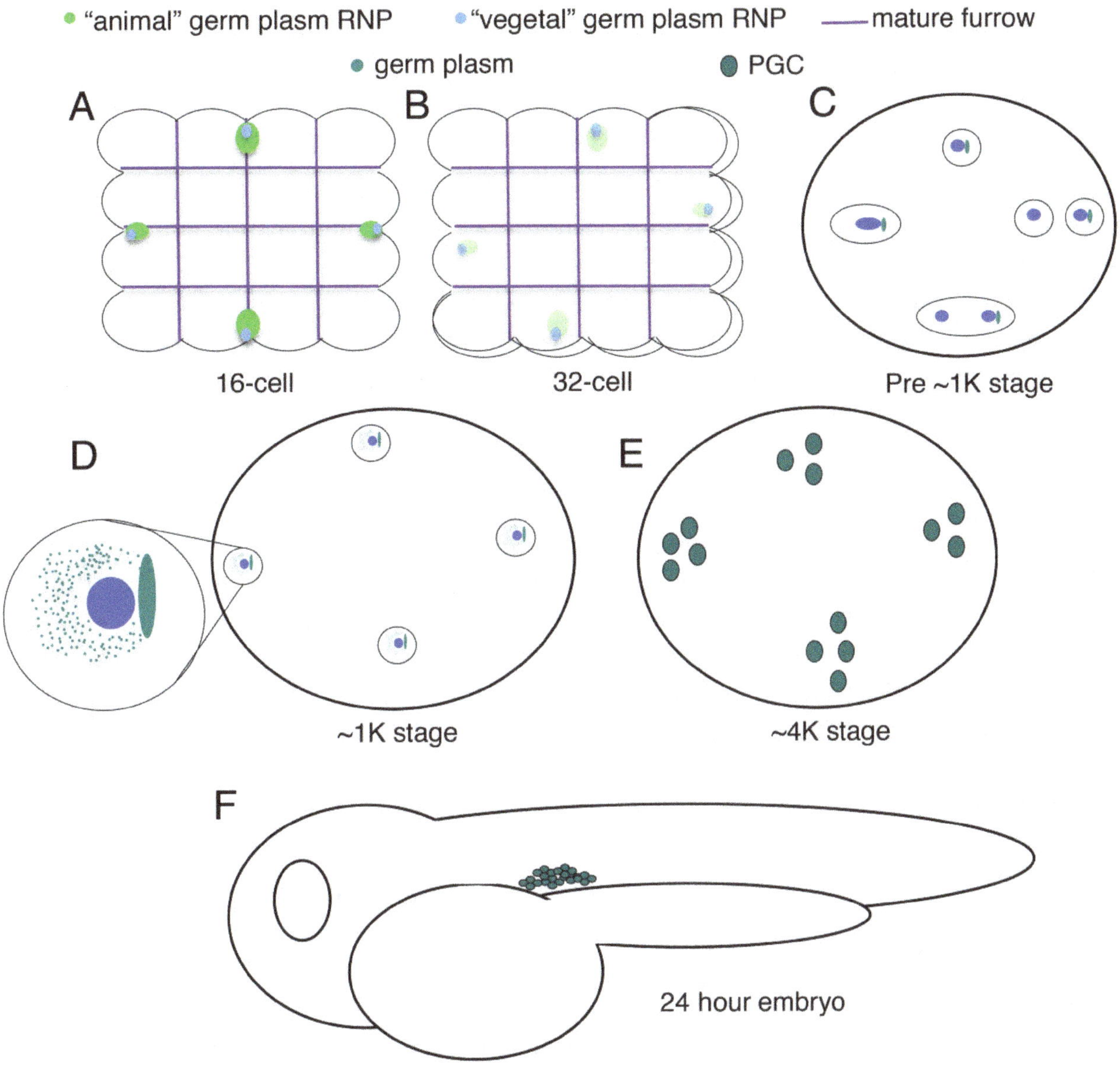

Figure 3. Germ plasm to primordial germ cell transition in zebrafish. A) Compact, yet separate animal and vegetal germ plasm aggregates at 16-cell stage are found at the edge of the first two furrows. B) At approximately the 32-cell stage, germ plasm aggregates ingress into four cells (aggregates are depicted with a lighter color to indicate that they are likely in bottom tier blastomeres, closest to their original location at the blastodisc–yolk cell boundary). C) Before PGC-specific expression, germ plasm asymmetrically segregates so that only one cell receives the aggregate (germ plasm masses are diagrammed at the various stages of the cell cycle that is represented as proceeding counterclockwise, indicating association of the germ plasm mass to one of the spindle poles to generate two asymmetric daughter cells, one with and one without the germ plasm mass—in this diagram, the counterclockwise progression is for the representation of the cell cycle only, and in reality each PGC is undergoing cell cycling independently). In (C) and (D), blue color depicts DNA. D) At approximately the 1K-cell stage, germ plasm aggregates break down into smaller particles, and daughter cells inherit these particles symmetrically. E) Cells from these previous founders begin to exhibit PGC-specific expression. F) After migration in the 24-hour embryo, PGCs accumulate in the prospective gonad just above the yolk.

localized, asymmetric segregation continues until about the 1000-cell stage, roughly coinciding with the activation of the genome at the midblastula transition (MBT).

During late blastula stage, the pattern of germ plasm segregation changes: now germ plasm distributes in a perinuclear arrangement and is inherited by both daughter cells [92]. At the same time, zygotic expression of the germ line-specific gene *vasa* begins. Due to the symmetric segregation of germ line RNAs during cell division, the number of PGCs increases, such that by gastrulation the number of germ plasm RNA-positive cells increases to around 30 cells [92].

Interestingly, DNA replication-inhibited embryos display a transition between asymmetric and symmetric segregation patterns that occurs at a developmental time similar to that in control embryos. This suggests that this transition in modes of germ cell determinant segregation does not rely on nucleo/cytoplasmic ratio, which has been proposed to regulate transcriptional activation at MBT [101, 102] or zygotic transcription initiation itself, but instead relies on a DNA-independent maternal temporal mechanism, possibly an intrinsic developmental timer or, alternatively, the counting of cell divisions [92].

2.4.2. Activation of the germ cell program

Little is known about the activation of PGC program in zebrafish. In mouse, activation of the PGC program involves the expression of three interdependent proteins: *Prdm1*/BLIMP1, *Tcfap2c*/AP2gamma and PRDM14 (reviewed in [103]). BLIMP1 induction of *Tcfap2c* allows for AP2gamma to initiate the expression of PGC-specific genes (*dnd1* and *nanos3*). In zebrafish, BLIMP1 activates *Tfap2a* (which encodes AP2alpha) during neural crest cell specification, but a role of these factors in PGC development has not been reported. Ziwi (zebrafish Piwi) protein, which also localizes to germ plasm in the cleavage furrows, is also expressed in PGCs in 24-hour embryos [104]. Piwi proteins act within mRNA storage particles involved in the translational control of mRNAs [105].

Even though Vasa protein does not colocalize to the germ plasm during the early cleavage period [92, 106, 107], it is found in perinuclear patches around the germinal vesicle during oogenesis and is uniformly distributed in all embryonic cells prior to MBT [92, 107]. At around 3-4 hours post-fertilization (hpf), when the zygotic genome is activated, Vasa protein levels increase [92]. The bulk of this increase in Vasa protein is dependent on the presence of a nucleus in the PGCs, suggesting that a large part of translated Vasa is derived from new zygotic expression. However, a small amount of Vasa does accumulate in embryos whose cells lack a nucleus, suggestive of translation of Vasa protein from maternal transcripts. This finding has led to the hypothesis that maternally inherited *vasa* mRNA in PGCs results in newly translated Vasa protein, which in turn triggers the activation of a PGC-specific gene expression program that includes zygotic *vasa*. After MBT, Vasa protein resumes perinuclear localization in PGCs. Interestingly, at this stage, Vasa protein does not colocalize with clusters of *vasa* RNA [92], suggesting that the bulk of the protein is not involved in the regulation of its own transcript.

An important hallmark of the activation of the germ cell program in animal systems is their subsequent migration (reviewed in [108, 109]). Two components found in the zebrafish germ plasm are required for PGC migration and maintenance in this organism: *nanos* [64] and *dnd* [67]. Morpholino knockdown of *dnd* illustrates a role of the protein product in the polarization and migration of PGCs, as morphants do not downregulate E-cadherin in PGCs, and cells remain in close contact with one another [67]. These results show that downregulation of E-cadherin levels in PGCs allows these cells to become motile in order to receive and interpret guidance cues for PGC migration to the gonad [110]. The reader is referred to Raz [111] and Paksa and Raz [109] for in-depth reviews of zebrafish PGC migration.

3. Comparative analysis of germ plasm aggregation in vertebrates: Independent yet similar solutions

3.1. Other fish species

The mode of PGC induction within teleost fish is not fully conserved. Fish in the ostariophysan lineage, such as carp, Fegrade's danio, tetra, and zebrafish, localize *vasa* RNA at the furrow. On the other hand, euteleost species, such as medaka, rainbow fish, and trout, lack *vasa* RNA localization, exhibiting instead diffusely distributed cytoplasmic *vasa* RNA [112, 113]. In the more basal teleost lineage containing butterfly fish, *vasa* RNA is localized [112]. These patterns of *vasa* RNA localization suggest that germ plasm-mediated PGC determination is an ancestral feature of the teleost lineage, which subsequently became lost in euteleosts. Consistent with a role of the 3′UTR in RNA localization, species exhibiting *vasa* RNA localization have highly conserved 3′UTRs in this mRNA compared to species with diffuse *vasa* mRNA distribution [112].

Embryos from sturgeon species, considered a primitive fish that acts as a basal outgroup for the teleost lineage, show many similarities to anurans including holoblastic cleavage, forming a distinct blastocoel and archenteron and undergoing primary neurulation. Two studies had varying conclusions on whether sturgeon embryos employ epigenesis or preformation [8, 114]. One group found that *vasa* and *dazl* RNAs failed to localize in oocytes [8], whereas the other group argued that sturgeon PGCs are specified in the vegetal hemisphere around the vegetal pole using a maternally derived germ plasm, as is the case in anurans. Interestingly, transplanting a single PGC from sturgeon to goldfish resulted in correct translocation of PGCs to the gonadal ridge [114], indicating that the mechanism for PGC migration is conserved across species in divergent lineages.

3.2. Amphibians (*X. laevis*)

Xenopus uses maternally inherited germ plasm for PGC determination and exhibits a number of other similarities to zebrafish. During oogenesis, the mitochondrial cloud in *Xenopus* embryos appears homologous to the zebrafish Bb, as these two structures are rich in mitochondria and have a role in germ plasm segregation [115]. *Xenopus* germ plasm contains homologs to many of the germ plasm components in zebrafish, including *Xdazl* and the *nanos* homolog *Xcat2*. In both *Xenopus* and zebrafish, germ plasm components originally localize to the vegetal pole of the egg during oogenesis in small islets (Figure 4). These two RNAs accumulate in the mitochondrial cloud in early oogenesis using a mitochondrial cloud localization element (MCLE) in their 3′UTR and subsequently disperse among the islets. Interestingly, germ line RNAs (*Xcat2*), injected after the time the early transport (METRO) mechanism is active, are able to localize to the vegetal pole using the late transport pathway. However, these RNAs do not enter the germ plasm [116], suggesting the presence of steps early in the RNP packaging process for these RNAs that are crucial for proper transcript localization.

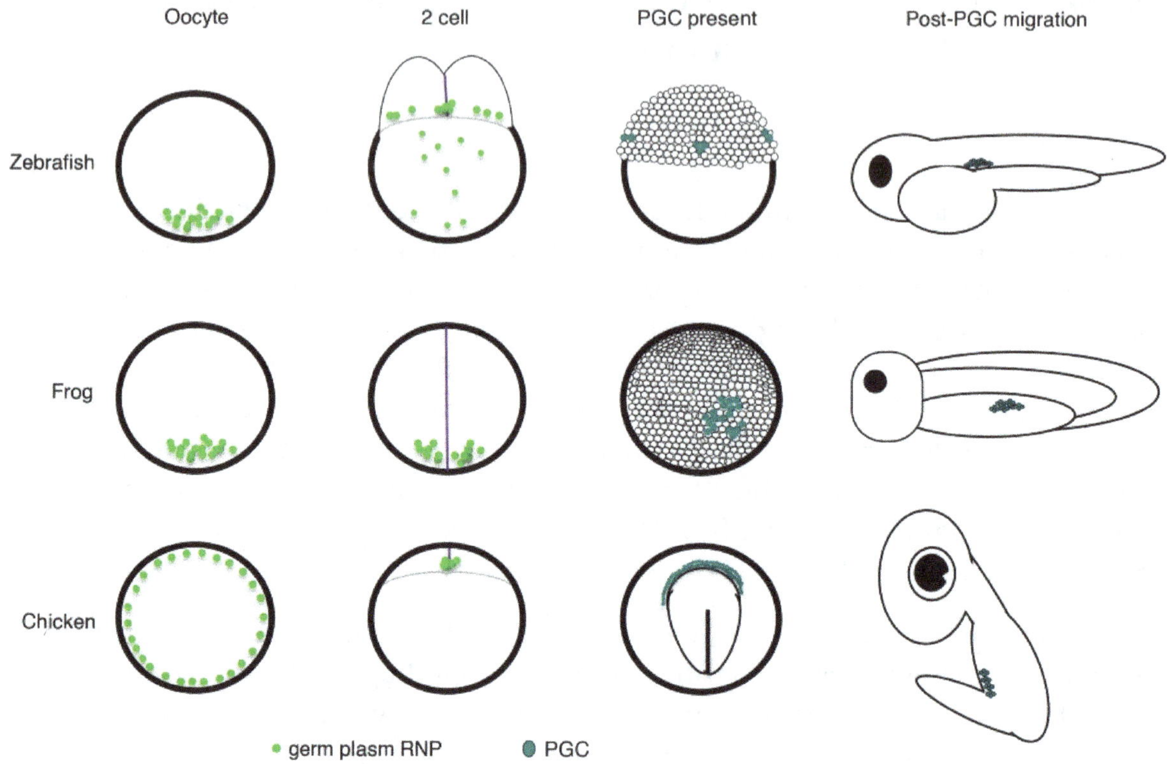

Figure 4. Comparison of vertebrate species employing germ plasm. All three species, zebrafish, frog and chicken, localize germ plasm during oogenesis. At the two-cell stage, two sites of localization of germ plasm (animal and vegetal) are found in zebrafish, only one vegetal site in frog and one site localizing at the basement of the first cleavage furrow in the chick. Both zebrafish and frog embryos have been shown to multimerize germ plasm RNPs into four large aggregates during the first several cycles, which during early embryogenesis are inherited by four separate sets of PGCs in zebrafish and one main PGC cluster in frogs. Chick PGCs are found in a crescent-shaped distribution at the anterior edge of the primitive streak. After PGC migration, each organism contains bilaterally situated groupings of PGCs at the site of the prospective gonads.

While the early/METRO pathway does not involve microtubules [117, 118], *Xenopus* germ plasm undergoes a second movement after fertilization, which relies on microtubules. Like zebrafish, *Xenopus* germ plasm undergoes local aggregation to multimerize islets into large patches of germ plasm in the first round of aggregation. The second aggregation process involves periodic surface contraction waves (SCWs) directing the germ plasm to the vegetal cortex to form large patches (Figure 4). SCWs can be inhibited by ultraviolet radiation at the vegetal pole [97] and require the kinesin family motor Xklp1 [119]. These aggregation waves result in the concentration of germ plasm at the cortex, forming four large aggregates that, as in zebrafish, are associated with the microtubule network, localize to cleavage furrows [83] and are later inherited by only four blastomeres [120].

The segregation of germ plasm in *Xenopus* continues to have parallels with that in zebrafish during the cleavage stages. During this period, germ plasm associates with the plasma membrane; during mitosis, it localizes to one of the spindle poles leading to asymmetric segregation until the gastrulation stages, when germ plasm acquires a perinuclear localization; and during cell division, symmetric expression leading to an increase in the number of germ plasm containing cells [121]. Although some experiments show that implanted PGCs can form functional gametes in *Xenopus* [122], others show explanted pre-PGCs, as well as isolated migrating PGCs, do not form germ cells when placed in an ectopic location, differentiating instead as somatic cells [123]. These observations suggest that, unlike *Drosophila* [124], *Xenopus* germ plasm specifies, but does not irreversibly determine, the germline.

Although there are many similarities in the pathways of germ plasm segregation in *Xenopus* and zebrafish, a major difference between the two species is that in zebrafish embryos germ plasm RNPs employ two distinct modes of transport within the early embryo: for animal and vegetal germ plasm RNPs; whereas in *Xenopus*, all germ plasm RNPs are located at the vegetal pole. The difference might be explained by the method each species employs for cellular cleavage. Teleost embryos undergo segregation of ooplasm away from the yolk followed by meroblastic cleavages only at one pole of the embryo. In *Xenopus* embryos, however, the yolk does not segregate away from dividing blastomeres, which as opposed to teleosts undergo holoblastic cleavage involving division across the entire embryo.

3.3. Other vertebrate species (chick)

Until relatively recently, studies suggested that the chick used epigenesis [125, 126], although these studies relied on in vitro culture [127]. However, Vasa protein (CVH in chick) was found to accumulate at the base of the membrane furrows in the early cleavage stage chick embryo, a location strikingly similar to that for *vasa* RNA localization in zebrafish embryos (Figure 4). CVH localizes to globular structures in chicken oocytes, which also contained mitochondrial cloud materials, suggesting preformation [6]. Considering the proposed independent appearance of germ plasm in various vertebrate lineages [8], it is possible that the parallels observed in teleosts, amphibians, and the chick reflect the cell division apparatus as a pre-existing intrinsic cellular mechanism that is readily co-opted for the local gathering and segregation in the embryo of maternally inherited germ plasm. Further analysis, involving common cellular and developmental themes in various animal lineages, will be required to better understand germ plasm evolution and its relationship with the basic embryonic cellular apparatus.

4. Conclusions

Epigenesis describes an inductive mechanism used notably by mammals in which tissues signal for a set of cells to become the PGCs. Preformation describes a mechanism using germ granules placed in the oocyte, which are collected into a set of cells to become the PGCs. Even though mechanisms differ, many of the RNAs and proteins that specify the germ line are conserved between animals using preformation and epigenesis.

Developmental biology studies have focused on genetic models to decipher the molecules and mechanisms for germ line establishment. Zebrafish use preformed germ granules known as germ plasm RNPs, which aggregate together, recruit to the furrow and distally compact into tight masses that ingress into only four cells. Throughout the remainder of maternal stage cell divisions, these four cells asymmetrically segregate the germ plasm aggregate so as only one of the dividing cells keeps the mass. When the zygotic genome is activated, these cells divide and generate the PGC population. It is tempting to speculate that the maternal process of germ plasm inheritance is designed to optimize the gathering of germ plasm material into large masses capable of influencing cell fate and that their subsequent asymmetric segregation during the cleavage stages preserves their full inductive potential until activation of the zygotic genome.

Understanding the mechanisms of germ cell determination will contribute to our ability to interpret cases of impaired fertility and will facilitate the promotion of healthy reproduction and assisted reproductive methods. In addition, recent studies in various biological systems have identified common links between germ cell gene expression programs, and those of stem and cancer cells [128–131], suggesting that a better understanding of germ cell biology will also contribute to the fields of regenerative and cancer biology. The zebrafish model system provides a tractable experimental system to gain mechanistic insights into these important topics relevant to human and animal health.

Acknowledgements

We thank current and former members of the Pelegri laboratory for useful discussions and participating on parts of the presented work. The research in our laboratory is funded by an National Institutes of Health (NIH) grant GM065303. C.E. has been additionally supported by NIH grant GM108449.

Author details

Celeste Eno and Francisco Pelegri*

*Address all correspondence to: fjpelegri@wisc.edu

Laboratory of Genetics, University of Wisconsin – Madison, Madison, WI, USA

References

[1] Wolpert L. Principles of development. London/Oxford: Current Biology/University Press; 1998.

[2] Swiers G, Chen Y-H, Johnson AD, Loose M. A conserved mechanism for vertebrate mesoderm specification in urodele amphibians and mammals. Dev. Biol. 2010;343:138-152.

[3] Johnson AD, Bachvarova RF, Drum M. Expression of Axolotl *DAZL* RNA, a marker of germ plasm: widespread maternal RNA and onset of expression in germ cells approaching the gonad. Dev. Biol. 2001;234:402-415.

[4] Smith LD. The role of a 'germinal plasm' in the formation of primordial germ cells in *Rana pipiens*. Dev. Biol. 1966;14:330-347.

[5] Bachvarova RF, Crother BI, Manova K, Chatfield J, Shoemaker CM, Crews DP, Johnson AD. Expression of Dazl and Vasa in turtle embryos and ovaries: evidence for inductive specification of germ cells. Evol. Dev. 2009;11:525-534.

[6] Tsunekawas N, Noito M, Sakai Y, Nishida T, Noce T. Isolation of chicken *vasa* homolog gene and tracing the origin of primordial germ cells. Development. 2000; 127: 2741-2750.

[7] Evans TC, Wade CM, Chapman FA, Johnson AD, Loose M. Acquisition of germ plasm accelerates vertebrate evolution. Science. 2014;343:200-203.

[8] Johnson AD, Richardson E, Bachvarova RF, Crother BI. Evolution of the germ line-soma relationship in vertebrate embryos. Reproduction. 2011;141:291-300.

[9] Shibata N, Umesono Y, Orii H, Sajurai T, Watanabe K, Agata K. Expression of *vasa(vas)*-related genes in germline cells and totipotent somatic stem cells of planarians. Dev. Biol. 1999;206:73-87.

[10] Anderson RA, Fulton N, Cowan G, Coutts S, Saunders PTK. Conserved and divergent patterns of expression of DAZL, VASA and OCT4 in the germ cells of the human fetal ovary and testis. BMC Dev. Biol. 2007;7:136.

[11] Magnusdottir E, Surani MA. How to make a primordial germ cell. Electroenceph. Clin. Neurophys. 1987;66:529-538.

[12] Boterenbrood EC, Nieuwkoop PD. The formation of the mesoderm in urodelean amphibians. V. Its regional induction by the endoderm. Wilhelm Roux' Arch. Dev. Biol. 1973;173:319-332.

[13] Johnson AD, Crother B, White ME, Patient R, Bachvarova RF, Drum M, Masi T. Regulative germ cell specification in axolotl embryos: a primitive trait conserved in the mammalian lineage. Philos. Trans. R. Soc. Lond. B Biol. Sci. 2003;358:1371-1379.

[14] Bachvarova RF, Masi T, Drum M, Parker N, Mason K, Patient R, Johnson AD. Gene expression in the axolotl germ line: *Axdazl, axvh, axoct-4,* and *Axkit*. Dev. Dyn. 2004;231:871-880.

[15] Lawson KA, Hage WJ. Clonal analysis of the origin of primordial germ cells in the mouse. Ciba Found. Symp. 1994;182:68-84.

[16] Ying Y, Qi X, Zhao GQ. Induction of primordial germ cells from murine epiblasts by synergistic action of BMP4 and BMP8B signaling pathways. Proc. Natl. Acad. Sci. USA. 2001;98:7858-7862.

[17] Ying Y, Zhao G-Q. Cooperation of endoderm-derived BMP2 and extraembryonic ectoderm-derived BMP4 in primordial germ cell generation in the mouse. Dev. Biol. 2001;232:484-492.

[18] Lawson KA, Dunn NR, Roelen BA, Zeinstra LM, Davis AM, Wright CV, Korving JP, Hogan BL. Bmp4 is required for the generation of primordial germ cells in the mouse embryo. Genes Dev. 1999;13:424-436.

[19] Saga Y. Mouse germ cell development during embryogenesis. Curr. Opin. Genet. Dev. 2008;18:337-341.

[20] Chen H-H, Welling M, Bloch DB, Muñoz J, Mientjes E, Chen X, Tramp C, Wu J, Yabuuchi A, Chou Y-F, Buecker C, Krainer A, Willemsen R, Heck AJ, Geijsen N. DAZL limits pluripotency, differentiation, and apoptosis in developing primordial germ cells. Stem Cell Rep. 2014;3:892-904.

[21] Schrans-Stassen PT, Saunders PT, Cooke HJ, de Rooij DG. Nature of the spermatogenic arrest in *Dazl-/-* mice. Biol. Reprod. 2001;65:771-776.

[22] Reijo R, Lee T-Y, Salo P, Alagappan R, Brown LG, Rosenberg M, Rozen S, Jaffe T, Strauss D, Hovatta O, de la Chapelle A, Silber S, Page DC. Diverse spermatogenic defects in humans caused by Y chromosome deletions encompassing a novel RNA binding protein gene. Nature Genetics. 1995;10:383-393.

[23] Kee K, Angeles VT, Flores M, Nguyen HN, Reijo Pera RA. Human *DAZL, DAZ* and *BOULE* genes modulate primordial germ-cell and haploid gamete formation. Nature. 2009;462:222-225.

[24] Extavour CG, Akam M. Mechanisms of germ cell specification across the metazoans: epigenesis and preformation. Development. 2003;130:5869-5884.

[25] Nakamura A, Seydoux G. Less is more: specification of the germline by transcriptional repression. Development. 2008;135:3817-3827.

[26] Rangan P, DeGennaro M, Jaime-Bustamante K, Coux R-X, Martinho RG, Lehmann R. Temporal and spatial control of germ-plasm RNAs. Curr. Biol. 2009;19:72-77.

[27] Huettner AF. The origin of the germ cells in *Drosophila melanogaster*. J. Morphol. 1923;37:385-423.

[28] Mahowald AP. Assembly of the Drosophila germ plasm. Int. Rev. Cytol. 2001;203:187-213.

[29] Geigy R. Action de l'ultra-violet sur le pole germinal dans l'oeuf de *Drosophila melanogaster* (Castration et mutabilite). Revue suisse Zool. 1931;38:187-288.

[30] Okada M, Kleinman IA, and Schneiderman HA. Restoration of fertility in sterilized Drosophila eggs by transplantation of polar cytoplasm. Dev. Biol. 1974;37:43-54.

[31] Mahowald AP. Fine structure of pole cells and polar granules in *Drosophila melanogaster*. J. Exp. Zool. 1962;151:201-216.

[32] Lerit DA, Gavis ER. Transport of germ plasm on astral microtubules directs germ cell development in *Drosophila*. Curr. Biol. 2011;21:439-448.

[33] Ephrussi A, Dickinson LK, and Lehmann R. oskar organizes the germ plasm and directs localization of the posterior determinant nanos. Cell. 1991;66:37-50.

[34] Lynch JA, Özüak O, Khila A, Abouheif E, Desplan C, Roth S. The phylogenetic origin of oskar coincided with the origin of maternally provisioned germ plasm and pole cells at the base of the holometabola. PLoS Genet. 2011;7:e1002029.

[35] Strome S, Wood WB. Generation of asymmetry and segregation of germ-line granules in early *C. elegans* embryos. Cell. 1983;35:15-25.

[36] Boyd L, Guo S, Levitan D, Stinchcomb DT, Kemphues KJ. PAR-2 is asymmetrically distributed and promotes association of P granules and PAR-1 with the cortex in *C. elegans* embryos. Development. 1996;122:3075-3084.

[37] Hird SN, White JG. Cortical and cytoplasmic flow polarity in early embryonic cells of *Caenorhabditis elegans*. J. Cell Biol. 1993;121:1343-1355.

[38] Hird SN, Paulsen JE, Strome S. Segregation of germ granules in living *Caenorhabditis elegans* embryos: cell-type-specific mechanisms for cytoplasmic localisation. Development. 1996;122:1303-1312.

[39] DeRenzo C, Reese KJ, Seydoux G. Exclusion of germ plasm proteins from somatic lineages by culin-dependent degradation. Nature. 2003;424:685-689.

[40] Deppe U, Schierenberg E, Cole T, Krieg C, Schmitt D, Yoder B, von Ehrenstein G. Cell lineages of the embryo of the nematode *Caenorhabditis elegans*. Proc. Natl. Acad. Sci. USA. 1978;75:376-380.

[41] Strome S, Wood WB. Immunofluorescence visualization of germ-line-specific cytoplasmic granules in embryos, larvae, and adults of *Caenorhabditis elegans*. Proc. Natl. Acad. Sci. USA. 1982;79:1558-1562.

[42] Sulston JE, Schierenberg E, White JG, Thomson JN. The embryonic cell lineage of the nematode *Caenorhabditis elegans*. Dev. Biol. 1983;100:64-119.

[43] Updike D, Strome S. P granule assembly and function in *Caenorhabditis elegans* germ cells. J. Androl. 2010;31:53-60.

[44] Brown FD, Tiozzo S, Roux MM, Ishizuka K, Swalla BJ, De Tomaso AW. Early lineage specification of long-lived germline precursors in the colonial ascidian *Botryllus schlosseri*. Development. 2009;136:3485-3494.

[45] Fujimura M, Takamura K. Characterization of an ascidian DEAD-box gene, Ci-DEAD: specific expression in the germ cells and its mRNA localization in the posterior-most blastomeres in early embryos. Dev. Genes Evol. 2000;210:64-72.

[46] Shirae-Kurabayashi M, Nishikata T, Takamura K, Tanaka KJ, Nakamoto C, Nakamura A. Dynamic redistribution of *vasa* homolog and exclusion of somatic cell determinants during germ cell specification in *Ciona intestinalis*. Development. 2006;133:2683-2693.

[47] Brown FD, Swalla BJ. Vasa expression in a colonial ascidian, *Botrylloides violaceus*. Evol. Dev. 2007;9:165-177.

[48] Takamura K, Fujimura M, Yamaguchi Y. Primordial germ cells originate from the endodermal strand cells in the ascidian *Ciona intestinalis*. Dev. Genes Evol. 2002;212:11-18.

[49] Bontems F, Stein A, Marlow F, Lyautey J, Gupta T, Mullins MC, Dosch R. Bucky ball organizes germ plasm assembly in zebrafish. Curr. Biol. 2009;19:414-422.

[50] Abrams EW, Mullins MC. Early zebrafish development: it's in the maternal genes. Curr. Opin. Genet. Dev. 2009;19:396-403.

[51] Kosaka K, Kawakami K, Sakamoto H, Inoue K. Spatiotemporal localization of germ plasm RNAs during zebrafish oogenesis. Mech. Dev. 2007;124:279-289.

[52] Ewen-Campen B, Schwager EE, Extavour CGM. The molecular machinery of germ line specification. Mol. Reprod. Dev. 2010;77:3-18.

[53] Rocak S, Linder P. DEAD-box proteins: the driving forces behind RNA metabolism. Nature Rev. Mol. Cell Biol. 2004;5:232-241.

[54] Lasko PF, Ashburner M. Posterior localization of vasa protein correlates with, but is not sufficient for, pole cell development. Genes Dev. 1990;4:905-921.

[55] Gruidl ME, Smith PA, Kuznicki KA, McCrone JS, Kirchner J, Roussell DL, Strome S, Bennett KL. Multiple potential germ-line helicases are components of the germ-line-specific P granules of *Caenorhabditis elegans*. Proc. Natl. Acad. Sci. USA. 1996;93:13837-13842.

[56] Sunanaga T, Watanabe A, Kawamura K. Involvement of *vasa* homolog in germline recruitment from coelomic stem cells in budding tunicates. Dev. Genes Evol. 2007;217:1-11.

[57] Ohashi H, Umeda N, Hirazawa N, Ozaki Y, Miura C, Miura T. Expression of *vasa* (*vas*)-related genes in germ cells and specific interference with gene functions by double-stranded RNA in the monogenean, *Neobenedenia girellae*. Int. J. Parasitol. 2007;37:515-523.

[58] Tanaka SS, Toyooka Y, Akasu R, Katoh-Fukui Y, Nakahara Y, Suzuki R, Yokoyama M, Noce T. The mouse homolog of Drosophila *vasa* is required for the development of male germ cells. Genes Dev. 2000;14:841-853.

[59] Braat AK, van de Water S, Korving J, Zivkovic D. A zebrafish Vasa morphant abolishes Vasa protein but does not affect the establishment of the germline. Genesis. 2001;30:183-185.

[60] Weidinger G, Wolke U, Köpprunner M, Klinger M, Raz E. Identification of tissues and patterning events required for distinct steps in early migration of zebrafish primordial germ cells. Development. 1999;126:5295-5307.

[61] Kobayashi S, Yamada M, Asaoka M, Kitamura T. Essential role of the posterior morphogen nanos for germline development in *Drosophila*. Nature. 1996;380:708-711.

[62] Forbes A, Lehmann R. Nanos and Pumilio have critical roles in the development and function of *Drosophila* germline stem cells. Development. 1998;125:679-690.

[63] Deshpande G, Calhoun G, Yanowitz JL, Schedl PD. Novel functions of *nanos* in downregulating mitosis and transcription during the development of the *Drosophila* germline. Cell. 1999;99:271-281.

[64] Köprunner M, Thisse C, Thisse B, Raz E. A zebrafish nanos-related gene is essential for the development of primordial germ cells. Genes Dev. 2001;15:2877-2885.

[65] Kedde M, Strasser MJ, Boldajipour B, Oude Vrielink JA, Slanchev K, le Sage C, Nagel R, Voorhoeve PM, van Duijse J, Ørom UA, Lund AH, Perrakis A, Raz E, Agami R. RNA-binding protein Dnd1 inhibits microRNA access to target mRNA. Cell. 2007;131:1273-1286.

[66] Mishima Y, Giraldez AJ, Takeda Y, Fujiwara T, Sakamoto H, Schier AF, Inoue K. Differential regulation of germline mRNAs in soma and germ cells by zebrafish miR-430. Curr. Biol. 2006;16:2135-2142.

[67] Weidinger G, Stebler J, Slanchev K, Dumstrei K, Wise C, Lovell-Badge R, Thisse C, Thisse B, Raz E. dead end, a novel vertebrate germ plasm component, is required for zebrafish primordial germ cell migration and survival. Curr. Biol. 2003;13:1429-1434.

[68] Blaser H, Eisenbeiss S, Neumann M, Reichman-Fried M, Thisse B, Thisse C, Raz E. Transition from non-motile behaviour to directed migration during early PGC development in zebrafish. J. Cell Sci. 2005;118:4027-4038.

[69] Strasser MJ, Mackenzie NC, Dumstrei K, Nakkrasae L-I, Stebler J, Raz E. Control over the morphology and segregation of zebrafish germ cell granules during embryonic development. BMC Dev. Biol. 2008;8:58.

[70] Arkov AL, Wang J-YS, Ramos A, Lehmann R. The role of Tudor domains in germline development and polar granule architecture. Development. 2006;133:4053-4062.

[71] Lachke SA, Alkuraya FS, Kneeland SC, Ohn T, Aboukhalil A, Howell GR, Saadi I, Cavallesco R, Yue Y, Tsai AC-H, Nair KS, Cosma MJ, Smith RS, Hodges E, AlFadhli

SM, Al-Hajeri A, Shamseldin HE, Bahbehani A, Hannon GJ, Bulyk ML, Drack AV, Anderson PJ, John SWM, Maas RL. Mutations in the RNA granule component TDRD7 cause cataract and glaucoma. Science. 2011;331:1571-1576.

[72] Nair S, Marlow F, Abrams E, Kapp L, Mullins M, Pelegri F. The chromosomal passenger protein Birc5b organizes microfilaments and germ plasm in the zebrafish embryo. PLoS Genetics. 2013;9:e1003448.

[73] Maegawa S, Yasuda K, Inoue K. Maternal mRNA localization of zebrafish DAZ-like gene. Mech. Dev. 1999;81:223-226.

[74] Suzuki H, Maegawa S, Nishibu T, Sugiyama T, Yasuda K, Inoue K. Vegetal localization of the maternal mRNA encoding an EDEN-BP/Bruno-like protein in zebrafish. Mech. Dev. 2000;93:205-209.

[75] Theusch EV, Brown KJ, Pelegri F. Separate pathways of RNA recruitment lead to the compartmentalization of the zebrafish germ plasm. Dev. Biol. 2006;292:129-141.

[76] Takeda Y, Mishima Y, Fujiwara T, Sakamoto H, Inoue K. DAZL relieves miRNA-mediated repression of germline mRNAs by controlling Poly(A) tail length in zebrafish. PLoS ONE. 2009;4:e7513.

[77] Hashimoto Y, Maegawa S, Nagai T, Yamaha E, Suzuki H, Yasuda K, Inoue K. Localized maternal factors are required for zebrafish germ cell formation. Dev. Biol. 2004;268:152-161.

[78] Welch E, Pelegri F. Cortical depth and differential transport of vegetally localized dorsal and germ line determinants in the zebrafish embryo. Bioarchitecture 2015;5:13-26.

[79] Yoon C, Kawakami K, Hopkins N. Zebrafish *vasa* homologue RNA is localized to the cleavage planes of 2- and 4-cell-stage embryos and is expressed in the primordial germ cells. Development. 1997;124:3157-3165.

[80] Tran LD, Hino H, Quach H, Lim S, Shindo A, Mimori-Kiyosue Y, Mione M, Ueno N, Winkler C, Hibi M, Sampath K. Dynamic microtubules at the vegetal cortex predict the embryonic axis in zebrafish. Development. 2012;139:3644-3652.

[81] Ge X, Grotjahn D, Welch E, Lyman-Gingerich J, Holguin C, Dimitrova E, Abrams EW, Gupta T, Marlow FL, Yabe T, Adler A, Mullins MC, Pelegri F. Hecate/Grip2a acts to reorganize the cytoskeleton in the symmetry-breaking event of embryonic axis induction. PLoS Genet. 2014;10:e1004422.

[82] Campbell PD, Heim AE, Smith MZ, Marlow FL. Kinesin-1 interacts with Bucky ball to form germ cells and is required to pattern the zebrafish body axis. Development. 2015;142:2996-3008.

[83] Robb DL, Heasman J, Raats J, Wylie C. A kinesin-like protein is required for germ plasm aggregation in Xenopus. Cell. 1996;87:823-831.

[84] Tarbashevich K, Dzementsei A, Pieler T. A novel function for KIF13B in germ cell migration. Dev. Biol. 2011;349:169-178.

[85] Pelegri F, Knaut H, Maischein H-M, Schulte-Merker S, Nüsslein-Volhard C. A mutation in the zebrafish maternal-effect gene *nebel* affects furrow formation and *vasa* RNA localization. Curr. Biol. 1999;9:1431-1440.

[86] Eno C, Pelegri F. Gradual recruitment and selective clearing generate germ plasm aggregates in the zebrafish embryo. Bioarchitecture. 2013;3:125-132.

[87] Kishimoto Y, Koshida S, Furutani-Seiki M, Kondoh H. Zebrafish maternal-effect mutations causing cytokinesis defects without affecting mitosis or equatorial *vasa* deposition. Mech. Dev. 2004;121:79-89.

[88] Yabe T, Ge X, Lindeman R, Nair S, Runke G, Mullins M, Pelegri F. The maternal-effect gene *cellular island* encodes Aurora B kinase and is essential for furrow formation in the early zebrafish embryo. PLoS Genet. 2009;5:e1000518.

[89] Danilchik MV, Funk WC, Brown E, Larkin K. Requirement for microtubules in new membrane formation during cytokinesis of Xenopus embryos. Dev. Biol. 1998;194:47-60.

[90] Jesuthasan S. Furrow-associated microtubule arrays are required for the cohesion of zebrafish blastomeres following cytokinesis. J. Cell Sci. 1998;111:3695-3703.

[91] Urven LE, Yabe T, Pelegri F. A role for non-muscle myosin II function in furrow maturation in the early zebrafish embryo. J. Cell Sci. 2006;119:4342-4352.

[92] Knaut H, Pelegri F, Bohmann K, Schwarz H, Nüsslein-Volhard C. Zebrafish *vasa* RNA but not its protein is a component of the germ plasm and segregates asymmetrically prior to germ line specification. J. Cell Biol. 2000;149:875-888.

[93] Pelegri F, Schulte-Merker S. A gynogenesis-based screen for maternal-effect genes in the zebrafish, *Danio rerio*. In: Detrich W, Zon LI, Westerfield M, editors. The Zebrafish: Genetics and Genomics. San Diego: Academic Press; 1999. p. 1-20.

[94] Bashirullah A, Halsell SR, Cooperstock RL, Kloc M, Karaiskakis A, Fisher WW, Etkin LD, Lipshitz HD. Joint action of two RNA degradation pathways controls the timing of maternal transcript elimination at the midblastula transition in *Drosophila melanogaster*. The EMBO J. 1999;18:2610-2620.

[95] Kashikawa M, Amikura R, Nakamura A, Kobayashi S. Mitochondrial small ribosomal RNA is present on polar granules in early cleavage embryos of *Drosophila melanogaster*. Dev. Growth Differ. 1999;41:495-502.

[96] Kloc M, Dougherety MT, Bilinski S, Chan AP, Brey E, King ML, Patrick CW Jr, Etkin LD. Three-dimensional ultrastructural analysis of RNA distribution within germinal granules of *Xenopus*. Dev. Biol. 2002;241:79-93.

[97] Savage R, Danilchik MV. Dynamics of germ plasm localization and its inhibition by ultraviolet irradiation in early cleavage *Xenopus* embryos. Dev. Biol. 1993;157:371-382.

[98] Braat AK, Zandbergen T, van de Water S, Goos HJT, Zivkovic D. Characterization of zebrafish primordial germ cells: morphology and early distribution of *vasa* RNA. Dev. Dyn. 1999;216:153-167.

[99] Whittington PM, Dixon KE. Quantitative studies of germ plasm and germ cells during early embryogenesis of *Xenopus laevis*. J. Embryol. Exp. Morph. 1975;33:57-74.

[100] Lambert JD, Nagy LM. Asymmetric inheritance of centrosomally localized mRNAs during embryonic cleavages. Nature. 2002;420:682-686.

[101] Kane DA, Kimmel CB. The zebrafish midblastula transition. Development. 1993;119:447-456.

[102] Lu X, Li JM, Tavazole S, Wieschaus EF. Coupling of zygotic transcription to mitotic control at the *Drosophila* mid-blastula transition. Development. 2009;136:2101-2110.

[103] Magnúsdóttir E, Surani MA. How to make a primordial germ cell. Development. 2014;141:245-252.

[104] Houwing S, Kamminga LM, Berezikov E, Cronembold D, Girard A, van den Elst H, Filippov DV, Blaser H, Raz E, Moens CB, Plasterk RHA, Hannon GJ, Draper BW, Ketting RF. A role for Piwi and piRNAs in germ cell maintenance and transposon silencing in zebrafish. Cell. 2007;129:69-82.

[105] Ku H-Y, Lin H. PIWI proteins and their interactors in piRNA biogenesis, germline development and gene expression. Natl. Sci. Rev. 2014; 1: 205-218.

[106] Braat AK, van de Water S, Goos H, Bogerd J, Zivkovic D. Vasa protein expression and localization in the zebrafish. Mech. Dev. 2000;95:271-274.

[107] Wolke U, Widinger G, Köprunner M, Raz E. Multiple levels of postranscriptional control lead to germ line-specific gene expression in the zebrafish. Curr. Biol. 2002;12:289-294.

[108] Richardson BE, Lehmann R. Mechanisms guiding primordial germ cell migration: strategies from different organisms. Nature Rev. Mol. Cell Biol. 2010;11:37-49.

[109] Paksa A, Raz E. Zebrafish germ cells: motility and guided migration. Curr. Opin. Cell Biol. 2015;36:80-85.

[110] Hartwig J, Tarbashevich K, Seggewiß J, Stehling M, Bandemer J, Grimaldi C, Paksa A, Groß-Thebing T, Meyen D, Raz E. Temporal control over the initiation of cell motility by a regulator of G-protein signaling. Proc. Natl. Acad. Sci. USA. 2014;113:11389-11394.

[111] Raz E. Guidance of primordial germ cell migration. Curr. Opin. Cell Biol. 2004;16:169-173.

[112] Knaut H, Steinbeisser H, Schwarz H, Nüsslein-Volhard C. An evolutionary conserved region in the vasa 3'UTR targets RNA translation to the germ cells in the zebrafish. Curr. Biol. 2002;12:454-466.

[113] Herpin A, Rohr S, Riedel D, Kluever N, Raz E, Schartl M. Specification of primordial germ cells in medaka (*Oryzias latipes*). BMC Dev. Biol. 2007;7:3.

[114] Saito TL, Psenicka M, Goto R, Adachi S, Inoue K, Arai K, Yamaha E. The origin and migration of primordial germ cells in sturgeons. PLoS ONE. 2014;9:e86861.

[115] Minakhina S, Steward R. Axes formation and RNA localization. Curr. Opin. Genet. Dev. 2005;15:416-421.

[116] Zhou Y, King ML. Localization of Xcat2 RNA, a putative germ plasm component, to the mitochondrial cloud in *Xenopus* stage 1 oocyte. Development. 1996;122:2947-2953.

[117] Kloc M, Etkin LD. Apparent continuity betwen the messenger transport organizer and late RNA pathways during oogenesis in *Xenopus*. Mech. Dev. 1988;73:95-106.

[118] Kloc M, Larabell C, Etkin LD. Elaboration of the messenger transport organizer pathway for localization of RNA to the vegetal cortex of *Xenopus* oocytes. Dev. Biol. 1996;180:119-130.

[119] Quaas J, Wylie C. Surface contraction waves (SCWs) in the Xenopus egg are required for the localization of the germ plasm and are dependent upon maternal stores of the kinesin-like protein Xklp1. Dev. Biol. 2002;243:272-280.

[120] MacArthur H, Houston DW, Bubunenko M, Mosquera L, King ML. DEADSouth is a germ plasm specific DEAD-box RNA helicase in *Xenopus* related to eIF4A. Mech. Dev. 2000;95:291-295.

[121] Whitingon PM, Dixon KE. Quantitative studies of germ plasm and germ cells during early embryogenesis of *Xenopus laevis*. J. Embryol. Exp Morphol. 1975;33:57-74.

[122] Ikenishi K, Okuda T, Nakazato S. Differentiation of presumptive primordial germ cells (pPGC)-like cells in explants into PGCs in experimental tadpoles. Dev. Biol. 1984;103:258-262.

[123] Wylie CC, Heasman J, Snape A, O'Driscoll M, Holwill S. Primordial germ cells of *Xenopus laevis* are not irreversibly determined early in development. Dev. Biol. 1985;112:66-72.

[124] Technau G. A single cell approach to problems of cell lineage and commitment during embryogenesis of *Drosophila melanogaster*. Development. 1987;100:1-12.

[125] Swift CH. Origin and early history of the primordial germ cells in the chick. Am. J. Anat. 1914;15:483-516.

[126] Eyal-Giladi H, Ginsburg M, Farbarov A. Avian primordial germ cells are of epiblastic origin. J. Embryol. Exp. Morphol. 1981;165:139-147.

[127] Karagenc L, Cinnamon Y, Ginsburg M, Petitte JN. Origin of primordial germ cells in the prestreak chick embryo. Dev. Genet. 1996;19:290-301.

[128] Juliano CE, Swartz SZ, Wessel GM. A conserved germline multipotency program. Development. 2010;137:4113-4126.

[129] Nagamatsu G, Kosaka T, Kawasumi M, Kinoshita T, Takubo K, Akiyama H, Sudo T, Kobayashi T, Oya M, Suda T. A germ cell-specific gene, Prmt5, works in somatic cell reprogramming. J. Biol. Chem. 2011;286:10641-10648.

[130] Yohn CB, Pusateri L, Barbosa V, Lehmann R. l(3)malignant brain tumor and three novel genes are required for Drosophila germ-cell formation. Genetics. 2003;165:1889-1900.

[131] Janiç A, Mendizabal L, Llamazares S, Rossell D, Gonzalez C. Ectopic expression of germline genes drives malignant brain tumor growth in *Drosophila*. Science. 2010;330:1824-1827.

2

The Effects of Extremely Low-Frequency Magnetic Fields on Reproductive Function in Rodents

Sang-Kon Lee, Sungman Park and Yoon-Won Kim

Abstract

Extremely low-frequency electromagnetic fields (ELF-EMF) are defined as those having frequencies up to 300 Hz, representing a non-ionising radiation having photon energy too weak to interact with biomolecular systems. Exposure to low-frequency electric field and magnetic field (MF) generally results in negligible energy absorption in the body. However, it is well established that ELF-MF induces biologic effects in various cellular functions. ELF-MF acting as a co-inducer can potentiate weak mutagenic signalling. The concern about possible adverse effects on human health of long-term exposure to ELF-MFs, especially at frequencies of 50 or 60 Hz generated from power lines and electric devices, has been increasing. Conversely, long-term effects of chronic exposure have been excluded from the scope of the guidelines of the International Commission on Non-Ionizing Radiation Protection (ICNIRP) because of insufficient consistent scientific evidence to fix the thresholds for such possible biological effects. The results regarding the adverse effects of ELF-MF on human or animal reproductive functions are contradictory or inconclusive. Overall conclusion of epidemiologic studies on ambient residential MF exposure consistently failed to establish a link between human adverse reproductive outcomes and chronic maternal or paternal exposure to low-frequency MFs. In animal studies, there is no compelling evidence for a causal relationship between disturbed prenatal development and ELF-MF exposure. Testicular spermatogenesis progresses through a complexly regulated cellular process involving mitosis and meiosis; this process seems to be vulnerable to external stressors, such as heat, MF exposure or chemical and physical agents. Exposure to ELF-MF did significant risk impaired implantation or the foetal development in animal studies. However, there is some consistency in the increase of minor skeletal alterations in animal experiments. The evidence derived from recent studies in male mice demonstrates that ELF-MF exposure is involved with an increase in the frequency of apoptosis in spermatogenic cells. Those results suggest that exposure to MF is related to possible cytogenetic effects on testicular germ cells and therefore may negatively affect reproduction. This chapter intends to present an overview on the effects of ELF-EMF exposure on the reproductive function and a plausible mechanism in rodent species.

Keywords: ELF-MF, 60 Hz, Reproduction, Germ cell apoptosis, Disruptors

1. Introduction

Life including human on earth has evolved in and adapted to the environment of various natural electromagnetic fields (EMFs) with relatively weak energy. In the last century, man-made EMFs with various spectrums were introduced into the natural environment. Long-term effects of man-made EMF on human health are not established. Human-made EMF is classified into three categories: low-frequency (LF) fields (1 Hz–100 kHz), high-frequency fields in the band of radiofrequency (100 kHz–3 GHz) and microwaves (above 3 GHz). Extremely low-frequency electromagnetic fields (ELF-EMF) are defined as those having frequencies up to 300 Hz. Ambient ELF-magnetic field (MF) is generally generated by the electric power transmission as alternating current at 50 or 60 Hz. The exposure to ELF-MF is increasing as a consequence of the wide use of electricity and electrical appliances at home or in the workplace. Therefore, it is a growing concern whether human-made EMF induces biological effects that might be harmful to human health.

For the induction of biological effect of ELF-MF, the MF is more deleterious than the electric field (EF) because MF induces an electric current in the body, while EF does not [1]. The direct biological effects of an electromagnetic field are divided into thermal effects by electromagnetic field energy absorption, stimulation function by induced electrical currents and non-thermal action by long-term exposure [2, 3]. The mechanisms of biological effects differ according to the varying frequency of EMF. Thermal effects mainly occur over 100 kHz radiofrequency. Despite statistical association between ambient residential MF exposure and childhood cancer in epidemiologic investigations [4–6] and scientific results of EMF relating to genotoxic effects [7–9], there is no plausible mechanism of cancer and no evidence for cancer in adults.

MF alone has generally not been related to genetic damage [10, 11]. However, MF exposure might enhance the effects of known DNA-damaging agents [12]. The International Agency for Research on Cancer (IARC) has classified it with 2B possibly carcinogenic to humans—based on the epidemiologic results on childhood leukaemia [10].

Available evidences are insufficient to confirm the effect of 50/60-Hz EMF generated by power lines and electric devices on human health. The magnitude and distribution of MF currents depend on the frequency, the size of the object and proximity between the objects and a conducting device [1]. In general, the conclusions of epidemiologic studies have not consistently demonstrated the association between human adverse reproductive outcome and maternal or paternal exposure to low-frequency magnetic fields. A meta-analysis failed to demonstrate an increased risk of spontaneous abortion or malformation in studies comparing pregnant woman using a video display terminal with those not using it [13]. The International Commission on Non-Ionizing Radiation Protection (ICNIRP) guidelines are based on short-term, immediate health effects such as peripheral nerve, muscle, burn and elevated tissue temperature [1]. Long-term effects of chronic exposure have been excluded from the scope of the ICNIRP guidelines because of insufficiently consistent scientific evidence to fix the thresholds for such putative biological effects.

In animal studies, exposure to ELF-MF does not significantly affect implantation and the development of a foetus [14–20] but may induce foetal death, congenital abnormalities, minor skeletal anomalies and a decrease in the number of foetus impregnated by exposed males [21–23]. There is increasing evidences from animal studies of adverse effects of exposure to ELF-MF on the male reproductive system, such as a decrease in sperm number and testis volume [24], an increase in the frequency of apoptosis of spermatogenic cells [25–27], a significant decrease in the diameter of seminiferous tubules [28] and an alteration of the pituitary–gonadal axis [29–31]. Conversely, other studies report that ELF-MF exposure has no adverse effects on the reproductive function in animal [32–34], although in those studies the daily exposure was relatively short.

Superficially located testes could be more affected by MF than the internal organs. The testis is the most sensitive tissue for thermal thresholds compared with other tissues, such as the spinal cord, intestines or skin [35]. Testicular spermatogenesis is a complex process comprising the transformation from spermatogonia, primary spermatocyte, secondary spermatocyte, round spermatid and elongated spermatid to sperm through a series of events involving mitosis, meiosis and cellular differentiation [36]. That makes the testis one of the most vulnerable organs in the whole body to external stressors, such as heat, MF or chemical agents. The thermal thresholds for tissue damage vary with the animal species and tissue; for example, tissue damage in mice is lower compared to humans and pigs [35]. The mechanisms involved in the reported adverse effects on reproductive function remain unclear. To date, the contradictory results of outcome regarding the biological effect of ELF-MF exposure seem to be related to the variability of exposure system, exposure conditions including intensity of MF and exposure duration and experimental animal including species and age. Thus, reproducibility of the study on EMF is almost unsuccessful in independent laboratories. Therefore, it is not easy to find the causal relationship between ELF-EMF exposure and experimental results. This chapter describes the overall effect of ELF-MF exposure on the reproductive function and biologic effect in mice or rats on the basis of reported scientific literatures.

2. ELF-MF exposure and epidemiologic study in human

Certain epidemiologic studies demonstrate that exposure to ELF-MF may lead to an increased risk of certain types of adult and childhood cancer, including leukaemia, cancer of the central nervous system and lymphoma [4–6]. However, others failed to find such an association [37–39]. The results of epidemiologic studies on the effects of ELF-EMF on reproductive function have been contradictory since 1986, when it was reported that electric blankets and heated water usage may increase the abortion rate and underweight delivery [4]. The possible effects of heat cannot be linked to those of EMF. The epidemiology study investigating the reproductive effect of residential exposure to ELF-MF has not found a relationship between MF and reproductive outcomes, such as foetal loss, pregnancy loss and miscarriage [40–43]. Two prospective studies show that no association exists between low birth weight or the rate of spontaneous abortion and the use of electric bed heaters [42, 43].

The limitation in most of the studies was that measurement of ELF field density was not included. Field strength of residential ELF-MF has been reported to vary between 0.05 and

0.11 μT in the USA and between 0.025 and 0.07 μT in Europe [44]. The results are inconclusive due to potential confounders and the low number of cases [40]. It was proposed that good practice for human studies should include a double-blind design, appropriate criteria for inclusion and exclusion of volunteer [45]. According to the ICNIRP guideline for limiting exposure to time-varying EMF (1 Hz to 100 kHz), the overall conclusion of epidemiologic studies shows no consistent association between human adverse reproductive outcomes and maternal or paternal exposure to low-frequency fields [1].

3. Effects of ELF-MF exposure on reproductive function

3.1. Effects on prenatal development: Teratologic studies

Various reports showed that male and female mice exposure to ELF-MF has no significant risk on fertility and on the reproductive function in mice. In mammals, prenatal exposure to ELF does not increase miscarriage and gross external, visceral or skeletal malformations using fields up to 20 mT strength [14, 16–18, 21]. No significant differences on testis volume and sperm parameters were observed in male offspring of pregnant rats exposed to a field density of 0.83 or 500 μT until 21 days of lactation [18]. On the contrary, others report that a significant decrease in the number of implantations and living foetus per litter was observed in male and female rats exposed to 50-Hz MF of 25 μT for 90 days before mating [23].

The increase in the development of minor skeletal anomalies has been consistently reported in mouse or rat experiments [15, 16, 21, 22] (Table 1). The lowest field density to induce skeletal alteration was 13 μT [22]. Since the skeletal alteration is a common finding for prenatal exposure and may result from statistical fluctuation, it may be considered biologically insignificant [46]. Interestingly, in a toxicity study of harmonics MF, exposure to 180-Hz MF in combination with 60-Hz MFs had no significant effects on litter size, litter weight or live birth rates but induced an increase in the incidence of rib variants. Nevertheless, the incidence was not significantly different from that in controls exposed to 60 Hz alone [16]. In rats, prenatally exposed to 60 Hz of field strength 1 mT, it was described the existence of certain alterations in testicular histology, such as decreased in height of seminiferous epithelium, an increased in size of Leydig cell and of connective tissue in the testis [47], suggesting that exposure to ELM-MF may be a risk factor for reproductive function. Still, prenatal exposure to ELF-MF in Wistar rat was not a biological significant risk factor for foetal development. Table 1 lists the reports regarding effects of ELM-MF on foetal development.

3.2. Multigenerational studies

Few multigenerational studies have been reported [19, 48]. In a Sprague-Dawley rats study encompassing three generations, continuous exposure to 60-Hz MF for 18.5 h per day at field strengths of 0, 2, 200 or 1000 μT or to an intermittent MF (1 h on/1 h off) at 1000 μT was performed. No significant exposure-related adverse effects were found in all three generations with respect to the reproductive function, namely in litter/breeding pair, percentage of fertile

MF (Hz)	Animal	Age	Intensity	Exposure	Duration	Skeleton anomalies	Effects on reproduction & anomaly	Result	Ref.
50	CD1 mice	Pregnant	20 mT	0–17 days	24 h	Non	Abortion rate congenital malformation	NS	14 (Kowalczuk et al, 1994)
50	Wistar rat	Pregnant	30 mT	1–29 days	24 h	Skeletal ossification ↑	No congenital malformation	S	15 (Mevissen et al, 1994)
60 or +180	SD rat	Mated female	0.2 mT	6–19 days	18.5 h/day	Rib variants	Litter size, litter weight or foetal development (⟷)	NS	16 (Ryan et al, 2000)
50	Swiss mice M, F	60 days, before mate	25 μT	90 days	24 h	Non	Implantation site, viable foetus, number of resorption, testis weight (⟷) and ovary weight ↑	NS	17 (Elbetieha et al, 2002)
60	SD rat	Pregnant female	0.83, 3, 500 μT	Gestation 6 days to lactation 21 days	21 h/day	Non	Litter size, anogenital distance, testis weight, sperm parameter (⟷)	NS	18 (Chung et al, 2003)
50	Wistar rat	Mated female	35 μT	0–20 days	24 h	Skeletal anomaly ↑	Pregnancy rates, incidences of resorptions, late foetal deaths (⟷)	S	21 (Huuskonen et al, 1993)
50	CBA/Ca mice	Mated female	13 μT	0–18 days	24 h	Skeletal anomaly ↑	Malformation/ resorption (⟷)	S	22 (Huuskonen et al, 1998)
50	B6C3F1 mice	Pregnant, F1	50 μT	7 days, F1 15.5 ms	12 h/day		Seminiferous tubule size ↓, in female, leukaemia ↑	S	28 (Qi et al, 2015)

NS, not significant; S, significant.

Table 1. Effects of ELF-MF on prenatal development.

pair, latency to parturition or litter size [48]. In contrast, another study using continuous exposure to 60-Hz MF at a field strength of 0.5 and 1.5 mT in three generations showed a consistent reduction of weight of the ovary and testis in F2 mice, although no significant effects were found on implantation. However, no significant difference of testis weight was observed in F3 male mice [19]. Interestingly, it was observed in F1 and F2, but not in F3 mice, an increased frequency of a certain type of tumours including lymphoma, adenocarcinoma or benign tumour compared with those in the control group. The results suggest that EMF exposure induces possible cytogenetic effects on living cells including gonadal cells and the biological adaptation for chronic exposure to ELF-MF may take place.

4. Effect on sperm count and testis weight

The testis volume reflects the activity of spermatogenesis in seminiferous tubules. The lumen diameter of the seminiferous tubule may be regulated by elongated spermatids in rats [49]. The reduction in the testicular volume generally indicates impairment of spermatogenesis.

It is well known that an ELF-MF with weak energy has a significant cytogenetic effect on spermatogenic cells in the testis. It has reported a significant decrease in the counts for mature spermatid or epididymal sperm and the alteration in sperm parameters in mice or rats chronically exposed to ELF-MF [24–27, 31, 50, 51] (Table 2). Recently, it was shown that chronic exposure to ELF-MF is related to a significant risk for chronic myeloid leukaemia in female and a decrease in the size of seminiferous tubules in male mice [28]. Testis weight was significantly lower than the control in accordance to a decreased sperm count [24, 29]. On the contrary, alteration of the sperm count may not reflect the testis weight [25–27].

Interestingly, testis weight increased in the exposed group at 14 μT MF for 16 weeks compared to that in sham control group, while it remained unaffected in mice exposed to the 200 μT, 0.1 and 0.5 mT MF for 8 weeks [25, 26]. No significant association between a decrease in mature spermatogenic cells and alteration of testis weight was observed for 8 weeks of ELF-MF exposure [25]. In another report, sperm counts decreased after MF exposure for 4 weeks without significant histopathological changes in the testis of mice, though the testicular weight was significantly lower than that of the control [24].

In rats, ELF-MF can impair spermatogenesis recovery after heat-induced reversible testicular damage [52]. Ultrastructural changes in spermatogonia and spermatocyte occurred earlier than degeneration of Sertoli cells, suggesting that spermatogenic cells may be more sensitive to EMF exposure than Sertoli cells.

Stressful conditions to the testis, such as MF exposure, which induces early germ cell degeneration and a reduction of spermatogenesis, may however not be reflected in a reduced sperm counts in the ejaculation until months later.

For long-term exposure up to 46 weeks to ELF-MF of 0.1 or 0.5 mT, testis weight decreased in mice of the first and the second generations. The reduction rate of testis weight on the second generation decreased significantly by about 60%, compared with 10% in the first generation, whereas testis weight was unaffected in the third generation [19]. Testicular histological findings failed to show significant changes in the first-generation mice, while an increase of phagocytic cells and active spermatogenesis were observed in the gonads of the second-generation mice but not in those of the third-generation mice. These results suggest that long-term continuous exposure may induce adaptive mechanisms, which protect the DNA from harmful influences.

Animal	Age	MF (Hz)	Density	Exposure	Duration	Testis	Functional	Growth rate	Result	Ref.
Mice		50	6.4 mT	4 weeks	Continuous	Testis weight $\downarrow$ sperm $\downarrow$ motility $\downarrow$ morphology $\downarrow$ mature sperm cell $\downarrow$			S	24 (Hong et al, 2003)
BALB/c	7 weeks	60	0.1 or 0.5 mT	8 weeks	Continuous	Testis weight $\leftrightarrow$			NS	25 (Lee et al, 2004)
BALB/c	18–26 gm	60	14 μT, 200 μT	16 weeks	Continuous	Testis weight $\uparrow$ at 14 μT, $\leftrightarrow$ at 200 μT, germ cell apoptosis $\uparrow$ in exposed group			S	26 (Kim et al, 2009)
SD rat	12 weeks	50	100 μT, 500 μT	10 ms	2 h/day	Sperm count, morphology, apoptosis	Oxidative stress parameter		NS	32 (Akdag et al, 2013)
CD-1 mice	12 weeks	60	2 mT	72 h, 10 days	8 h/day	Meiotic chromosome aberration, sperm morphology			NS	33 (Heredia-Rojas et al, 2004)
SD rat	8 weeks	50	500 μT	4, 8 weeks	4 h/day	Testis weight, histology, oxidative status	T ($\leftrightarrow$)		NS	34 (Duan et al, 2014)
Wistar rat	Foetal	60	1 mT	13th gestation to 21st postnatal	30 min × 3/day	Histomorphologic parameter: seminiferous epithelium $\downarrow$, Leydig cell, connective tissue $\uparrow$	T ($\leftrightarrow$)	NS	S	47 (Tenorio et al, 2011)
CFLP mice	8–9 weeks	50	100 μT	14 days	23.5 h/day		HCG-stimulated T-response higher		S	89 (Forgács et al, 2004)
BALB/c	8 weeks		3 mT	8 weeks	4 h/day	1 spermatocyte, Leydig cell $\uparrow$	M1*/M2 $\downarrow$		S	103 (Ebrahimi-Kalan et al, 2011)
Wistar rat	180–200 gm	SMF	128 mT	30 days	1 h/day	Spermatogenesis $\leftrightarrow$	T $\downarrow$, DNA oxidation (+)		S	104 (Amara et al, 2006)

NS, not significant; S, significant; *M, methionein.

Table 2. Effects of ELF-MF on histomorphology and function of testis in rodents.

4.1. Intermittent exposure

Intermittent exposure of ELF-MF may lead to chromosomal damage in dividing cells [53]. A negative result was reported in a study regarding the genotoxicity of ELF-MF performed at continuous exposure [54, 55].

Several possible mechanisms may explain why intermittent ELF-MF can induce genotoxicity, including micronuclei formation [56], chromosomal aberrations in human amniotic cells [57], induction of DNA strand breakage in cultured human fibroblast [7] or dose-dependent DNA damage [58]. In contrast to continuous ELF-MF exposure, the application of intermittent MF results in a significant increase of DNA damage. Nonetheless, a major limitation is that the most results suggesting a genotoxic effect of intermittent MF were obtained from *in vitro* studies.

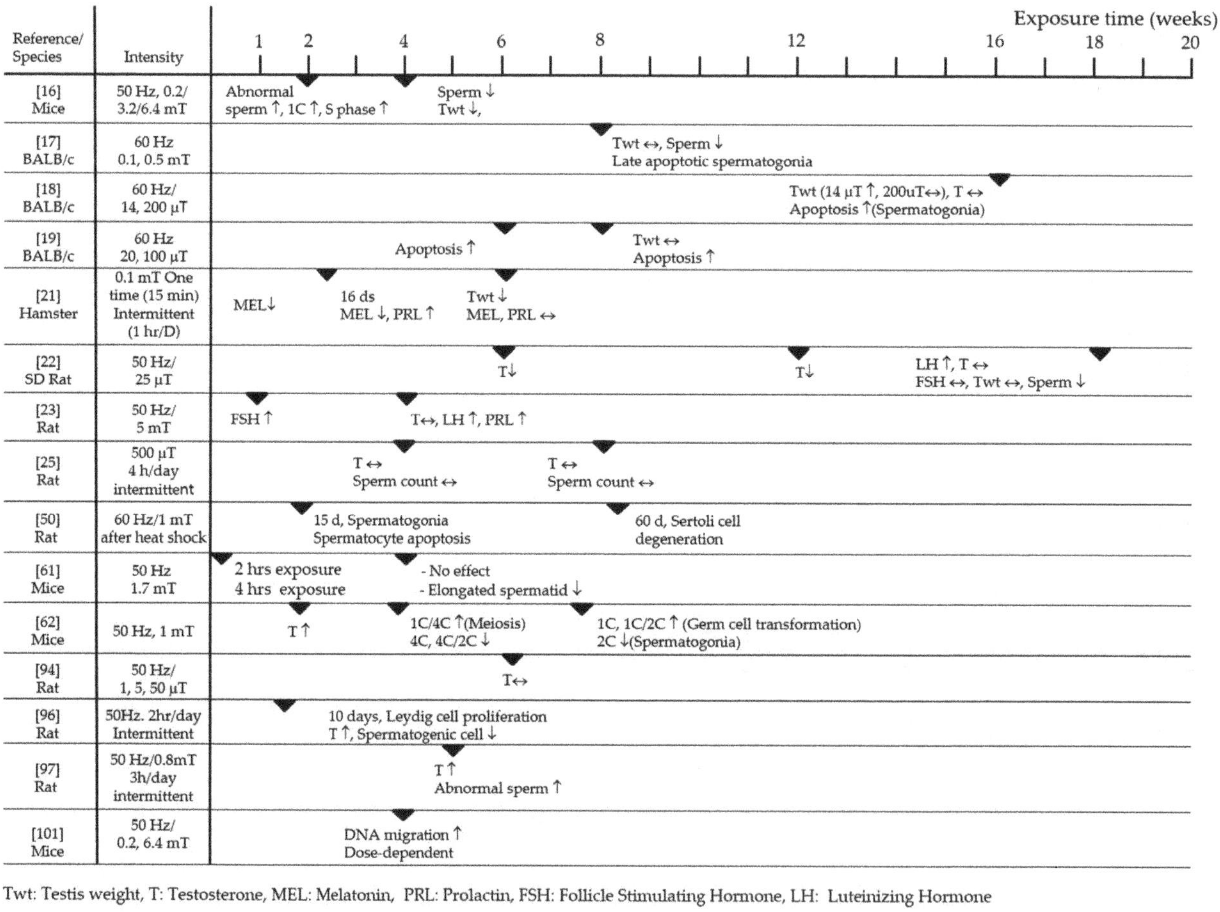

Figure 1. Biological effects of ELF-EMF exposure in rodent male reproductive function.

Moreover, the results of intermittent exposure to ELF-MF are inconsistent. Cultured human diploid fibroblasts exposed intermittently to ELF-MF of 50 Hz at 1 mT presented a significant increase of DNA damage, in contrast to the recorded in a continuous ELF-MF exposure [7]. The highest level of induced DNA damage occurred at 5-min fields-on/10-min fields-off, among various intermittent exposure conditions. The results suggest that more than 10-min extended off-time may give time for recovery. However, in rat studies, the intermittent exposure to 50-Hz ELF-MF of 500 μT (the European reference level for occupational exposure) had no adverse effects on spermatogenesis applied 4 h per day for 4 or 8 weeks [34]. In addition, there were no significant differences between ELF-MF–exposed rats and sham controls regarding parameter for oxidative stress. Other studies showed that exposure to intermittent ELF-MF reports no significant effects on sperm morphology, meiotic chromosome aberration after 2 mT MF for 72 h or 10 days, nor on sperm parameters and germ cell apoptosis after 100 or 500 μT, 2 h per day for 10 months [32, 33]. It suggests that relatively low intensity and short-term exposure to EMF would not be significant risk factors on spermatogenesis. Figure 1 lists the reported biologic effects on testis function in animals exposed to ELF-MF.

5. Germ cell apoptosis and ELF-MF exposure

Apoptosis, also called programmed cell death, is a key phenomenon in the control of sperm production. It is suggested that surplus cells and genetically abnormal cells are spontaneously eliminated by apoptosis as a defense mechanism during spermatogenesis [36]. The regulation of germ cell apoptosis during spermatogenesis is mediated by Sertoli cell–derived signals over each germ cell to which it is closely associated. Spontaneous apoptosis and pathological increase of germ cell death are induced by various external stimuli including exposure to heating, deprivation of gonadotropin and testosterone and chemotherapeutic agents [59–62]. The mechanisms of germ cell apoptosis triggered by exposure to ELF-MF are not well understood; however, it is considered to be different from those induced by aging [60], heat or hormonal deprivation [61, 62].

In mice, spontaneous apoptosis is most commonly observed in spermatocytes, including dividing spermatocytes, whereas the apoptotic rate in spermatogonia is significantly lower [63]. Histological characteristics of the seminiferous epithelium correlated with aging in rats indicate a decrease in the proliferation of spermatogonia and an increase in spermatogonia apoptosis [60]. Apoptosis associated to heat or testosterone treatment occurs mainly in round spermatids and pachytene spermatocytes [64, 65]. In mouse testis irradiated with single doses of γ rays, ionising radiation of up to 5 Gy, marked changes of testicular histology were induced by even 0.5 Gy. An apoptosis was characterized by a rapid onset of degeneration of spermatogonia and preleptotene spermatocyte [66]. However, the typical morphological characteristics of apoptosis, such as margination of chromatin or nuclear fragmentation, are rarely seen. Apoptosis related to androgen withdrawal predominantly affects spermatocytes and round spermatids [67].

Prominent histopathological alteration in testes exposed to ELF-MF showed an increased frequency of germ cell apoptosis and a decrease of mature spermatogenic cells, especially sperm [25–27]. In ELF-EMF–exposed mice, main TUNEL-positive cells are spermatogonia [25, 26] (Fig. 2).

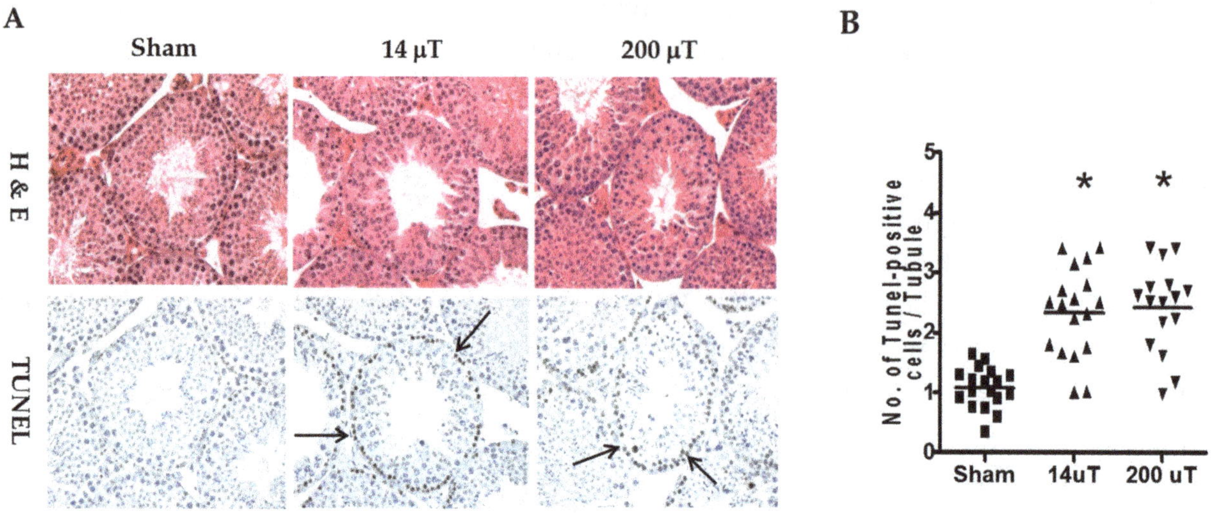

Figure 2. Effects of exposure to 60-Hz EMF on testicular germ cells apoptosis. A. Routine haematoxylin and eosin (upper row) and TUNEL (bottom row) staining of cross-sectioned seminiferous tubules. Most TUNEL-positive cells (arrow) were spermatogonia. (Magnification: ×400). B. Frequency of apoptosis was increased in mice of exposed groups. The data are mean ± S.E. *P < 0.001 *vs.* sham control. Adapted from Kim YW, Kim HS, Lee JS, Kim YJ, Lee SK, Seo JN, et al. Effects of 60 Hz 14 μT magnetic field on the apoptosis of testicular germ cell in mice. Bioelectromagnetics. 2009;30:66–72.

The continuous exposure to a 60-Hz MF may affect biological processes including apoptotic cell death and spermatogenesis in the male reproductive system of mice in duration- and dose-dependent manner [27]. The continuous exposure to ELF-MF of 0.1 or 0.5 mT for 8 weeks induced testicular germ cell apoptosis in BALB/c mice [25]. A significant increase in the incidence on testicular germ cell death was referred, although non-significant body or testis weight was recorded. The continuous exposure to a 60-Hz MF at 100 μT for 8 weeks or at 14 μT for 16 weeks induced testicular germ cell apoptosis in mice [27] (Fig. 3). The minimum dose to induce apoptosis in testicular germ cell in mice was less than 20 μT at continuous exposure to a 60-Hz MF for 8 weeks and the minimum duration was 6 weeks at continuous exposure of field strength 100 μT.

5.1. Flow cytometric analysis

Flow cytometric analysis showed that in mice exposed to 60-Hz MF of 0.1 mT or 0.5 mT for 8 weeks, an increase in late apoptosis of testicular germ cells was originated [25]. Moreover, the testicular biopsy score showed a significant decrease in mature spermatogenic cells or spermatozoa in exposed mice without concurrent significant effect on the testis weight. It has been accepted that there was a high correlation between the testicular biopsy score and sperm count [68]. Flow cytometric studies in mice exposed to ELF-MF of 6.4 mT for 2 weeks revealed a significant decrease in mature spermatogenic cells (round spermatids, 1C) compared to the controls [24], whereas the differentiating spermatogonia cells (S phase) were significantly increased. After 4-week exposure, the testis weight of exposed mice was significantly lower compared with control, although no significant changes in the percentage of spermatogonia (2C) or primary spermatocyte (4C) were observed. These results suggest a possible cytotoxic

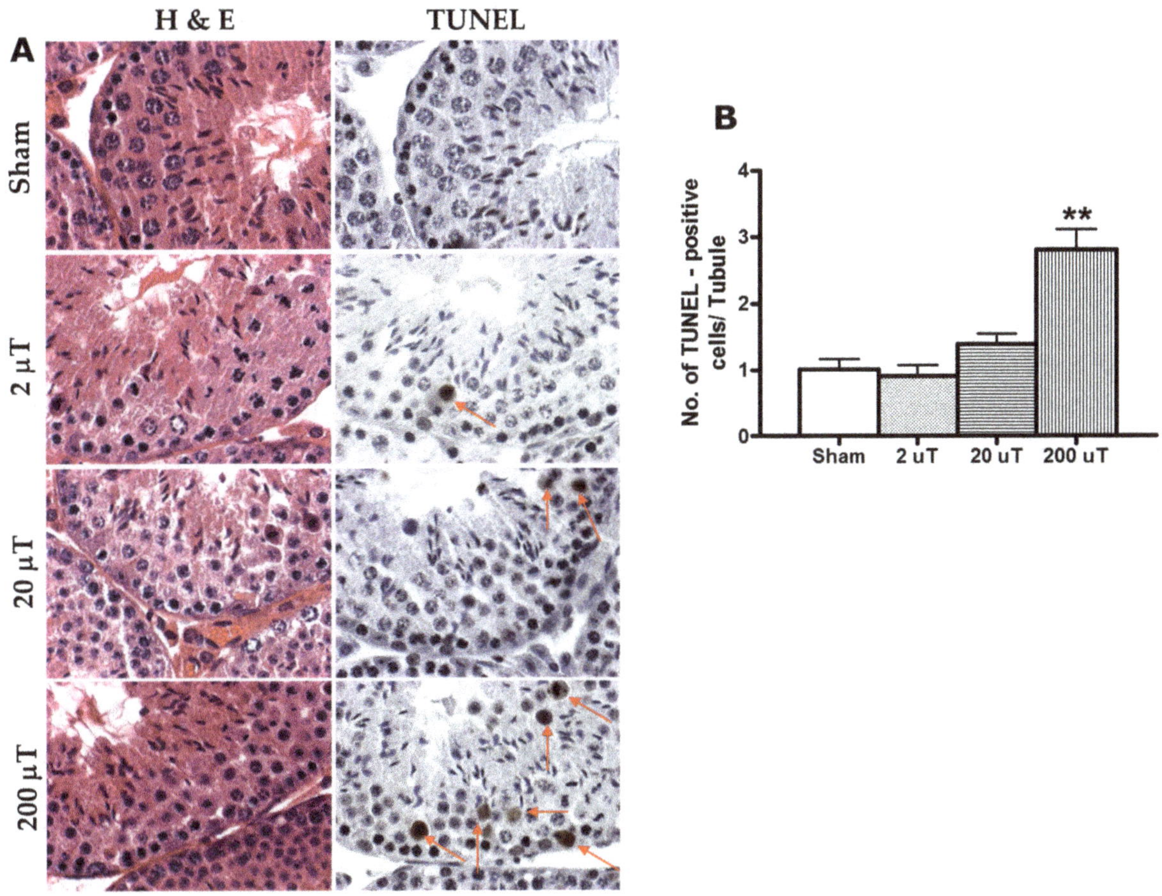

Figure 3. Effects of a 60-Hz MF on testicular germ cells apoptosis in mice exposed for 8 weeks. A. Routine haematoxylin and eosin (left) and TUNEL (right) staining of cross-sectioned seminiferous tubules. (Magnification: ×400). B. Frequency of apoptosis dose-dependent increased in exposed groups. The data are means ± S.E. **P < 0.001 *vs.* sham control. Adapted from Kim HS, Park BJ, Jang HJ, Ipper NS, Kim SH, Kim YJ, et al. Continuous exposure to 60 Hz magnetic fields induces duration- and dose-dependent apoptosis of testicular germ cells. Bioelectromagnetics. 2014;35:100–107.

effect on differentiating spermatogonia. Moreover, a decrease in testis weight could be related to early loss of mature spermatogenic cells. A 28 days of exposition to 50-Hz EMF of 1.7 mT, in mice, had no effects when exposition was limited to 2 h, but when exposition was lengthened to 4 h, a significant decrease in elongated spermatids was observed [50]. Mice exposed to 50-Hz MF of 1.0 mT for 52 days presented a significantly higher total germ cell transformation and lower spermatogonia population compared to the corresponding control groups [69]. In summary, flow cytometric analysis shows that long-term exposure to EMF-MF has a possible effect on apoptosis of mature spermatogenic cells and a differentiation of spermatogonia.

6. Genotoxic effect of ELF-MF exposure

MF of very high intensity clearly induces adverse biological effects. However, time-varying ELF-MF is too weak to break DNA strands. Still, literature review on the genotoxic potential of electric and magnetic fields demonstrated that ELF-MF might cause genotoxic effects [55]. The International Agency for Research on Cancer (IARC) concluded that ELF-MF might be carcinogenic to humans based on evidences associating residential exposure to MF with twice the risk for childhood leukaemia in children exposed to more than 0.4 μT [10, 70].

In regards to the mechanism by which ELF-MF induce DNA damage, it was suggested that MF could act as a co-inducer of DNA damage rather than as a genotoxic agent *per se*. Genotoxic effects of EMF may occur indirectly by the generation of oxygen radicals or impairment of a radical scavenging mechanism [71]. *In vivo* studies showed a dose-dependent increase in DNA strand breaks following acute (2 h) 60-Hz MF exposure at exposure levels ranging from 0.1 to 0.5 mT in rat brain cells [72]. In human diploid fibroblasts *in vitro*, intermittent exposure (5 min on/10 min off) to ELF-MF results in a significant increase of DNA strand breakage in contrast to continuous exposure [7]. The results suggest that intermittent MF exposure may defer the adaptive mechanism indicating that ELF-MF–induced damage could be removed by the DNA repair mechanism. At extended off-times, no significant differences compared to the control were observed.

A recent review of on the topic including *in vivo* and *in vitro* studies related to the induction of DNA strand breakage by ELF-MF exposure highlighted conflicting evidences [73]. Proposed biological and biochemical responses of EMF effects are variable, including those on cell proliferation [6, 7], cell shape, modification of cell membrane structure [8], alteration of gene expression [9], induction of apoptosis [10] or production of reactive oxygen species (ROS) [70, 74, 75]. Reported changes in cell proliferation patterns [76, 77], alteration of Leydig cells and pineal gland function [78, 79] may also be related to genotoxic effects.

6.1. Free radicals and EMF exposure

Results on genotoxic effects of ELF-MF are contradictory. Several mechanisms have been proposed to explain DNA damage by indirect actions related to ELF-MF exposure. The possible biological mechanism of interaction involves the alteration of the cell membrane as the target for field interaction [80], disruption of the membrane protein, which may be affected by MF [81], or free radical–mediated damages on macromolecules [73, 82]. The changes of redox state induced by disturbed oxidative stress are related to cell cycle disturbance [83].

In biological systems, free radicals are produced by normal metabolism and electron transfer reaction in the cell membranes, mainly in the mitochondria membrane [80]. The balance between ROS production and antioxidants capacity can be disturbed by external stressors, such as exposure to MF or chemical agents. Modulation of antioxidants by ELF-MF can impair the intracellular defense mechanism inducing the development of DNA damage, which may be related to cancer development. The investigation for a correlation between exposure to ELF-MF and an increased incidence of tumours is however contradictory. The modulation of cellular redox balance is affected by the enhancement of an oxidative intermediate, or the inhibition or reduction of antioxidants. Those may be influenced by environmental factors such as ELF-MF [84]. EMF-MF might compromise the intracellular defense activity promoting the development of DNA damage. Exposure of the cell to 50-Hz MF and simultaneous treatment with an oxidant may affect the DNA damage [85]. As DNA damage is not repaired, a nuclear enzyme triggers apoptosis. Moderate oxidative stress induced apoptosis, whereas a higher dose of ROS initiated cell necrosis [86]. It seems that EMF enhances the physiologic functions such as activation of certain cell types. ELF-MF affects gene transcription, cell growth and apoptosis, as well as the membrane-mediated signal transduction process [9]. Therefore, although ELF-MF may be weak, it may affect various biological functions of living organs.

7. Cell proliferation and EMF exposure

There are several *in vitro* studies related to cell differentiation under EMF exposure conditions [76, 77, 87]. Cell proliferation and DNA synthesis are related to the initial induction of mutation that eludes the physiologic defense system and DNA repair. In contrast to such biological effects of ELF-MF exposure, there are beneficial effects on tissue healing in bone fracture regarding ELF-MF exposure [88]. It has been shown that ELF-MF influence proliferation and DNA damage in both normal and tumour cells *in vitro* through the action of free radical species [77].

It was reported a dose-dependent increase in the proliferation rate in certain cell types, namely the HL-60 leukaemia cells and rat fibroblast, exposed to ELF-EMF, followed by the simultaneous increase in DNA strand breakage and in 8-hydroxy-2'-deoxyguanosine (8-OHdG) formation, one of the prominent forms at lesion of radical-induced DNA damage. The effects of ELF-MF on cell proliferation and DNA damage were prevented by antioxidant treatment [77].

Another *in vitro* study refers to mutations in the hypoxanthine–guanine phosphoribosyl transferase gene as induced in human osteosarcoma cell line and in Chinese hamster ovary cells (CHO) after exposure to ELF-MF [76]. An increase of human chorionic gonadotropin (HCG)-stimulated testosterone production was observed after exposure of 50-Hz MF in a mouse primary Leydig cell culture [89]. Other study revealed that EMF exposure originated the stimulation of Ca^{2+} influx in rat pituitary cells, as part of the regulatory process in the Leydig cell steroidogenesis [90]. Modulation of calcium signalling by EMF was pointed as a possible candidate for activation of biochemical reactions [91]. On the other hand, it could also be possible that an alteration in cAMP contents and in intracellular communication might be associated with its effects on steroidogenesis [92].

Exposition of cryptorchid rats to intermittent EMF stimulation for 10 days induced Leydig cell proliferation, along with an increase in plasma testosterone and in testis weight [93]. ELF-MF exposure increased the HCG-stimulated capacity for testosterone production in mice Leydig cells *ex vivo* [89]. Since they may not be hormonally mediated, it was hypothesised that the possible biological effects on Leydig cells would involve direct cytotoxic effects [79]. In another study, a significant increase in the size and weight of testes was related to an increase in the amount of interstitial tissue. Elevated testosterone levels after a 10-week exposure to 50 Hz of 100 μT MF was also observed [94].

The percentage of cells in S phase significantly increased in mice exposed to 6.4 mT, with a subsequent decrease in sperm count and testis weight [95]. Mice exposed to 50 Hz for 52 days also evidenced increased total germ cell transformation, the spermatogonia cell population being significantly lower than in the corresponding control [69]. These results suggest that long-term exposure to ELF-MF has possible effects on the proliferation and differentiation of spermatogonia.

8. Hypothalamic–pituitary–gonadal axis and EMF exposure

Spermatogenesis is controlled by the hypothalamic–pituitary–gonadal (HPG) axis. In the model of hormonal deprivation, hypophysectomy-associated cell loss in the testis results from germ cell apoptosis [61]. Several studies report a suppression of melatonin production in the pineal gland of rats as a consequence of EMF exposure [29, 80, 96, 97]. Acute MF exposure can result in altered pineal gland and HPG function. One-time or intermittent exposure to 60-Hz MF at 0.1 mT is associated with a reduction in melatonin concentration. Daily intermittent exposures for 16 days increases prolactin levels and suppresses normal nocturnal rise in pineal melatonin production. However, at 42 days, there is no significant effect in melatonin or prolactin levels [29]. It suggests that pituitary–gonadal axis may adapt to chronic exposure to EMF. A reduced circulating concentration of melatonin may result in an increased prolactin release but the pituitary stimulated estrogen and testosterone levels by the gonads [98]. It is proposed that melatonin might be considered essential to both spermatogenesis and folliculogenesis [99].

Testosterone is crucial for the spermatogonia differentiation into round or elongated spermatids. Deprivation of gonadotropin or testosterone induces germ cell apoptosis [61]. In rats or mice, despite a decrease in sperm count or increase in frequency of germ cell apoptosis, exposure to ELF-MF did not affect serum testosterone level [26, 30, 34, 100].

In rats, follicle-stimulating hormone (FSH) increased within 1 week and luteinizing hormone (LH) increased in 4 weeks after exposure to 50 Hz of 5 mT without significant changes in the peripheral testosterone levels [30]. Since FSH levels affects spermatogenesis, an elevated FSH level suggests the disturbance of the spermatogenic process. In rats, the seminiferous tubules with the maximal response to FSH are also those presenting higher spontaneous apoptosis in spermatogonia [101]. It was also consistently observed that mature spermatogenic cells, such as spermatid and sperm, decrease in a relatively early phase of EMF exposure [25, 26]. It has been hypothesised that the early histologic findings in testis after ELF-MF exposure—a decrease in mature spermatogenic cells, such as round and elongated spermatid—may stimulate FSH secretion in the hypophysis by a positive feedback. This would be followed by a decrease in testosterone synthesis due to possibly damage in Leydig cells. Testosterone levels are supposed to be partially recovered by up-regulation of pituitary gonadotropin. On the other hand, 13 days of short-period exposure to MF in mice lead to a rise in testosterone levels [69]. *In vivo* experiments showed that 2 h per day exposure for 10 days of EMF stimulation results in Leydig cell proliferation, an increase in testosterone level and testis weight but a decrease in germ cell population [93]. In rats, ELF-MF exposure for 5 weeks, 3 h per day in 50

Hz of 0.8 T MF originated an increase in testosterone levels [102]. It seems that the testosterone transiently increases in the early phase of MF exposure due to Leydig cell proliferation followed by a decline of testosterone production related to damage in Leydig cells induced by MF exposure.

In mice, exposure to 50 Hz of 100 μT for 48 h originates a markedly increase in the steroidogenic capacity of Leydig cells without alterations in the serum testosterone level or in the testicular histology [79]. The results suggest that the effect of MF exposure on mouse Leydig cells and an alteration in testosterone level might not be mediated by gonadotropin. In Sprague-Dawley rats exposed to 50 Hz at field strength 25 μT, testosterone levels were significantly decreased only after 6 and 12 weeks of exposition, which was followed by a significant increase in the serum levels of LH after 18 weeks of exposure, the FSH levels remaining unaffected [31]. It was proposed that an MF-induced decline in testosterone level would stimulate the HPG axis with positive feedback.

Testosterone level in mice exposed to EMF for 16 weeks was not modified despite the marked increased germ cell apoptosis [26]. Differentiating spermatogenic cell apoptosis may occur in the early phase of ELF-MF exposure without alteration in the peripheral testosterone level [25, 27, 52], supporting the idea that the biological effect of MF exposure on germ cell apoptosis may not be hormonally mediated.

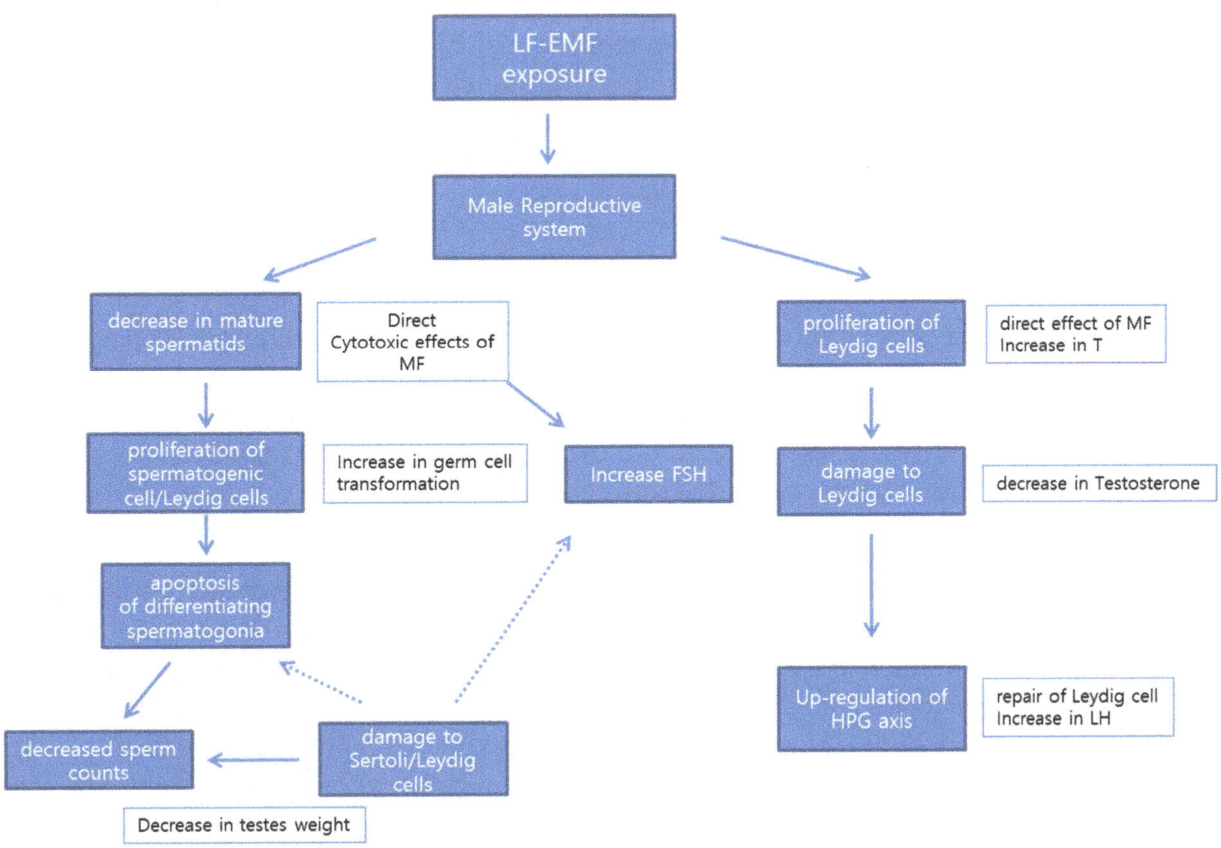

Figure 4. Diagram of the proposed mechanism for apoptosis in testis exposed to ELF-EMF.

Summarizing, cellular proliferation of Leydig cells may be induced at a relatively early phase after ELF-MF exposure. By consequence, testosterone production transiently increased, that is afterwards followed by a decrease in testosterone production due to disturbance of Leydig cell function, which in turn may stimulate LH production [19, 35]. However, the damaged Leydig cells induced by MF exposure may be repaired, in spite of germ cell death. The susceptibility to biological action of ELF-MF may differ according to the cell type [59] and Leydig cells may be more resistant to EMF exposure than germ cells (Fig. 4).

9. Summary and conclusion

A high-intensity MF with thermal effects is clearly teratogenic in laboratory and animal studies. A 50/60-Hz ELM-MF generated by power lines or an electric appliance is too weak, however, to induce DNA strand breakage. Nevertheless, studies regarding genotoxicity demonstrate that ELF-MF with a non-thermal exposure level is related to DNA damage in biologic systems. It is suggested that ELF-MF may act as a co-inducer to potentiate a suboptimal mutagenic signal.

Gonadal tissue involving germ cell differentiation and development is sensitive to external stressors, such as radiation, heat and exposure to chemical or physical agents. Putative harmful effects of MF on reproductive function have emerged as a major concern.

No consistent evidences exist on the adverse effects of ELF-MF on reproduction. To date, scientific literature reveals no significant risk on implantation or foetal development, but to minor skeletal alterations, an increased frequency of germ cell apoptosis in ELF-MF–exposed mice or rats has been consistently described. It was also suggested that a field of up to 20 mT does not increase a gross external or skeletal anomaly.

Accumulated evidence showed that ELF-MF exposure is cytogenetic to gonadal cells in rodents in a dose- and duration-dependent manner. Spermatogenic cells may be more sensitive to MF exposure than Leydig or Sertoli cells in testes. The pathway of testicular germ cell apoptosis following ELF-MF exposure is not well established. Based on relevant studies, germ cell apoptosis may be directly triggered by ELF-MF and not be hormonally mediated. However, chronic ELF-MF exposure disturbed the HPG axis. Testicular histology revealed alterations on Leydig cells producing testosterone and Sertoli cells supporting spermatogenesis in long-term MF-exposed mice.

Continuous exposure to ELF-MF in mice induces apoptosis of spermatogenic cells especially in mature spermatid, in a dose- and duration-dependent manner. For inducing apoptosis of testicular germ cells in mice, the minimum dose is represented by a field strength of 20 μT at continuous exposure to 60-Hz MF for 8 weeks or by a minimum duration of 16 weeks at continuous exposure to 100 μT, whereas intermittent exposure to ELF-MF, as low as 70 μT, induces genotoxic effects *in vitro* [7]. Furthermore, intermittent exposure to MF might be more harmful to living cells than continuous exposure.

The magnitude of the biologic effects depends on the density of the magnetic fields, the duration of exposure and the time of recovery. It may be a dynamic compensatory mechanism of spermatogenesis during germ cell apoptosis responding to exposure to ELF-EMF according to the intensity of EMF, the exposure pattern and duration. Long-term effects of chronic exposure have been excluded from the ICNIRP guidelines because of insufficient consistent scientific evidence to set a threshold for such possible adverse effects. According to the latest ICNIRP guideline for low-frequency MF (1 Hz to 100 kHz), the safety levels at short-term exposure are 1 mT for occupational exposure and 200 μT for the general population [11]. Those safety levels are two-fold higher in the field density than those in the previous guideline. Until now, safety levels for long-term exposure are not determined.

Adverse effects of ELM-MF are mainly from animal experiments. The experimental MF exposure condition may generally differ from those found in the environment in real life. The stochastic probability of the occurrence of the biologic effects to a certain ELM-MF exposure level should be determined for risk assessment. For a better understanding of the mechanism regulating biological effects of ELM-MF, molecular signalling pathway needs to be elucidated.

Author details

Sang-Kon Lee[1], Sungman Park[2] and Yoon-Won Kim[2,3*]

*Address all correspondence to: ywkim@hallym.ac.kr

1 Department of Urology, Sacred Heart Hospital, School of Medicine, Hallym University, Gangwon, Republic of Korea

2 Institute of Medical Science, School of Medicine, Hallym University, Chuncheon, Gangwon, Republic of Korea

3 Department of Microbiology, School of Medicine, Hallym University, Chuncheon, Gangwon, Republic of Korea

References

[1] ICNIRP (International Commission on Non-Ionizing Radiation protection). Guidelines for limiting exposure to time varying electric and magnetic fields (1 Hz to 100 kHz). Health Phys. 2010;99:818–836.

[2] Foster KR, Glaser R. Thermal mechanisms of interaction of radiofrequency energy with biological systems with relevance to exposure guidelines. Health Phys. 2007;92:609–620.

[3] Gaestel M. Biological monitoring of non-thermal effects of mobile phone radiation: recent approaches and challenges. Biol Rev. 2010;85:489–500.

[4] Wertheimer N, Leeper E. Possible effects of electric blankets and heated waterbeds on fetal development. Bioelectromagnetics. 1986;7:13–22.

[5] Savitz DA, Wachtel H, Barnes FA, John EM, Tvrdik JG. Case control study childhood cancer and exposure to 60 Hz magnetic fields. Am J Epidemiol. 1988;128:21–38.

[6] Feychting M, Forssen U, Floderus B. Occupational and residential magnetic field exposure and leukemia and central nervous system tumor. Epidemiology. 1997;8:384–389.

[7] Ivancsits S, Diem E, Pilger A, R°udiger HW, Jahn O. Induction of DNA strand breaks by intermittent exposure to extremely-low-frequency electromagnetic fields in human diploid fibroblasts. Mutat Res. 2002;519:1–13.

[8] Simko M. Induction of cell activation processes by low frequency electromagnetic fields. Scientific World J. 2004;4 Suppl 2:4–22.

[9] Kovacic P, Somanathan R. Electromagnetic fields: mechanism, cell signaling, other bioprocesses, toxicity, radical, antioxidants and beneficial effects. J Recept Signal Transduct Res. 2010;30:214–226.

[10] International Agency for Research on Cancer (IARC). Nonionizing radiation Part I: static and extremely low frequency (ELF) electric and magnetic fields. Monographs. 2002;80:1–395.

[11] World Health Organization (WHO). Extremely Low Frequency Fields. Environmental Health Criteria 238. Geneva: World Health Organization. 2007. p. 347–355.

[12] Juutilainen J, Kumlin T, Naarala J. Do extremely low frequency magnetic fields enhance the effects of environmental carcinogens? A meta-analysis of experimental studies. Int J Radiat Biol. 2006;82:1–12.

[13] Shaw GW, Croen LA. Human adverse reproductive outcomes and electromagnetic fields exposure: review of epidemiologic studies. Environ Health Perspect. 1993;101:107–119.

[14] Kowalczuk CI, Robbins L, Thomas JM, Butland BK, Saunders RD. Effects of prenatal exposure to 50 Hz magnetic fields on development in mice: I. Implantation rate and fetal development. Bioelectromagnetics. 1994;15:349–361.

[15] Mevissen M, Buntenkotter S, Loscher W. Effects of static and time-varying (50-Hz) magnetic fields on reproduction and fetal development in rats. Teratology. 1994;50:229–237.

[16] Ryan BM, Polen M, Gauger JR, Mallett E Jr, Kearns MB, Bryan TL, et al. Evaluation of the developmental toxicity of 60 Hz magnetic fields and harmonic frequencies in Sprague-Dawley rats. Radiat Res. 2000;153:637–641.

[17] Elbetieha A, Al-Akhras MA, Darmanl H. Long-term exposure of male and female mice to 50 Hz magnetic field: effects on fertility. Bioelectromagnetics. 2002;23:168–172.

[18] Chung MK, Kim JC, Myung SH, Lee DI. Developmental toxicity evaluation of ELF magnetic fields in Sprague Dawley rats. Bioelectromagnetics. 2003;24:231–240.

[19] Kim YW, Lee JS, Jang IE, Choi YH, Kang SH, Jung KC, et al. Effects of continuous exposure of 60 Hz magnetic fields on the mice through the third-generation. IEEK. 2001;28:220–233.

[20] Negishi T, Imai S, Itabashi M, Nishimura I, Sasano T. Studies of 50 Hz circularly polarized magnetic fields of up to 350 microT on reproduction and embryo-fetal development in rats: exposure during organogenesis or during preimplantation. Bioelectromagnetics. 2002;23:369–389.

[21] Huuskonen H, Juutilainen J, Komulainen H. Effects of low-frequency magnetic fields on fetal development in rats. Bioelectromagnetics. 1993;14:205–213.

[22] Huuskonen H, Juutilainen J, Julkunen A, M°aki-Paakkanen J, Komulainen H. Effects of low-frequency magnetic fields on fetal development in CBA/Ca mice. Bioelectromagnetics. 1998;19:477–485.

[23] Al-Akhras MA, Elbetieha A, Hasan MK, Al-Omari I, Darmani H, Albiss B. Effects of extremely low frequency magnetic field on fertility of adult male and female rats. Bioelectromagnetics. 2001;22:340–344.

[24] Hong R, Liu Y, Yu YM, Hu K, Weng EQ. Effects of extremely low frequency electromagnetic fields on male reproduction in mice. Zhonghua Lao Dong Wei Sheng Zhi YeBing Za Zhi. 2003;21:342–345.

[25] Lee JS, Ahn SS, Jung KC, Kim YW, Lee SK. Effects of 60 Hz electromagnetic field exposure on testicular germ cell apoptosis in mice. Asian J Androl. 2004;6:29–34.

[26] Kim YW, Kim HS, Lee JS, Kim YJ, Lee SK, Seo JN, et al. Effects of 60 Hz 14 μT magnetic field on the apoptosis of testicular germ cell in mice. Bioelectromagnetics. 2009;30:66–72.

[27] Kim HS, Park BJ, Jang HJ, Ipper NS, Kim SH, Kim YJ, et al. Continuous exposure to 60 Hz magnetic fields induces duration- and dose-dependent apoptosis of testicular germ cells. Bioelectromagnetics. 2014;35:100–107.

[28] Qi G, Zuo X, Zhou L, Aok E, Okamula A, Watanabe M, et al. Effects of extremely low-frequency electromagnetic fields (ELF-EMF) exposure on B6C3F1 mice. Environ Health Prev Med. 2015;20:287–293.

[29] Wilson BW, Matt KS, Morris JE, Sasser LB, Miller DL, Anderson LE. Effects of 60Hz Magnetic field exposure on the pineal and hypothalamic-pituitary-gonadal axis in the Siberian hamster (Phodopus sungorus). Bioelectromagnetics. 1999;20:224–232.

[30] Mostafa RM, Moustafa YM, Ali FM, Shafik A. Sex hormone status in male rats after exposure to 50 Hz, 5 mTesla magnetic field. Arch Androl. 2006;52:363–369.

[31] Al-Akhras MA, Darmani H, Elbetieha A. Influence of 50 Hz magnetic field on sex hormones and other fertility parameters of adult male rats. Bioelectromagnetics. 2006;27:127–131.

[32] Akdag MZ, Dasdag S, Uzunlar AK, Ulukaya E, Oral AY, Celik N, et al. Can safe and long-term exposure to extremely low frequency (50 Hz) magnetic fields affect apoptosis, reproduction, and oxidative stress? Int J Radiat Biol. 2013;89:1053–1060.

[33] Heredia-Rojas JA, Caballero-Hernandez DE, Rodriguez-de la Fuente AO, Ramos-Alfano G, Rodriguez-Flores LE. Lack alteration on meiotic chromosomes and morphological characteristics of male germ cells in mice exposed to a 60 Hz and 2.0 mT magnetic field. Bioelectromagnetics. 2004;25:63–68.

[34] Duan W, Liu C, Wu H, Chen C, Zhang T, Gao P, et al. Effects of exposure to extremely low frequency magnetic fields on spermatogenesis in adult rats. Bioelectromagnetics. 2014;35:58–69.

[35] Dewhirst MW, Viglianti BL, Lora-Michiels M, Hanson M, Hoopes PJ. Basic principles of thermal dosimetry and thermal thresholds for tissue damage from hyperthermia. Int J Hyperthermia. 2003;19:267–294.

[36] Blanco-Rodriguez J. A matter of death and life: the significance of germ cell death during spermatogenesis. Int J Androl. 1998;29:13–31.

[37] Verkasalo PK, Pukkala E, Hongisto MY, Valjus JE, Järvinen PJ, Heikkilä KV, et al. Risk of cancer in Finnish children living close to power lines. BMJ. 1993;307(6909): 895–899.

[38] Tomenius L. 50Hz electromagnetic environment and the incidence of childhood tumor in Stockholm County. Bioelectromagnetics. 1986;7:191–207.

[39] Schreiber GH, Swaen GM, Meijers JM, Slangen JJ, Sturmans F. Cancer mortality and residence near electricity transmission equipment: a retrospective cohort study. Int J Epidemiol. 1993;22:9–15.

[40] Juutilainen J, Matilainen P, Saarikoski S, Läärä E, Suonio S. Early pregnancy loss and exposure to 50-Hz magnetic fields. Bioelectromagnetics. 1993;14:229–236.

[41] Savitz DA, Ananth CV. Residential magnetic fields, wire codes, and pregnancy outcome. Bioelectromagnetics. 1994;15:271–273.

[42] Bracken MB, Belanger K, Hellenbrand K, Dlugosz L, Holford TR, McSharry JE, et al. Exposure to electromagnetic fields during pregnancy with emphasis on electrically heated beds: association with birth weight and intrauterine growth retardation. Epidemiology. 1995;6:263–270.

[43] Lee GM, Neutra RR, Hristova L, Yost M, Hiatt RA. The use of electric bed heaters and the risk of clinically recognized spontaneous abortion. Epidemiology. 2000;11:406–415.

[44] WHO. Environmental Health Criteria 238 – Extremely Low Frequency Fields. 2007. p. 1–543.

[45] Repacholi MH, Cardis E. Criteria for EMF health risk assessment. Radiol Prot Dosim. 1997;72:305–312.

[46] Juutilainen J. Developmental effects of electromagnetic fields. Bioelectromagnetics. 2005;26:S107–S115.

[47] Tenorio BM, Jimenez GC, Morais RN, Torres SM, Nogueira RA, Silva Junior VA. Testicular development evaluation in rats exposed to 60 Hz and 1 mT electromagnetic field. J Appl Toxicol. 2011;31:223–230.

[48] Ryan BM, Symanski RR, Pomeranz LE, Johnson TR, Gauger JR, McCormick DL. Multigeneration reproductive toxicity assessment of 60 Hz magnetic fields using a continuous breeding protocol in rats. Teratology. 1999;59:156–162.

[49] Sharpe RM. Possible role of elongated spermatids in control of stage-dependent changes in the diameter of the lumen of the rat seminiferous tubule. J Androl. 1989;10:304-311.

[50] De Vita R, Cavallo D, Raganella L, Eleuteri P, Grollino MG, Calugi A. Effects of 50 Hz magnetic fields on mouse spermatogenesis monitored by flow cytometric analysis. Bioelectromagnetics. 1995;16:330–334.

[51] Ramadan LA, Abd-Allah AR, Aly HA, Saad-el-Din AA. Testicular toxicity effects of magnetic field exposure and prophylactic role of coenzyme Q10 and L-carnitine in mice. Pharmacol Res. 2002;46:363–370.

[52] Tenorio BM, Ferreira Filho MB, Jimenez GC, de Morais RN, Peixoto CA, Nogueira Rde A, et al. Extremely low-frequency magnetic fields can impair spermatogenesis recovery after reversible testicular damage induced by heat. Electromagn Biol Med. 2014;33:139–146.

[53] Winker R, Ivancsits S, Pilger A, Adlkofer F, Rudiger HW. Chromosomal damage in human diploid fibroblasts by intermittent exposure to extremity low-frequency electromagnetic fields. Mutat Res. 2005;585:43–49.

[54] McCann J, Dietrich F, Rafferty C, Martin AO. A critical review of the genotoxic potential of electric and magnetic fields. Mutat Res. 1993;297:61–95.

[55] McCann J, Dietrich F, Rafferty C. The genotoxic potential of electric and magnetic fields: an update. Mutat Res. 1998;411:45–86.

[56] Simko M, Kriehuber R, Lange S. Micronucleus formation in human amnion cells after exposure to 50 Hz MF applied horizontally and vertically. Mutat Res. 1998;418:101–111.

[57] Nordenson I, Mild KH, Andersson G, Sandstrom M. Chromosomal aberrations in human amniotic cells after intermittent exposure to fifty hertz magnetic fields. Bioelectromagnetics. 1994;15:293–301.

[58] Ivancsits S, Diem E, Jahn O, Rüdiger HW. Intermittent extremely low frequency electromagnetic fields cause DNA damage in a dose-dependent way. Int Arch Occup Env Health. 2003;76:431–436.

[59] Cai L, Hales BF, Robaire B. Induction of apoptosis in the germ cells of adult male rats after exposure to cyclophosphamide. Biol Reprod. 1997;56:1490–1497.

[60] Morales E, Paster LM, Hom R, Zuasti A, Ferrer C, Calvo A, et al. Effect of aging on the proliferation and apoptosis of testicular germ cells in the Syrian hamster mesocricetus auratus. Reprod Fertil Dev. 2003;15:89–98.

[61] Sinha Hikim AP, Rajavashisth TB, Sinha Hikim I, Lue Y, Bonavera JJ, Leung A, et al. Significance of apoptosis in the temporal and stage-specific loss of germ cells in the adult rat after gonadotropin deprivation. Biol Reprod. 1997;57:1193–1201.

[62] Woolveridge I, de Boer-Brouwer M, Talor MF, Teerds KJ, Wu FCW, Morris ID. Apoptosis in the rat spermatogenic epithelium following androgen withdrawal: change in apoptosis-related genes. Biol Reprod. 1999;60:461–470.

[63] Allan DJ, Harmon BV, Roberts SA. Spermatogonial apoptosis has three morphologically recognizable phases and shows no circadian rhythm during normal spermatogenesis in the rat. Cell Prolif. 1992;25:241–250.

[64] Lue Y, Sinha Hikim AP, Wang C, Im M, Leung A, Swerdloff RS. Testicular heat exposure enhances the suppression of spermatogenesis by testosterone in rats: the "two-hit" approach to male contraceptive development. Endocrinology. 2000;141:1414–1424.

[65] Tananainen JS, Tilly JL, Vihko KK, Hsueh AJ. Hormonal control of apoptotic cell death in the testis: gonadotrophins and androgens as testicular survival factors. Mol Endocrinol. 1993;7:643–650.

[66] Hasegawa M, Wilson G, Russel LD, Meistrich ML. Radiation-induced cell death in the mouse testis: relationship to apoptosis. Radiat Res. 1997;147:457–467.

[67] Russel LD, Clermont Y. Degeneration of germ cells in normal, hypophysectomized and hormone treated hypophysectomized rats. Anat Rec. 1977;187:347–366.

[68] Johnsen SG. Testicular biopsy score count – a method for registration of spermatogenesis in human testes: normal values and results in 335 hypogonadal males. Hormones. 1970;1:2–25.

[69] Furuya H, Aikawa H, Hagino T, Yoshida T, Sakabe K. Flow cytometric analysis of the effects of 50 Hz magnetic fields on mouse spermatogenesis. Japanese J Hygiene. 1998;53:420–425.

[70] Johansen C. Electromagnetic fields and health effects-epidemiologic studies of cancer, diseases of the central nervous system and arrhythmia related heart disease. Scan J Work Environ Health. 2004;30:1–80.

[71] Consales C, Merla C, Marino C, Benassi B. Electromagnetic fields, oxidative stress, and neurodegeneration. Int J Cell Biol. 2012;2012:683897. doi:10.1155/2012/683897.

[72] Lai H, Singh NP. Melatonin and a spin-trap compound block radiofrequency electromagnetic radiation-induced DNA strand breaks in rat brain cells. Bioelectromagnetics. 1997;18:446–454.

[73] Ruiz-Gómez MJ, Martínez-Morillo M. Electromagnetic fields and the induction of DNA strand breaks. Electromagn Biol Med. 2009;28:201–214.

[74] Svedenstal BM, Johanson KJ, Mattsson MO, Paulsson LE. DNA damage, cell kinetics and ODC activities studied in CBA mice exposed to electromagnetic fields generated by transmission lines. In Vivo. 1999;13:507–513.

[75] Simko M, Droste S, Kriehuber R, Weiss DG. Stimulation of phagocytosis and free radical production in murine macrophages by 50 Hz electromagnetic field. Eur J Cell Biol. 2001;80:562–566.

[76] De Mattei M, Caruso A, Traina GC, Pezzetti F, Baroni T, Sollazzo V. Correlation between pulsed electromagnetic fields exposure time and cell proliferation increase in human osteosarcoma cell lines and human normal osteoblast cells in vitro. Bioelectromagnetics. 1999;20:177–182.

[77] Wolf FI, Torsello A, Tedesco B, Fasanella S, Boninsegna A, D'Ascenzo M, et al. 50-Hz extremely low frequency electromagnetic fields enhance cell proliferation and DNA damage: possible involvement of a redox mechanism. Biochimica et Biophysica Acta. 2005;1743:120–129.

[78] Wilson BW, Anderson LE, Hilton DI, Phillips RD. Chronic exposure to 60 Hz electric fields: effects on pineal function in the rat. Bioelectromagnetics. 1981;7:371–380.

[79] Forgács Z, Thuróczy G, Paksy K, Szabó LD. Effect of sinusoidal 50 Hz magnetic field on the testosterone production of mouse primary Leydig cell culture. Bioelectromagnetics. 1998;19:429–431.

[80] Bonhomme-Faivre L, Marion S, Bezie Y, Auclair H, Fredj G, Hommeau C. Study of human neurovegetative and hematologic effects of environmental low-frequency (50 Hz) electromagnetic fields produced by transformers. Arch Environ Health. 1998;53:87–92.

[81] Bersani F, Marinelli F, Ognibene A, Matteucci A, Cecchi S, Santi S, et al. Intramembrane protein distribution in cell cultures is affected by 50 Hz pulsed magnetic fields. Bioelectromagnetics. 1997;18:463–469.

[82] Scaiano JC, Mohtat N, Cozens FL, McLean J, Thansandote A. Application of the radical pair mechanism to free radicals in organized systems: can the effects of 60 Hz be predicted from studies under static fields? Bioelectromagnetics. 1994;15:549–554.

[83] Suthanthiran M, Anderson ME, Sharma VK, Meister A. Glutathione regulates activation-dependent DNA synthesis in highly purified normal human T lymphocytes stimulated via the CD2 and CD3 antigens. Proc Natl Acad Sci U S A. 1990;87:3343–3347.

[84] Simko M. Cell type specific redox status is responsible for diverse electromagnetic field effects. Curr Med Chem. 2007;14:1141–1152.

[85] Zmyslony M, Palus J, Jajte J, Dziubaltowska E, Rajkowska E. DNA damage in rat lymphocytes treated *in vitro* with iron cations and exposed to 7 mT magnetic fields (static or 50 Hz). Mutat Res. 2000;453;89–96.

[86] Lennon SV, Martin SJ, Cotter TG. Dose-dependent induction of apoptosis in human tumor cell lines by widely diverging stimuli. Cell Prolif. 1991;24:203–214.

[87] Rodemann HP, Bayreuther K, Pfleiderer G. The differentiation of normal and transformed human fibroblasts in vitro is influenced by electromagnetic fields. Exp Cell Res. 1989;182:610–621.

[88] Luben RA, Cain CD, Chen MCY, Rosen DM, Adeyt WR. Effects of electromagnetic stimuli on bone and bone cells *in vitro*: inhibition of responses to parathyroid hormone by low-energy low-frequency fields. Proc Natl Acad Sci U S A. 1982;79:4180–4184.

[89] Forgács Z, Somosy Z, Kubinyi G, Sinay H, Bakos J, Thuróczy G, et al. Effects of wholebody50-Hz magnetic field exposure on mouse Leydig cells. Scientific World J. 2004;4:83–90.

[90] Barier E, Dufy B, Veyret B. Stimulation of Ca2+ influx in rat pituitary cells under exposure to a 50 Hz magnetic field. Bioelectromagnetics. 1996;17:303–311.

[91] Liburdy RP. Calcium signaling in lymphocytes and ELF fields. Evidence for an electric field metric and a site of interaction involving the calcium ion channel. FEBS Lett. 1992;301:53–59.

[92] Schimmelpfeng J, Stein JC, Dertinger H. Action of 50 Hz magnetic fields on cycle AMP and intracellular communication in monolayers and spheroids of mammalian cells. Bioelectromagnetics. 1995;16:381–386.

[93] Ozguner IF, Dindar H, Yagmurlu A, Savas C, Gokcora IH, Yucesan S. The effect of electromagnetic field on undescended testis after orchiopexy. Int Urol Nephrol. 2002;33:87–93.

[94] Picazo ML, De Miguel MP, Leyton V, Franco P, Varela L, Paniagua R, et al. Long-term effects of ELF magnetic fields on the mouse testis and serum testosterone levels. Electro- and Magnetobiology. 1995;14:127–134.

[95] Hong R, Zhang Y, Liu Y, Weng EQ. Effects of extremely low frequency electromagnetic fields on DNA of testicular cells and sperm chromatin structure in mice. Zhonghua Lao Dong Wei Sheng Zhi Ye Bing Za Zhi. 2005;23:414–417.

[96] Kato M, Honma K, Shigemitsu T, Shiga Y. Effects of exposure to a circularly polarized 50 Hz magnetic field on plasma and pineal melatonin levels in rats. Bioelectromagnetics. 1993;14:97–106.

[97] Yellon SM. Acute 60 Hz magnetic field exposure effects on the melatonin rhythm in the pineal gland and circulation of the adult Djungarian hamster. J Pineal Res. 1994;16:136–144.

[98] Reiter RJ. Effects of light and stress on pineal function. In: Wilson BW, Stevens RG, Anderson LE, editors. Extremity Low Frequency Electromagnetic Fields: The Question of Cancer. Columbus, Ohio: Battelle; 1990. p. 87–107.

[99] Wójtowicz M, Jakiel G. Melatonin and its role in human reproduction. Ginekol Pol. 2002;73:1231–1237.

[100] Kato M, Honma K, Shigemitsu T, Shiga Y. Circularly polarized, sinusoidal, 50 Hz magnetic field exposure does not influence plasma testosterone levels of rats. Bioelectromagnetics. 1994;15:513–518.

[101] Parvinen M, Marana R, Robertson DM, Hansson V, Ritzen EM. Functional cycle of rat Sertoli cells: differential binding and action of follicle-stimulating hormone at various stages of the spermatogenic cycle. In: Steinberger A, Steinberger E, editors. Testicular Development, Structure and Function. Raven Press; 1980. p. 425–432.

[102] Sert C, Akdag MZ, Basham H, Buyukbayram H, Dasdag S. ELF magnetic field effects on fatty acid composition of phospholipid fraction and reproduction of rat's testis. Electromagn Biol Med. 2002;21:19–29.

[103] Ebrahimi-Kalan A, Roudkenar MH, Halabian R, Milan BP, Zarrintan A, Roushandeh AM. Down-regulation of Methionein 1 and 2 after exposure to electromagnetic field in mouse testis. Int Biomedical J. 2011;15:151–156.

[104] Amara S, Abdelmelek H, Garrel C, Guiraud P, Douki T, Ravanat JL, et al. Effects of subchronic exposure to static magnetic field on testicular function in rats. Arch Med Res. 2006;37:947–952.

3

Chromosome Abnormalities in Domestic Animals as Causes of Disorders of Sex Development or Impaired Fertility

Izabela Szczerbal and Marek Switonski

Abstract

Cytogenetic evaluation is an important step in the diagnosis of infertile or sterile animals. Moreover, the analysis of sex chromosomes is crucial for a proper classification of disorders of sex development (DSD). For many years, chromosome studies mainly addressed the livestock species, while recently, increasing interest in such analysis in companion animals is observed. New molecular and cytogenetic tools and techniques have given opportunities for a precise identification of chromosome mutations. Among them, fluorescence *in situ* hybridization, besides chromosome banding, has become a gold standard. In this chapter, recent advances in the cytogenetic diagnosis of cattle, pigs, horses, dogs and cats are presented.

Keywords: Aneuploidy, translocation, intersexuality, sex reversal, freemartinism, trisomy, monosomy, centric fusion, inversion, chimerism, mosaicism

1. Introduction

Genetic background of disorders of sex development (DSD) and impaired fertility (IF) accompanied with normal sex development are important issues in animal breeding. The identification of the causative gene and chromosome mutations is especially important in these species, in which artificial insemination is a common reproduction technology. In such species, a routine cytogenetic evaluation of sires is conducted in numerous countries [1–5]. Chromosome studies are also recommended in the case of phenotypically normal males and females with IF, since it can be caused by balanced chromosomal translocations or inversions. Finally, the cytogenetic analysis is considered to be the first diagnostic step for DSD animals with ambiguous genitalia or abnormal sexual behaviour. The importance of chromosome analysis in the case of DSD was highlighted in livestock species [6, 7] and companion animals [8, 9].

In this chapter, recent studies on chromosome abnormalities detected in domestic mammals, including livestock (cattle, pig and horse) and companion animals (dog and cat), are discussed. We focused on novel reports, which were not reviewed earlier [2, 7, 9].

1.1. Historical overview of domestic animal cytogenetics

Chromosome analyses of domestic mammals were started over 50 years ago. Their importance in animal breeding was pointed out by Ingemar Gustavsson, who identified a centric fusion (Robertsonian translocation) between chromosomes 1 and 29 in the Swedish Red and White cattle [10]. Five years later, an adverse influence on fertility of the carriers was documented [11]. Since this discovery, the development of veterinary cytogenetics of livestock species has been very rapid. It was found that 1/29 centric fusion is widespread in cattle breeds, with an exception of Holstein-Friesian cattle [12]. In horses, X monosomy appeared to be the most frequent chromosome abnormality [6], while in pigs reciprocal translocations were predominant [13].

The discovery of chromosome banding techniques was a crucial step for mammalian cytogenetics. The first banding technique was developed with the use of a fluorochrome, quinacrine mustard, on plant (broad bean, *Vicia faba*) chromosomes [14]. This type of banding was named Q-banding. Later, several other banding techniques were developed: G-banding, C-banding, R-banding, silver staining (Ag-I), etc. The use of these techniques in domestic animal cytogenetics was reviewed by Iannuzzi and Di Berardino [15].

Chromosome banding techniques facilitated the establishment of international standard karyotypes for domestic mammals. The standards are required for a precise diagnosis of chromosome aberrations. The first G-banded standards were published by Ford et al. [16] with the following species included in this report: the cattle, pig, horse, sheep, goat, cat and rabbit. In the following years, updated or novel standards were proposed for several species, including the pig [17], cattle, sheep, goat [18], horse [19], dog [20, 21] and cat [22].

Cytogenetic studies were revolutionized in the late 1980s by the use of a molecular technique named FISH (fluorescent *in situ* hybridization), based on the hybridization between labelled locus specific or whole chromosome painting probes with chromosomes on a microscopic slide [15]. This technique has become a gold standard in cytogenetic diagnosis, and it is commonly used for a precise and reliable detection of chromosomal structural abnormalities and aneuploidies. A wide range of molecular probes are commercially available for the human and the mouse, but not for domestic animal species. Therefore, a homemade preparation of the probes for these species is required. To obtain whole chromosome painting probes, very useful in cytogenetic diagnosis, microdissection of a given chromosome [23] or its flow sorting facilitate the amplification or molecular cloning of chromosome specific sequences [24, 25]. Also, locus-specific probes are useful in cytogenetic diagnosis, and such probes are derived from genome libraries, usually cloned in bacterial artificial chromosome (BAC) vectors (https://bacpac.chori.org/). An example of both types of probes, applied for the detection of X and Y chromosomes, is presented in Fig. 1.

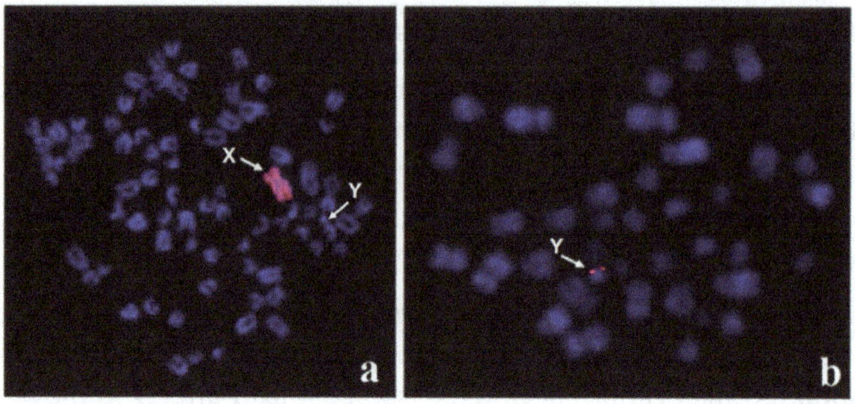

Figure 1. The use of FISH technique for the detection of sex chromosomes: (a) identification of the canine X chromosome (red), using a whole chromosome painting probe, (b) detection of the feline Y chromosome by the *SRY*-locus specific probe (red signal).

The latest achievements in developing molecular tools applied to cytogenetic diagnosis include SNP (single nucleotide polymorphism) microarrays and arrays for comparative genome hybridization (aCGH). They may be used for a precise identification of break points leading to structural chromosome rearrangements. Recently, a new molecular karyotyping approach, based on the FISH technique, performed on a single microscopic slide, with a set of subtelomere-specific probes for both the p arm and the q arm of each chromosome has been developed [26]. It is foreseen that this tool will soon be commercialized for cattle and pig chromosomes.

1.2. Chromosome abnormalities

The diploid chromosome number (2n) is one of the characteristic features of a species (Table 1). In some species (e.g. the red fox, raccoon dog), a variable number of chromosomes, due to the presence of B chromosomes, is observed. Chromosomes are classified into four morphological categories, based on the arm ratio, which is a quotient of the long arm length divided by the short arm length.

Species	2n	Morphology of sex chromosomes*	
		X	Y
Cattle	60	SM	M
Sheep	54	A	M
Goat	60	A	M
Pig	38	M	M
Horse	64	SM	A
Dog	78	SM	M
Cat	38	M	M

* M – metacentric, SM – submetacentric, A – acrocentric

Table 1. Chromosome number and sex chromosome morphology in selected domestic mammals

The effect of chromosome abnormalities, classified into three main categories (Fig. 2), is not uniform. Euploidies, autosomal aneuploidies and large unbalanced structural rearrangements usually cause early embryonic death. An exception is found in sex chromosome aneuploidies, which usually are not lethal but are responsible for disorders of sex development. Balanced chromosome rearrangements cause impaired fertility due to the mortality of embryos with an unbalanced chromosome complement.

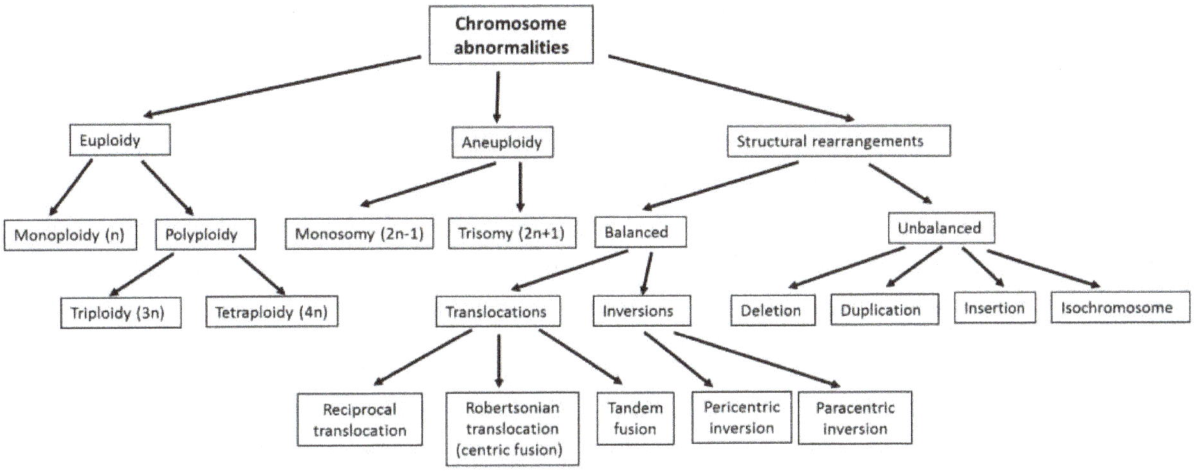

Figure 2. Classification of chromosome abnormalities.

An analysis of the sex chromosome complement is required to diagnose disorders of sex development. The use of conventional Giemsa-stained chromosomes facilitates the identification of X and Y chromosomes in the cattle and dog (Fig. 3), while in others, it is not an easy task. In the cat and horse, the morphology of both sex chromosomes is similar to some autosomes, and their identification requires the application of the banding or FISH techniques. In the pig, only the Y chromosome is easily recognized.

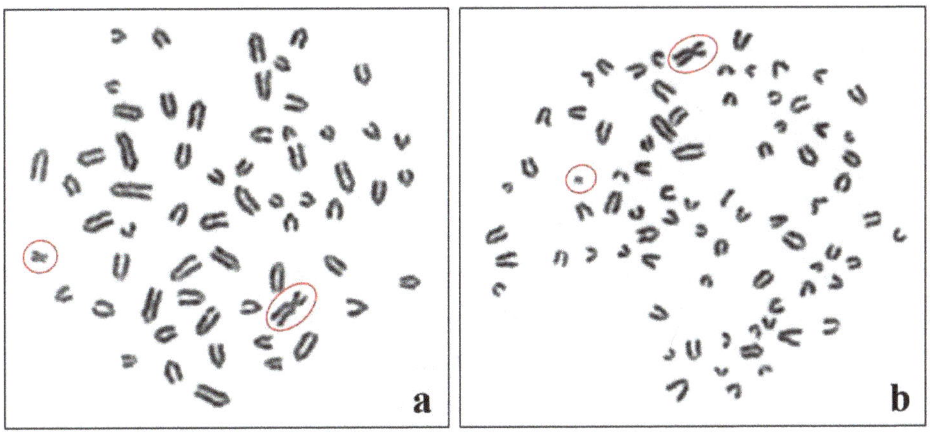

Figure 3. Giemsa-stained metaphase spreads: (a) bull, 2n = 60,XY, (b) male dog, 2n= 78,XY. Bi-armed sex chromosomes are indicated: X (large) and Y (small).

1.3. Classification of disorders of sex development

The sex determination of mammals is a very complex process, in which sex chromosomes and dozens of genes are involved [27]. A majority of these genes are localized on autosomes, and only a few of them reside on the sex chromosomes, including *SRY* (sex determining region Y), triggering male differentiation of the gonads and located on the Y chromosome, and *AR* (androgen receptor), involved in the development of Wolffian ducts and external male genitalia, having its locus on the X chromosome. Interestingly, crucial genes for ovary differentiation, i.e. *RSPO1* (R-spondin 1), *CTTNB1* (catenin beta 1), *WNT-4* (wingless-type mouse mammary tumor virus integration site family, member 4) and *FOXL2* (forkhead box L2), are localized on autosomes.

Due to the crucial role of sex chromosomes (especially the Y chromosome in male determination), mammalian DSDs are classified into three main categories, based on sex chromosome complement: (1) sex chromosome DSD, (2) XX DSD and (3) XY DSD. This classification was originally developed for human DSDs [28], later adopted for companion animals (the dog and cat)[8, 9], and it may also be applied to mammalian livestock.

The sex chromosome DSD comprises sex chromosome aneuploidies, structural rearrangements involving sex chromosomes and lymphocyte chimerism (XX/XY) observed in freemartins. In XX DSD animals, a major abnormality, diagnosed in several species (goat, pig, horse and dog), is manifested by the presence of testicles or ovotestes, as well as the uterus with oviducts and duct deferens in spite of the lack of the *SRY* gene. The XY DSD includes monogenic abnormalities, e.g. androgen insensitivity syndrome (AIS) and persistent Mullerian duct syndrome (PMDS), as well as syndromes in which the background is elusive or complex (e.g. cryptorchidism and hypospadias).

1.4. Meiosis in carriers of chromosome abnormalities

Meiosis is a crucial step of gametogenesis, and its alteration may cause impaired fertility. Unbalanced chromosome segregation at meiotic anaphase I or anaphase II may occur in animals with a normal chromosome set. As a consequence, among normal, haploid gametes a low proportion of unbalanced gametes can be present. The fertilization of such gametes leads to the development of embryos with an abnormal chromosome number. A majority of embryos with an autosomal aneuploidy die during early prenatal development, and decreased fertility of the carrier is noticed. In contrast, sex chromosome aneuploidies, especially trisomies, are not lethal.

The risk of unbalanced segregation is increased in gametogenesis of carriers of structural chromosome aberrations. It is well known that in gametogenesis of such carriers, chromosome pairing at meiotic prophase I is altered and, as a consequence, an unbalanced chromosome segregation at anaphase I may occur. Since the carriers produce normal and unbalanced gametes, the chromosome abnormality is inherited by 50% of the offspring.

Chromosome pairing at the pachytene substage of prophase I is usually studied by visualization of synaptonemal complexes under light or electron microscopies [29]. A postnatal study of chromosome pairing is possible only in males, because female meiosis starts during foetal life. Recently, the use of immunofluorescent visualization of proteins building synaptonemal complexes (e.g. SCP1 – synaptonemal protein 1) or engaged in crossing over (e.g. MLH1) has become more common [30].

Meiotic chromosome segregation at anaphase I can be studied postnatally in females and males. To study the segregation in females, *in vitro* maturation of oocytes is usually applied to obtain chromosome spreads at metaphase II. The spreads can be conventionally stained with Giemsa dye or analysed with the use of *in situ* hybridization approach, e.g. comparative genomic hybridization (CGH) [31]. The segregation analysis in spermatogenesis is usually carried out on sperms hybridized with fluorescently labelled locus-specific probes [32].

1.5. Mosaicism and chimerism

Chromosome studies of an individual sometimes reveal the presence of two or more cell lines with different chromosome complements. Such a situation may be classified as chromosome mosaicism or chimerism. Mosaicism is a condition when cell lines are derived from one embryo and are produced by unbalanced sister chromosome segregation during the mitotic anaphase. The incidence of mosaicism is frequent in animals with sex chromosome aneuploidies, e.g. a majority of infertile mares with X monosomy are mosaics with two cell lines – normal (64,XX) and monosomic (63,X). The presence of cell lines derived from different embryos is called chimerism. A classic example is the lymphocyte chimerism (with XX and XY chromosome sets) in heterosexual twins or multiplets, which developed placental anastomoses.

Mosaicism and chimerism are commonly diagnosed in animals with disorders of sex development. Thus, the identification of two or more cell lines is an important diagnostic issue. Since the incidence of the second cell line can be low, its detection requires analyses of a large number of metaphase spreads. For example, to exclude the presence of another cell line at the level of, e.g. 3% (with 95% confidence), 100 metaphase spreads should be analysed [33]. Moreover, the molecular analysis of genetic markers, e.g. microsatellites, is required to distinguish between lymphocyte and whole body chimerism [34].

2. Sex chromosome DSD

The first step of DSD diagnosis, following clinical investigation, should be cytogenetic analysis focused on the identification of sex chromosomes. This analysis facilitates the identification of DSDs caused by chromosome abnormalities: XX/XY chimerism present in freemartins or sex chromosome aneuploidies (X monosomy and trisomies: XXY, XYY and XXX). It is well known that sex chromosome DSDs usually cause infertility or sterility [7], but it should be pointed out that some of them are associated with an increased risk of gonad malignancy [35].

2.1. XX/XY lymphocyte chimerism

Freemartinism, observed in females originating from heterosexual pregnancies, is the most common and extensively studied DSD in cattle and other ruminants [36, 37]. The frequency of freemartinism depends on the incidence of twin (multiple) pregnancies associated with the development of placental anastomoses between embryos. The anastomoses in bovine twin pregnancies are developed very often (over 90%). In contrast, the incidence of twinning is rather low (2–5%); however, in recent years, an increasing trend is observed in dairy cattle breeds [38]. The cytogenetic detection of the XY cell line, originating from a male co-twin, can be difficult in a female co-twin, if it occurs with a low frequency. It is not exceptional for the frequency of the XY cell line to be below 5%. Therefore, the use of molecular detection of genes localized on the Y chromosome (*SRY, ZFY* or *AMELY*), along with cytogenetic analysis, is recommended [39]. Chimerism was also observed in heifers originating from a single birth, the so-called singletons [38]. The authors analysed cytogenetically and molecularly 12 heifer singletons with underdeveloped internal genitalia and found 7 with lymphocyte XX/XY chimerism among them. It indicated that during the foetal life, a male co-twin embryo died after the development of placental anastomoses. This report showed that the incidence of bovine freemartins is underestimated. In the bovine male co-twin, chimerism is detected, but these males are normally developed with no sign of DSD. Among other ruminants, freemartinism was quite extensively studied in sheep, and the incidence of this DSD, diagnosed by a chromosome study, was around 5% XX/XY [40, 41]. Studies in goats also indicate that the frequency of freemartins is similar to that in sheep [42].

The lymphocyte XX/XY chimerism was also reported in non-ruminant females. This condition was identified in horses [43, 44], pigs [45] and dogs [34]. A normal phenotype was reported in chimeric mares, but they were not clinically examined, thus the conclusion concerning their sex development and fertility is still a matter of controversy. In contrast, in chimeric female dogs, the sex development is usually altered [9], and in some cases, ambiguous external genitalia were observed [34, 46].

Sex development and fertility of XX/XY chimeric males are usually unaffected [7]. Such males are diagnosed during the routine cytogenetic examination of sires, mainly bulls and boars, used in artificial insemination (AI). Recently, an extensive study on fertility of seven chimeric AI boars was carried out by Barsac et al. [45]. The authors concluded that this condition has a non-significant effect on their reproductive performance. An unusual case of lymphocyte chimerism was reported by Demyda-Peyras et al. [47], who identified 64,XY (approx. 96%) and 63,X (approx. 4%) cell lines in a phenotypically normal colt.

2.2. Sex chromosome aneuploidies

The most deleterious effects on sex development, leading to sterility, are caused by X monosomy and XXY trisomy (Table 2). The incidence of sex chromosome aneuploidies is not uniform across species. The X monosomy is most common in mares, rather rare in dogs and cats, and very rare in cattle and pigs. In turn, the XXY trisomy is diagnosed rather rarely, with an exception of tortoiseshell male cats. Two other sex chromosome aneuploidies (XXX and XYY) were reported incidentally. These rough estimations should be taken with caution, because mainly males are cytogenetically analysed, and their numbers are not comparable across species – very large in cattle, large in pigs and rather small in dogs and cats. In contrast, in horses, the number of studied mares is much larger than that of stallions.

Mutation	The most common clinical features	Effects on fertility	Decreasing incidence in different species
X monosomy	Usually normal external genitalia; abnormal oestrus or no clinical signs of oestrus; small or hypoplastic ovaries	Sterility	Horse > cat and dog > cattle and pig
XXX trisomy	Normal external genitalia; usually lack of clinical signs, sometimes irregular oestrus	Infertility	Dog, horse
XXY trisomy	Small and hypoplastic testes; abnormal spermatogenesis (oligospermia or azoospermia)	Sterility	Cat and cattle > dog and horse > pig
XYY trisomy	Usually normal phenotype; small testes	Infertility	Cattle

Table 2. Aneuploidies and their impact on fertility

The incidence of X monosomy was evaluated in a screening study of young and phenotypically normal horses [48]. The authors analysed karyotypes of 500 horses, including 262 mares and 228 stallions. Among them, eight mares (3.1%) with X monosomy were identified, while all males had a normal karyotype. This study strongly confirmed that X monosomy is the most important cause of equine DSD. In cattle, this aneuploidy is very rare, and a second case of this abnormality was recently described in a 3-year-old Longhorn heifer with normal external genitalia, an underdeveloped uterus and gonads unidentifiable by rectal ultrasonography [49]. Several cases of X monosomy were diagnosed in dogs and cats [9], and the most recent report concerns an unusual case of feline X monosomy – Fig. 4 [50]. This animal presented a virilized phenotype, manifested by a rudimentary penis and scrotum-like structure. Interestingly, the histological examination of the gonad revealed the presence of corpora lutea and primordial follicles.

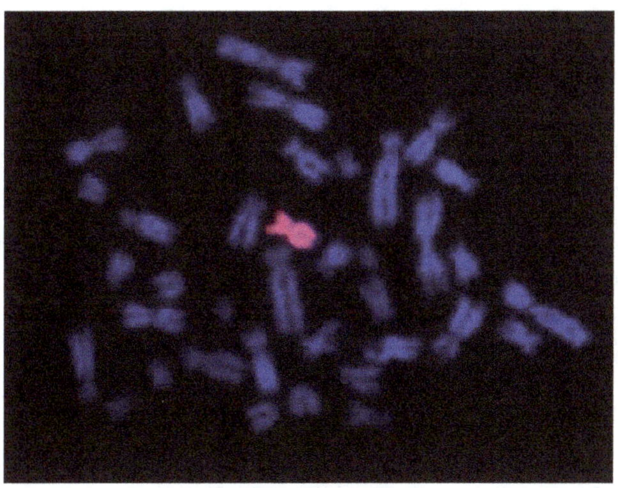

Figure 4. Metaphase spread with X monosomy (37,X) in a cat with a virilised phenotype. Application of whole X chromosome painting probe facilitated identification of a single X chromosome. For details, see Szczerbal et al. [50].

The largest number of XXY trisomy cases was reported in tortoiseshell male cats, and recently such a case was described [51]. In cattle, more than 20 cases of this abnormality were diagnosed in infertile bulls [2, 3]. However, it was rarely diagnosed in pigs, horses and dogs. A recent report concerned a mosaic 39,XXY/38,XY azoospermic young boar, recruited for artificial insemination [52]. The authors analysed the chromosome constitution in lymphocytes and fibroblasts culture derived from the gonads. In both tissues, a prevalence of the trisomic cell line (approx. 95%) was observed. It is known that XXY trisomy in men is associated with an elevated risk of malignancy. Such a case was also reported in a XXY trisomic dog, suffering from testicular cancer [53].

Two other common sex chromosome aneuploidies (XXX and XYY) were rarely reported in domestic mammals. Several such cases were described in horses, cattle and dogs. The carriers usually are infertile, e.g. an XXX mare [54] and an XXX female dog [55].

3. Decreased fertility in carriers of balanced chromosome aberrations

Balanced structural rearrangements (centric fusions, tandem fusions, reciprocal translocations and inversions) are inheritable mutations, which affect the fertility of carriers, without a visible alteration of sex development. The most pronounced decrease of fertility is associated with reciprocal translocations, including sterility of males carrying the X-autosome or Y-autosome rearrangements. In contrast, the fertility of carriers of other balanced aberrations (centric fusions, tandem fusions and inversions) is only slightly decreased.

3.1. Centric fusions (Robertsonian translocations)

Centric fusions are the most extensively studied chromosome aberrations in domestic mammals due to the discovery of a wide distribution of the 1/29 centric fusion in cattle. It was recently shown with the use of advanced molecular techniques that the 1/29 mutation is not a simple centric fusion, but it is a complex rearrangement including deletion, inversion and centric fusion [56].

The 1/29 Robertsonian translocation was identified in numerous cattle breeds, but it seems to be absent in Holstein-Friesian cattle, the most important dairy breed. In some beef breeds, its incidence exceeds 10% [2], while in some local breeds, e.g. in Portugal, the frequency of the carriers is above 50% [57]. Also, other types of centric fusions, involving other autosomes, were described in cattle [58]. In sheep, goats, pigs and dogs, centric fusions were rather rarely identified, while in horses and cats, there are no such reports.

The fertility of centric fusion carriers is slightly altered due to the formation of a trivalent [59] and its abnormal segregation at meiotic anaphase I [60], leading to the formation of aneuploid embryos, which die at early embryonic stages of development. In the case of bull carriers, the non-return ratio (NRR) is decreased by approx. 5%, but the aberration is inherited by 50% of the offspring. Interestingly, in pig carriers of the 13/17 Robertsonian translocation, the rate of unbalanced gametes due to the abnormal segregation of the trivalent was low in sperms (approx. 3%) and much higher in secondary oocytes (approx. 19%) [61].

3.2. Reciprocal translocations

Reciprocal translocations are the most common chromosome rearrangements in pigs. It was estimated that an overall frequency of these rearrangements among 7,700 young boars cytogenetically evaluated in France was 0.47% [1]. Altogether, over 130 different translocations were identified in this species by 2012 [62], and novel cases were published in recent years (Table 3, Fig. 5).

Breed	Description of the translocation	Effect on fertility	Reference
Synthetic line (Large White × Pietrain)	t(1;6)(q17;p11)	Smaller litter size	[70]
Large White	t(1;2;7)(q11q17;p16;q13) complex rearrangements, including translocations, insertions and inversions	Low litter size	[71]
Polish Large White	t(6;16)(p13;q23)	Reduced litter size (20% decrease)	[67]
Not described	t(Y;1)(q1:q16)	Azoospermia	[72]
Not described*	t(4;7)(p+;q-)	Not described	[73]
Not described*	t(2;8)(q−;q+)	Not described	[73]
Not described**	t(1;5)	Not described	[74]
Not described**	t(3;4)	Reduced litter size (24%)	[74]
Not described**	t(8;13)	Reduced litter size (24%)	[74]
Not described**	t(7;15)	Reduced litter size (38%)	[74]

*derived from intracytoplasmatic sperm injection (ICSI) and embryo transfer (ET)

** identified among 300 young boars from Duroc, Landrace, Pietrain and Yorkshire breeds

Table 3. Reciprocal translocations in pigs reported in 2011–2015 (not included in the review by Raudsepp and Chowdhary [62]).

The incidence of reciprocal translocations in cattle, the most extensively cytogenetically studied species, is much lower. In the literature, there are reports on 21 different cases of reciprocal translocations, including two cases of the X-autosome and another two cases of Y-autosome translocations [63]. A low incidence of the translocation carriers seems to be a result of the limited resolution of the commonly used Giemsa staining This approach is sufficient to detect aneuploidies and centric fusions but does not facilitate the detection of a majority of reciprocal translocations. Interesting statistical and simulation estimations were performed by de Lorenzi et al. [64], who suggested that approx. only 16% of reciprocal translocations can be detected using Giemsa staining. Taking into consideration a real number of the identified

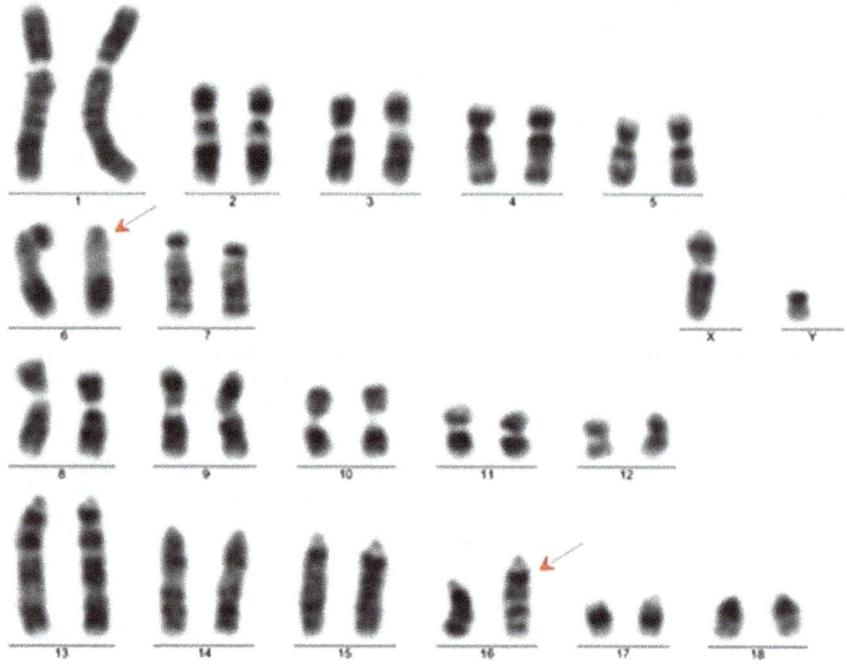

Figure 5. G-banded karyotype of a boar carrier of reciprocal translocation – t(6;16)(p13;q23). For details, see Kociucka et al. [67].

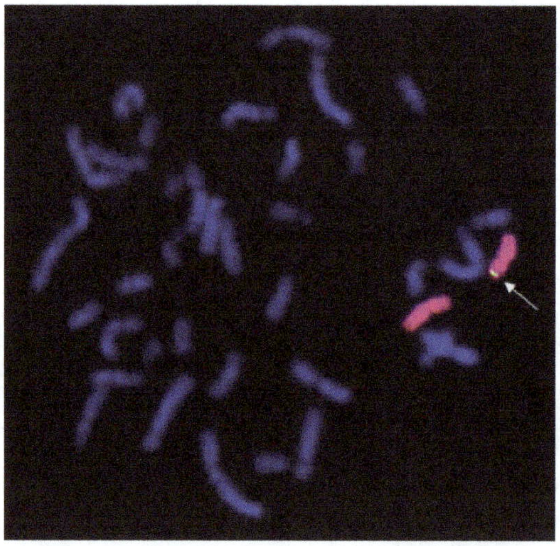

Figure 6. Metaphase spread with a translocation between X and Y chromosomes – t(X;Y)(p22;p12), in a cat with disorder of sex development. FISH technique facilitated the localization of *SRY* gene (arrow) on X chromosome. For details, see Szczerbal et al. [66].

translocations and this estimation, they suggested that the frequency of this chromosome is <0.14%. It would be three-fold less than in pigs.

In other domestic mammals, reciprocal translocations were rarely identified. In mares suffering from repeated early embryonic loss, three translocations, including two reciprocal, were identified between autosomes: t(1;21), t(16;22) and t(4;13) [65]. An interesting case was reported by Szczerbal et al. [66], who for the first time in domestic mammals diagnosed a translocation between the X and Y chromosomes, t(X;Y)(p22;p12), in a female cat (38,XX) with a disorder of sex development, manifested by the presence of a rudimentary penis (Fig. 6).

Similarly to centric fusions, decreased fertility is caused by the abnormal segregation of a tetravalent during anaphase I. However, the proportion of chromosomally unbalanced gametes is much higher in carriers of reciprocal translocations and varies between 20 and 50% [67]. As a consequence, a large proportion of the embryos die at an early embryonic stage, and thus the decrease of fertility is much more pronounced. In the case of reciprocal translocations involving sex chromosomes, the carriers are usually infertile, with an exception of the female carriers, which are infertile (Table 4).

Type of mutation	Most common features	Reduction of fertility	Incidence, according to the number of reported cases
Robertsonian translocation	Normal phenotype; normal libido; normal semen picture; lower non-return rate (cattle)	5%	Cattle – frequently in local breeds and some beef breeds Pig, dog – rare
Reciprocal translocation autosome/autosome	Normal phenotype; decreased litter size (pig), decreased non-return rate (cattle)	20–50%	Pig – frequently Cattle, horse – rare
Reciprocal translocation X/autosome	Females: normal phenotype; decreased litter size (pig)	20–50%	Pig, cattle, dog, horse – rare
	Males: abnormal spermatogenesis (oligospermia or azoospermia)	Sterility	
Reciprocal translocation Y/ autosome	Males: abnormal spermatogenesis (oligospermia or azoospermia)	Sterility	Pig, cattle – rare
Translocation X/Y	Disorder of sex development, including abnormal external genitalia	Sterility	Only one case in a cat has been diagnosed to date

Table 4. An overview of chromosome translocations in domestic mammals: phenotype, fertility and incidence

3.3. Other balanced chromosome rearrangements

Chromosome inversions were diagnosed in cattle and pigs, but their impact on fertility of the carriers is poorly recognized. Among different cases of inversions, one is of special interest. It was a pericentric inversion of the pig chromosome 4 – inv(4)(p1.4;q2.3), which was diagnosed in several AI boars in France. Analyses of sperms and oocytes of the male and female carriers showed the incidence of unbalanced gametes to be approx. 4% [68].

Tandem fusions were very rarely diagnosed in domestic mammals. In recent years, such a rearrangement was diagnosed in a young boar with a normal phenotype and semen characteristics [69]. In the tandem fusion, two acrocentrics were involved: t(14;17)(q29;q10). The carrier was not accepted for use in artificial insemination.

4. Conclusions

The long history of cytogenetic evaluation of domestic mammals, mainly applied to males used in artificial insemination, showed that heritable, balanced chromosome abnormalities do not cause visible phenotypic effects but are responsible for impaired fertility. Therefore, young males considered as candidates for use in artificial insemination stations are commonly cytogenetically evaluated. It is important to point out that the first step for the classification of sex development disorders is cytogenetic analysis. Eventually, the use of chromosome studies plays a crucial role in searching for causes of infertility of phenotypically normal animals, as well as individuals with an abnormal sex development.

Acknowledgements

This study was financed by the National Science Centre in Poland - grant 2012/05/B/NZ9/00907

Author details

Izabela Szczerbal and Marek Switonski*

*Address all correspondence to: switonsk@up.poznan.pl

Department of Genetics and Animal Breeding, Poznan University of Life Sciences, Poznan, Poland

References

[1] Ducos A, Berland HM, Bonnet N, Calgaro A, Billoux S, Mary N, Garnier-Bonnet A, Darré R, Pinton A. (2007). Chromosomal control of pig populations in France: 2002-2006 survey. Genet Sel Evol. 39: 583–597.

[2] Ducos A, Revay T, Kovacs A, Hidas A, Pinton A, Bonnet-Garnier A, Molteni L, Slota E, Switonski M, Arruga MV, van Haeringen WA, Nicolae I, Chaves R, Guedes-Pinto H, Andersson M, Iannuzzi L. (2008). Cytogenetic screening of livestock populations in Europe: an overview. Cytogenet Genome Res. 120: 26–41.

[3] Citek J, Rubes J, Hajkova J. (2009). Robertsonian translocations, chimerism, and aneuploidy in cattle. J Dairy Sci. 92: 3481–3483.

[4] Krumrych W (2009). Overview of the cytogenetic evaluation of young bulls bred in the north-western part of Poland in 1999-2008. Bull Vet Inst Pulawy. 53: 669–672.

[5] Quach TA, Villagómez DA, Coppola G, Pinton A, Hart EJ, Reyes ER, Basrur PK, King WA. (2009). A cytogenetic study of breeding boars in Canada. Cytogenet Genome Res. 126: 271–280.

[6] Lear TL, McGee RB. (2012). Disorders of sexual development in the domestic horse, *Equus caballus*. Sex Dev. 6: 61–71.

[7] Favetta LA, Villagómez DA, Iannuzzi L, Di Meo G, Webb A, Crain S, King WA. (2012). Disorders of sexual development and abnormal early development in domestic food-producing mammals: the role of chromosome abnormalities, environment and stress factors. Sex Dev. 6: 18–32.

[8] Poth T, Breuer W, Walter B, Hecht W, Hermanns W. (2010). Disorders of sex development in the dog-adoption of a new nomenclature and reclassification of reported cases. Anim Reprod Sci. 121: 197–207.

[9] Meyers-Wallen VN. (2012). Gonadal and sex differentiation abnormalities of dogs and cats. Sex Dev. 6: 46–60.

[10] Gustavsson I, Rockoborn G. (1964). Chromosome abnormality in three cases of lymphatic leukaemia in cattle. Nature 203: 990.

[11] Gustavsson I. (1969). Cytogenetics, distribution and phenotypic effects of a translocation in Swedish cattle. Hereditas. 63: 68–169.

[12] Switonski M. (2014). 50 years of studies on bovine 1/29 Robertsonian translocation - from Giemsa staining to genomic analysis. Chromosome Res. 22: 395.

[13] Long SE. (1991). Reciprocal translocations in the pig (*Sus scrofa*): a review. Vet Rec. 128: 275–278.

[14] Caspersson T, Zech L, Modest EJ, Foley GE, Wagh U, Simonsson E. (1969). Chemical differentiation with fluorescent alkylating agents in *Vicia faba* metaphase chromosomes. Exp Cell Res. 58: 128–140.

[15] Iannuzzi L, Di Berardino D. (2008). Tools of the trade: diagnostics and research in domestic animal cytogenetics. J Appl Genet. 49: 357–366.

[16] Ford CE, Pollock DL, Gustavsson I. (1980). Proceedings of the First International Conference for the standardisation of banded karyotypes of domestic animals. University of Reading Reading, England. 2nd–6th August 1976. Hereditas. 92: 145–162.

[17] Gustavsson I. (1988). Standard karyotype of the domestic pig. Committee for the Standardized Karyotype of the Domestic Pig. Hereditas. 109: 151–157.

[18] Cribiu EP, Di Berardino D, Di Meo GP, Eggen A, Gallagher DS, Gustavsson I, Hayes H, Iannuzzi L, Popescu CP, Rubes J, Schmutz S, Stranzinger G, Vaiman A, Womack J. (2001). International System for Chromosome Nomenclature of Domestic Bovids (ISCNDB 2000). Cytogenet Cell Genet. 92: 283–299.

[19] Bowling AT, Breen M, Chowdhary BP, Hirota K, Lear T, Millon LV, Ponce de Leon FA, Raudsepp T, Stranzinger G. (1997). International system for cytogenetic nomenclature of the domestic horse. Report of the Third International Committee for the Standardization of the domestic horse karyotype, Davis, CA, USA, 1996. Chromosome Res. 5: 433–443.

[20] Switonski M, Reimann N, Bosma AA, Long S, Bartnitzke S, Pieńkowska A, Moreno-Milan MM, Fischer P. (1996). Report on the progress of standardization of the G-banded canine (*Canis familiaris*) karyotype. Committee for the Standardized Karyotype of the Dog (*Canis familiaris*). Chromosome Res. 4: 306–309.

[21] Breen M, Bullerdiek J, Langford CF. (1999). The DAPI banded karyotype of the domestic dog (*Canis familiaris*) generated using chromosome-specific paint probes. Chromosome Res. 7: 401–406.

[22] Cho KW, Youn HY, Watari T, Tsujimoto H, Hasegawa A, Satoh H. (1997). A proposed nomenclature of the domestic cat karyotype. Cytogenet Cell Genet. 79: 71–78.

[23] Bugno M, Słota E, Pieńkowska-Schelling A, Schelling C. (2009). Identification of chromosome abnormalities in the horse using a panel of chromosome-specific painting probes generated by microdissection. Acta Vet Hung. 57: 369–381.

[24] Graphodatsky A, Ferguson-Smith MA, Stanyon R. (2012). A short introduction to cytogenetic studies in mammals with reference to the present volume. Cytogenet Genome Res. 137: 83–96.

[25] Rubes J, Musilova P, Mastromonaco GF. (2008). Cytogenetics of wild and captive bred non-domestic animals. Cytogenet Genome Res. 120: 61–68.

[26] O'Connor RE, Fonseka G, D.K. Griffin DK (2014). Identification of chromosomal translocations in pigs using FISH with subtelomeric probes and the development of a novel screening tool for their application. Chromosome Res. 22: 396–397.

[27] Eggers S, Sinclair A. (2012). Mammalian sex determination—insights from humans and mice. Chromosome Res. 20: 215–238.

[28] Pasterski V, Prentice P, Hughes IA (2010). Impact of the consensus statement and the new DSD classification system. Best Practice Res Clin Endocrinol Metab. 24: 187–195.

[29] Switonski M, Stranzinger G. (1998). Studies of synaptonemal complexes in farm mammals - a review. J Hered. 89: 473–480.

[30] Mary N, Barasc H, Ferchaud S, Billon Y, Meslier F, Robelin D, Calgaro A, Loustau-Dudez AM, Bonnet N, Yerle M, Acloque H, Ducos A, Pinton A. (2014). Meiotic re-

combination analyses of individual chromosomes in male domestic pigs (*Sus scrofa domestica*). PLoS One. 9(6): e99123.

[31] Hornak M, Jeseta M, Musilova P, Pavlok A, Kubelka M, Motlik J, Rubes J, Anger M. (2011). Frequency of aneuploidy related to age in porcine oocytes. PLoS One. 6(4): e18892.

[32] Massip K, Berland H, Bonnet N, Calgaro A, Billoux S, Baquié V, Mary N, Bonnet-Garnier A, Ducos A, Yerle M, Pinton A. (2008). Study of inter- and intra-individual variation of meiotic segregation patterns in t(3;15)(q27;q13) boars. Theriogenology. 70: 655–661.

[33] Hook EB. (1977). Exclusion of chromosomal mosaicism: tables of 90%, 95% and 99% confidence limits and comments on use. Am J Hum Genet. 29: 94–97.

[34] Szczerbal I, Nowacka-Woszuk J, Nizanski W, Salamon S, Ochota M, Dzimira S, Atamaniuk W, Switonski M. (2014a). A case of leucocyte chimerism (78,XX/78,XY) in a dog with a disorder of sexual development. Reprod Domest Anim. 49: e31–e34.

[35] Dzimira S, Nizanski W, Ochota M, Madej JA. (2015). Histopathological pattern of gonads in cases of sex abnormalities in dogs: an attempt of morphological evaluation involving potential for neoplasia. Pathol Res Pract. 211: 772–775.

[36] Peretti V, Ciotola F, Albarella S, Paciello O, Dario C, Barbieri V, Iannuzzi L. (2008). XX/XY chimerism in cattle: clinical and cytogenetic studies. Sex Dev. 2: 24–30.

[37] Esteves A, Bage R, Payan-Carreira R. (2012). Freemartinism in cattle. In: Ruminants – Anatomy, Behavior and Diseases (ed. RE Mendes), Nova Science Publishers Inc., New York, USA, 99–120.

[38] Szczerbal I, Kociucka B, Nowacka-Woszuk J, Lach L, Jaskowski JM, Switonski M. (2014b). A high incidence of leukocyte chimerism (60,XX/60,XY) in single born heifers culled due to underdevelopment of internal reproductive tracts. Czech J Anim Sci. 59: 445–449.

[39] Nowacka J, Switonski M, Mackowski M, Slota E, Radko A, Zabek T, Urbaniak K. (2004). The use of cytogenetic and molecular techniques reveals ambiguity of the freemartinism diagnosis in cattle. Czech J Anim Sci. 49: 239–243.

[40] Szatkowska I, Switonski M (1996). Evidence on hereditary occurrence of placental anastomoses in heterosexual twins in sheep. Hereditas. 124: 107–110.

[41] Brace MD, Peters O, Menzies P, King WA, Nino-Soto MI. (2008). Sex chromosome chimerism and the freemartin syndrome in Rideau Arcott sheep. Cytogenet Genome Res. 120: 132–139.

[42] Szatkowska I, Zych S Udala A, Dybus A, Blaszczyk P, Sysa P, Dabrowski T. (2004). Freemartinism: three cases in goats. Acta Vet Brno. 73: 375–378.

[43] Juras R, Raudsepp T, Das PJ, Conant E, Cothran EG (2010). XX/XY blood lymphocyte chimerism in heterosexual dizygotic twins from an American Bashkir curly horse. Case Report. J Equine Vet Sci. 30: 575–580.

[44] Demyda-Peyrás S, Membrillo A, Bugno-Poniewierska M, Pawlina K, Anaya G, Moreno-Millán M. (2013). The use of molecular and cytogenetic methods as a valuable tool in the detection of chromosomal abnormalities in horses: a case of sex chromosome chimerism in a Spanish purebred colt. Cytogenet Genome Res. 141: 277–283.

[45] Barasc H, Ferchaud S, Mary N, Cucchi MA, Lucena AN, Letron IR, Calgaro ABonnet N, Dudez AM, Yerle M, Ducos A, Pinton A. (2014). Cytogenetic analysis of somatic and germinal cells from 38,XX/38,XY phenotypically normal boars. Theriogenology. 81: 368–372.e1.

[46] Beccaglia M, Luvoni GC. (2012). Prediction of parturition in dogs and cats: accuracy at different gestational ages. Reprod Domest Anim. 47 Suppl 6: 194–196.

[47] Demyda-Peyrás S, Anaya G, Bugno-Poniewierska M, Pawlina K, Membrillo A, Valera M, Moreno-Millán M. (2014). The use of a novel combination of diagnostic molecular and cytogenetic approaches in horses with sexual karyotype abnormalities: a rare case with an abnormal cellular chimerism. Theriogenology. 81: 1116-1122.

[48] Bugno M, Słota E, Kościelny M. (2007). Karyotype evaluation among young horse populations in Poland. Schweiz Arch Tierheilkd. 149: 227–232.

[49] Romano JE, Raussdepp T, Mulon PY, Villadóniga GB. (2015). Non-mosaic monosomy 59,X in cattle: a case report. Anim Reprod Sci. 156: 83–90.

[50] Szczerbal I, Nizanski W, Dzimira S, Nowacka-Woszuk J, Ochota M, Switonski M. (2015a). X monosomy in a virilized female cat. Reprod Domest Anim. 50: 344–348.

[51] Pedersen AS, Berg LC, Almstrup K, Thomsen PD. (2014). A tortoiseshell male cat: chromosome analysis and histologic examination of the testis. Cytogenet Genome Res. 142: 107–111.

[52] Pinton A, Barasc H, Raymond Letron I, Bordedebat M, Mary N, Massip K, Bonnet N, Calgaro A, Dudez AM, Feve K, Riquet J, Yerle M, Ducos A. (2011). Meiotic studies of a 38,XY/39,XXY mosaic boar. Cytogenet Genome Res. 133: 202–208.

[53] Reimann-Berg N, Murua Escobar H, Nolte I, Bullerdiek J. (2008). Testicular tumor in an XXY dog. Cancer Genet Cytogenet. 183: 114–116.

[54] De Lorenzi L, Molteni L, Zannotti M, Galli C, Parma P. (2010). X trisomy in a sterile mare. Equine Vet J. 42: 469–470.

[55] O'Connor CL, Schweizer C, Gradil C, Schlafer D, Lopate C, Prociuk U, Meyers-Wallen VN, Casal ML. (2011). Trisomy-X with estrous cycle anomalies in two female dogs. Theriogenology. 76: 374–380.

[56] De Lorenzi L, Genualdo V, Gimelli S, Rossi E, Perucatti A, Iannuzzi A, Zannotti M, Malagutti L, Molteni L, Iannuzzi L, Parma P. (2012a). Genomic analysis of cattle rob(1;29). Chromosome Res. 20: 815–823.

[57] Iannuzzi A, Di Meo GP, Caputi Jambrenghi A, Vonghia G, Iannuzzi L, Rangel-Figueiredo T. (2008). Frequency and distribution of rob(1;29) in eight Portuguese cattle breeds. Cytogenet Genome Res. 120: 147–149.

[58] De Lorenzi L, Molteni L, Denis C, Eggen A, Parma P. (2008). A new case of centric fusion in cattle: rob(21;23). Anim Genet. 39: 454–455.

[59] Switonski M., Gustavsson I., Plöen L. (1987). The nature of the 1;29 translocation in cattle as revealed by synaptonemal complex analysis using electron microscope. Cytogeneics and Cell Genetics 44: 103–111.

[60] Bonnet-Garnier A, Pinton A, Berland HM, Khireddine B, Eggen A, Yerle M, Darré R, Ducos A. (2006). Sperm nuclei analysis of 1/29 Robertsonian translocation carrier bulls using fluorescence in situ hybridization. Cytogenet Genome Res. 112: 241–247.

[61] Pinton A, Calgaro A, Bonnet N, Ferchaud S, Billoux S, Dudez AM, Mary N, Massip K, Bonnet-Garnier A, Yerle M, Ducos A. (2009). Influence of sex on the meiotic segregation of a t(13;17) Robertsonian translocation: a case study in the pig. Hum Reprod. 24: 2034–2043.

[62] Raudsepp T, Chowdhary BP. (2011). Cytogenetics and Chromosome Maps, In: The Genetics of the Pig, 2nd edition (eds. M.F. Rothschild and A. Ruvinsky), CAB International, Wallingford, UK and Cambridge, MA, USA.

[63] Iannuzzi A, Genualdo V, Perucatti A, Pauciullo A, Varricchio G, Incarnato D, Matassino D, Iannuzzi L. (2015). Fatal outcome in a newborn calf associated with partial trisomy 25q and partial monosomy 11q, 60,XX,der(11)t(11;25)(q11;q14~21). Cytogenet Genome Res. 146: 222–229.

[64] De Lorenzi L, Morando P, Planas J, Zannotti M, Molteni L, Parma P. (2012b). Reciprocal translocations in cattle: frequency estimation. J Anim Breed Genet. 129: 409–416.

[65] Lear TL, Lundquist J, Zent WW, Fishback WD Jr, Clark A. (2008). Three autosomal chromosome translocations associated with repeated early embryonic loss (REEL) in the domestic horse (*Equus caballus*). Cytogenet Genome Res. 120: 117–122.

[66] Szczerbal I, Stachowiak M, Dzimira S, Sliwa K, Switonski M. (2015b). The first case of 38,XX (SRY-positive) disorder of sex development in a cat. Mol Cytogenet. 8: 22.

[67] Kociucka B, Szczerbal I, Bugaj S, Orsztynowicz M, Switonski M. (2014). A high incidence of adjacent-1 meiotic segregation pattern, revealed by multicolor sperm FISH, in a carrier boar of a new reciprocal translocation t(6;16)(p13;q23). Cytogenet Genome Res. 142: 21–27.

[68] Massip K, Yerle M, Billon Y, Ferchaud S, Bonnet N, Calgaro A, Mary N, Dudez AM,

Sentenac C, Plard C, Ducos A, Pinton A. (2010). Studies of male and female meiosis in inv(4)(p1.4;q2.3) pig carriers. Chromosome Res. 18: 925–938.

[69] Danielak-Czech B, Slota E. (2008). Tandem fusion-translocation: a unique karyotype rearrangement in the domestic pig. Annals of Animal Science 8: 343–348.

[70] Martin-Lluch M, De la Cruz-Vigo P, Ortuno V, Gomez-Fidalgo E, Carrascosa C, Sanchez-Sanchez R. (2014). Cytogenetic study of a reciprocal translocation (1;6)(q17; p11) in a subfertile boar. Chromosome Res. 22: 399.

[71] De la Cruz-Vigo, Martin-Lluch M, Gonzalez-Bulnes A, Gomez-Fildago E, Carrascosa C, Sanchez-Sanchez R. (2014). A complex structural translocation with three chromosomes involved in a boar. Chromosome Res. 22: 400.

[72] Barasc H, Mary N, Letron R, Calgaro A, Dudez AM, Bonnet N, Lahbib-Mansais Y, Yerle M, Ducos A, Pinton A. (2012). Y-autosome translocation interferes with meiotic sex inactivation and expression of autosomal genes: a case study in the pig. Sex Dev. 6: 143–150.

[73] García-Vázquez FA, Hernández-Caravaca I, Martín M, Gómez E, Rodríguez A, Sánchez-Sánchez R, Gadea J. (2011). Two cases of reciprocal chromosomal translocation (4;7)(p+;q-) (2;8)(q-;q+) in piglets produced by ICSI. Reprod Domest Anim. 46: 728–730.

[74] Quach A, Revay T, Macedo M, Wyss S, Sullivan B, King WA. (2014). The incidence of translocations in young breeding boars in Canada. Chromosome Res. 22: 397.

4

Assisted Reproductive Techniques on South American Wild Mammals

Alexandre R. Silva, Alexsandra F. Pereira, Gabriela L. Lima, Gislayne C. X. Peixoto and Ana Liza P. Souza

Abstract

For South American wild mammals, assisted reproductive techniques (ARTs) have been established in accordance with the procedures used for domestic mammals. However, often, it is not possible to infer the experimental conditions from one species to another. In general, the use of these ARTs aims the conservation of the animal genetic material, using gamete manipulation, embryo technology by *in vivo* or *in vitro* methods, and cloning. In all animals, the previous knowledge of reproductive physiology becomes the starting point for the use of ARTs in the expansion of all species of interest. In this sense, the purpose of this chapter is to develop an approach of theoretical, technical, and applied aspects of the ARTs in South American wild mammals, with emphasis on the state-of-the-art and its progress and perspectives.

Keywords: Wildlife, gamete, embryo technology, germplasm bank

1. Introduction

South America, the fourth largest continent, presents an extreme geographic variation that contributes to its large number of biomes, which includes the marine rainforests, alpine, deserts, savannahs, grasslands, chaparral, desert scrub, fresh water, marine, and deciduous desert. Due to such biome diversity, the South American fauna consists of various unique animals.

In the past decade, a rapid and continuous decline in mammalian species has been documented, indeed affecting South American countries. Therefore, the worry put on the development of conservation strategies, as well as on the rapid expansion of the commercial interest on wild species around the world, requires an equally rapid development and adoption of assisted reproductive techniques (ARTs). These techniques allow the conservation and multiplication of genetic valuable individuals, as well as facilitate the transport of the germplasm among distant regions. However, the application of ARTs is still a challenge, since data related to basic reproductive information is scarce, and it is indispensable for the genetic management of rare species. Because of the high diversity in reproductive mechanisms among mammals, it is not possible to directly apply protocols developed from one species to another, hindering the extrapolation of developed ARTs. In this sense, this chapter highlights the importance of applying ARTs for the South American wild mammals, showing the most recent studies in this area and the perspectives for its use in conservation programs.

2. Manipulation of spermatozoa

Technology of male gamete consists of various forms for sperm processing, since its collection until its storage by cooling or cryopreservation. In wild animals, the gamete recovery is generally conducted by electroejaculation (EEJ) [1]; nevertheless, for animals unable to ejaculate, or immediately *postmortem*, one possibility is to obtain epididymal sperm or testicular tissue, which can be cultivated and cryopreserved [2]. After collection, samples could be stored in cryobanks, which maintains the sperm quality for an indefinite time, providing valuable genetic material to be used for ARTs destined to the wild mammals' conservation and multiplication.

2.1. Collection of ejaculated sperm

The EEJ is a safe and effective method for semen collection, especially in wild animals, as it allows handling live animals under anesthesia. It consists of a controlled electrical stimulation of the ejaculatory reflex. A transrectal probe coupled with a specific voltage-producing unit [1] is used to apply the stimulus. The types and disposition of the electrodes, the electric stimulation protocol, and the anatomical characteristics of each species exert important influences on the efficiency of the method [3]. Due to the large species diversity, it is necessary to establish adequate species-specific stimulation protocols using appropriate anesthetic procedures, according to animal responses and well-being [4]. Spermatozoon from different species is presented in Figure 1.

In general, most researchers use adaptations of the original protocol developed for felid collection [5], consisting of a total of 80 electrical stimulations divided into three series: 30 stimuli (10 stimuli, 2 to 4 V, series 01), 30 stimuli (10 stimuli, 3 to 5 V, series 02), and 20 stimuli (10 stimuli, 5 and 6 V, series 03), with a 5-minute interval among series. This serial EEJ protocol was adapted for various South American felids such as the ocelot (*Leopardus pardalis*), the margay (*L. wiedii*), the tigrina (*L. tigrinus*) [6], the jaguar (*Panthera onca*) [7], and other carnivores such as coatis (*Nasua nasua*) [8] (Table 1) and maned wolves (*Chrysocyon brachyurus*) [9]. In addition, this protocol has been adapted for other animals such as sloths [10], giant anteaters

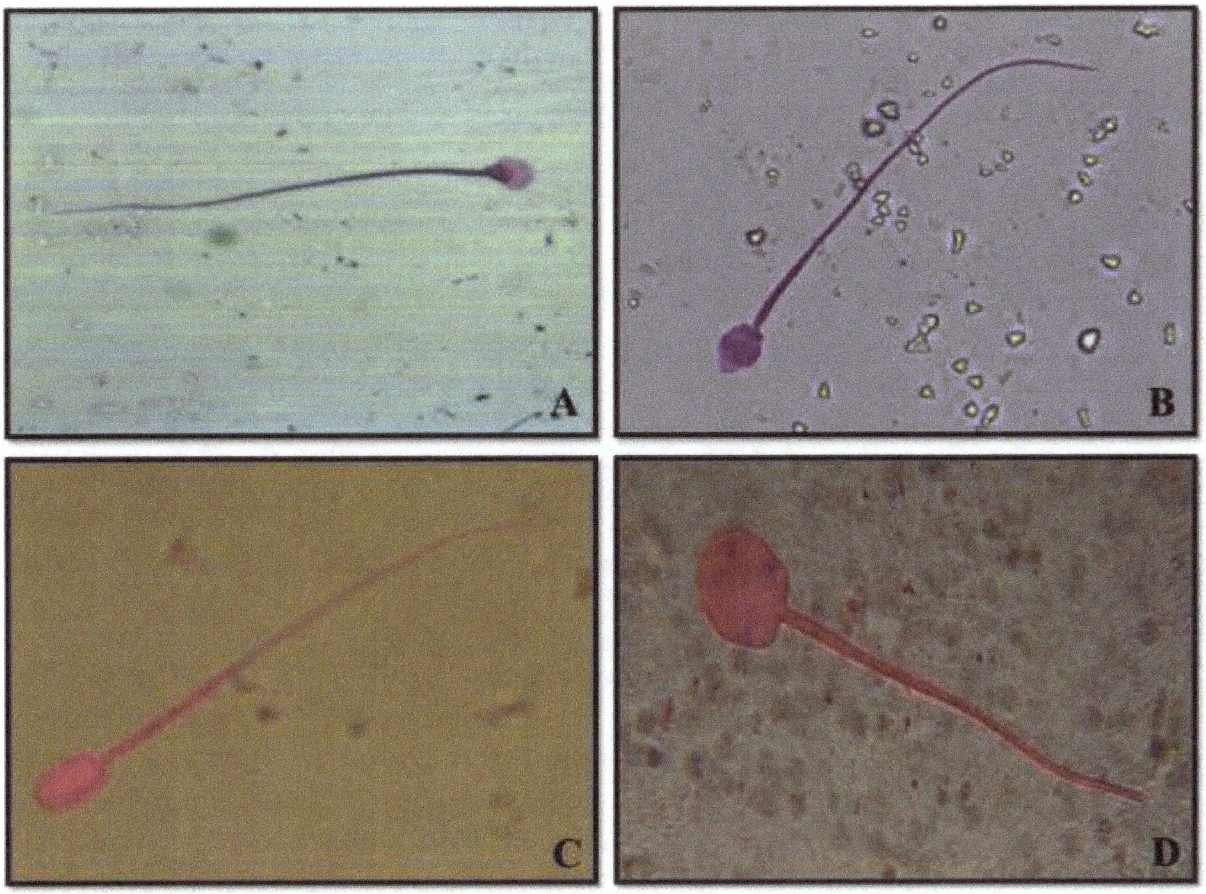

Figure 1. Sperm images of some animals in Latin America, (A) Spix's yellow-toothed cavy—*Galea spixii*, (B) Coati—*Nasua nasua*, (C) Collared peccary—*Pecari tajacu*, (D) Six-banded armadillos—*Euphractus sexcinctus*.

[11], six-banded armadillos (*Euphractus sexcinctus*) [12], tapirs (*Tapirus bairdii*) [13], and agoutis (*Dasyprocta leporina*) [14] (Table 1).

A totally different serial protocol was described for primate species. For spider monkey (*Ateles geoffroyi*), stimulation series begins at 1 V, followed by 1 V increments up to a maximum of 10 V. Each series lasts 5 minutes, with 3-minute rest periods between series [15]. For capuchin monkey (*Sapajus apella*), stimulation session is comprised of six series, 10, 15, 20, 25, 30, and 35 electrical stimuli (12.5–100 mA), with a 30-second interval between consecutive series [16].

In some species, however, a better response is achieved using continuous electrical stimulation. For collared peccaries (*Pecari tajacu*) EEJ (Table 1), the stimulatory cycle consists of 10 stimuli in each voltage, starting from 5 V, followed by voltage increases of 1 V up to 12 V, then, the voltage remains being applied during 10 minutes, continuously [17].

2.2. Epididymal sperm retrieval

Obtaining viable sperm directly from the epididymis tail is an additional option for the conservation of genetic material of South American wild animals, particularly for those accidentally killed. However, different factors such as the epididymis size and vas deferens

Parameters	*Collared peccary* [18]	*Agouti* [14]	*Ring-tailed coati* [8]	*Six-banded armadillo* [12]
Sperm concentration (x10^6)	765.0 ± 313.7	357.0 ± 61.2	197.5 ± 204.9	450.0 ± 14.0
Motility (%)	86.7 ± 2.6	80 ± 29.6	91.3 ± 9.2	61.0 ± 7.0
Vigor (0-5)	4.4 ± 0.3	3.5 ± 1.9	4.5 ± 0.6	2.0 ± 0.2
Viability (%)	92.3 ± 1.6	65.5 ± 26.1	73.1 ± 14.5	55.0 ± 7.0
Membrane functional integrity (%)	75.3 ± 2.3	68.5 ± 17.6	74.3 ± 12.0	46.0 ± 6.0
Normal morphology (%)	83.2 ± 2.2	83.9 ± 8.6	81.2 ± 11.8	86.0 ± 2.0

Table 1. Characteristics of semen obtained by electroejaculation from various South American species.

diameter [18] as well as the possibility for contamination of samples with blood or epithelial cells are considered as limiting factors [19].

Epididymal sperm can be retrieved by flotation or retrograde flushing. In the flotation method, the epididymis is cut into small pieces in a diluent solution for the removal of sperm; this is one of the preferred techniques for small species due to the reduced size of the epididymis [20]. On the other hand, the retrograde washing is accomplished by injecting a buffered solution at the vas deferens, as it was described for the agouti [21]. In some species, such as collared peccaries [22] and yellow-toothed cavies (*Galea spixii*) [23], both recovery methods can be conducted with no notorious differences on sperm quality.

2.3. Sperm preservation

The research on gamete preservation needs to focus on the species-specific variations in sperm physiology, as well as the changes that gametes undergo during sperm processing [24]. The factors most often cited as crucial for success include the sperm processing and packaging, type and composition of extender, duration of equilibrium time, freezing rate, storage, and thawing rate [25].

The short-term preservation of sperm on a liquid form is rarely applied for South American species. In capuchin monkey, for instance, dilution of semen in coconut water–based extender allowed its preservation for 24 hours at 33°C [16]. Moreover, the use of a TRIS extender containing egg yolk or *Aloe vera* extract was efficient in preserving the collared peccary semen for 36 hours at 5°C [26].

Nowadays, intensive research focuses on the development of effective methods for sperm cryopreservation. In general, methods are adapted from the most closely related phylogenetic domestic species; for example, the domestic pig serves as an experiment model for wild pigs; however, such adaptation is not always effective. This fact is verified for six-banded armadillos that present unique semen characteristics, such as high viscosity, large sperm dimensions, and *rouleaux* formation, which hinders the success of freezing protocols [27].

Extenders based on TRIS plus egg yolk and glycerol were reported for collared peccary semen cryopreservation [17]. Other compounds have been described, such as the INRA® extender for tapir [13], the lactose-egg yolk extender for jaguars [28], and a combination of TES and TRIS extenders for capuchin monkey [16]. Recently, a coconut water–based extender was reported for semen cryopreservation in collared peccaries [29], agoutis [30], and squirrel monkeys (*Saimiri collinsi*) [31].

Although most researchers use egg yolk and glycerol as main cryoprotectant, a recent study demonstrated that the dimethyl sulfoxide (DMSO) would be more appropriate for the preservation of maned wolf semen than that of glycerol. Moreover, low-density lipoprotein [32] and *Aloe vera* extract [33] could effectively substitute the egg yolk for the cryopreservation of collared peccary semen.

Regarding preservation of epididymal sperm from South American wild mammals, studies are scarce. For collared peccaries [34] and cavies [35], epididymal sperm was efficiently cryopreserved using TRIS extender added to egg yolk and glycerol. In agoutis, however, epididymal sperm is better cryopreserved in coconut water–based extenders [36].

2.4. Artificial insemination

Artificial insemination (AI) is the single most important technique ever devised for genetic improvement of animals. It refers to the artificial process of sperm deposition into the female genital system, at the appropriate time, seeking the fertilization of the oocyte. The main factors that determine the fertility rates derived from AI include the individual fertility of players, the way in which sperm is collected and manipulated, the ability of the inseminator, the female management (the time of insemination), the type of sperm used (fresh, chilled, or frozen), the insemination dose, and the site for the semen deposition [37].

In spite of AI popularity among domestic animals breeders, its use remains limited in South American wild animals. Initial studies include the successful production of offspring derived from laparoscopic intrauterine AI in ocelots. One female was inseminated with 7.5×10^6 frozen-thawed spermatozoa after receiving 500 IU equine chorionic gonadotropin (eCG) and 225 IU human chorionic gonadotropin (hCG), giving birth to a male offspring after 78 days of gestation [38]. Moreover, deposition of fresh or cryopreserved semen, as well as cryopreserved epididymal sperm, into the cervix of marmoset (*Callithrix jacchus*) resulted in pregnancy and the production of offspring [39].

2.5. Testicular tissue processing

The preservation of testicular tissue for prolonged periods is one big challenge in the field of cryobiology, because spermatogonia, Sertoli, and Leydig cells contain large amounts of water, therein increasing the risk for intracytoplasmic crystal formation [40]. If the testicular tissue has preserved active spermatogenesis, it can be used for the extraction of sperm or elongated spermatids that could be used for oocyte fertilization using intracytoplasmic sperm injection (ICSI) [31]. Moreover, testicle freezing can be used in cases where the animal dies suddenly [41].

Although this is a promising area, studies on the collection of sperm or even on the preservation of testicular tissues are very scarce. In South American individuals, the unique published results can be found in collared peccaries, in which the first successful xenotransplantation of fresh testicular tissue was also documented [42]. In addition, its testicular tissue was efficiently vitrified using ethylene glycol (EG) as cryoprotectant at 3.0 or 6.0 M [43].

3. Estrous and ovulation control

For wild females exhibiting irregular or regular cycles of sexual activity, estrus synchronization support assisted breeding procedures particularly to maximize the chances of conception and the use of fresh or cryopreserved semen [44]. The administration of exogenous hormones, whether or not in association, can artificially synchronize the estrous and ovulation of wild females, altering its endogenous endocrine environment [45]. Research has dramatically increased the number of synchronization protocols for wild animals, being the prostaglandins, progestins, or gonadotropins the most available, each with specific functions and peculiar mechanism of action.

One of the most ancient methods for estrus synchronization is the use of a luteolytic agent such as the prostaglandin F2α (PGF2α) or its analogues, which provokes the *corpus luteum* degeneration [46]. Another approach is by using gonadotropin-releasing hormone (GnRH), which represents the final common pathway where by internal and external relevant stimuli converge to control reproduction. GnRH neurons, thus, stimulate pituitary gonadotropin secretion to appropriately regulate gametogenesis and sex steroid secretion [47]. On the other hand, progestins inhibit the synthesis and secretion of GnRH that subsequently inhibits follicular growth and development as well as the ovulation [48]. Additionally, gonadotropin-based protocols include the use of equine and human chorionic gonadotropins (eCG and hCG, respectively), as well as the application of follicle stimulating hormone (FSH) and luteinizing hormone (LH), as a synthetic or extract form [49]. The action of eCG is similar to that of FSH at stimulating the ovary follicles to produce mature oocyte, thus promoting the outward signs of estrus [50]. In contrast, the hCG action is similar to that of LH, causing the release of mature oocyte at ovulation, and promoting corpora lutea formation [51]. Currently, the majority of studies focus on the use of hormonal associations. This is acceptable because the results of studies have shown that such protocols increased rate, occurrence, and ovulation speed, therefore increasing the fertility rate [52].

Various attempts have been made to establish effective protocols for estrous control in South American wild felines, mainly due to the variety of manifestations of reproductive cycles. For example, the margay present spontaneous ovulations during normal estrous cycles, but ocelot and tigrina are induced ovulators. Luteal control in feline is not a feasible option, due to several factors: (1) although some individuals ovulate spontaneously at unpredictable intervals, many felids are obligated to induce ovulations and (2) the diestrus *corpus luteum*, when present, is refractory to prostaglandins and dopamine agonists up to day 40 postovulation [50].

The use of porcine FSH (pFSH) with minimal LH activity is reported for inducing follicular growth in wild felines [53]. This protocol was effectively applied for pumas (*Puma concolor*) [54], jaguars (*Panthera onca*) [55, 56], and jaguarundis (*Puma yagouaroundi*) [49]. Recently, the ability of an oral progestin, altrenogest (0.192 mg/kg, 14 days), to suppress the ovarian follicular activity in tigrina was demonstrated. This protocol was also effective in reducing hyperstimulation and hyperestrogenism after hCG and eCG administration. However, the authors affirmed that not all females responded uniformly [45]. In fact, studies with both conventional gonadotropins showed diversity of response in different feline species. For example, the ovarian response to a combined treatment with eCG injection followed 80-84 hours later by a hCG administration resulted in a conception rate lower than 30% in ocelots (400 IU eCG and 200 IU hCG) [38] and tigrina (200 IU eCG and 150 IU hCG) [51]. Exogenous gonadotropins are immunologically complex foreign peptides, and thereby, the development of antibodies against gonadotropin is expected sequelae in wild feline submitted to short-term repeated treatments [57]. For this motive, the isolation, characterization, and production of recombinant gonadotropins has been suggested to avoid these complications. On the other hand, it is important to remember that felid species present a large variation in body sizes, a factor that directly influences the synchronization protocols, as well as the individual ovarian response to gonadotropin stimulation [58].

A considerable effort has been made to develop a method to induce estrus and ovulation in the maned wolf, the biggest South American canids. An implant of deslorelin (2.1 mg), a GnRH analogue, was 100% effective for induction of ovary activity, with ovulation occurring between days 9 and 16 (mean: 12.5 ± 1.4 days) after implant when paired with a male, a physiological condition required for the effective ovulation. Alternatively, a single injection of equine recombinant LH (reLH), associated with the deslorelin removal effectively induced ovulation in unpaired females, without apparent adverse impact on fertility in subsequent breeding seasons [59]. Regarding the coati, a carnivore from the Procyonidae family, the unique attempt of estrous control was made by using melengestrol acetate, a progestin, but treatments induced purulent vaginal discharge and uterine adenocarcinoma, and consequently the authors recommend caution when using this drug [60].

In ungulates, as the brown brocket deer (*Mazama gouazoubira*), a monovulatory nonseasonal breeder, estrous synchronization can be achieved using an intravaginal progesterone device (CIDR®) for 8 days, associated with an injection of 265 μg cloprostenol at the removal of the device [61]. In collared peccaries, on the other hand, the use of two injections of 60 μg cloprostenol administered at a 9-day interval was effective for estrous synchronization [62].

In primates, it is known that PGF2α can act on luteal cells to inhibit the luteotrophic actions of LH or hCG *in vitro* [63]. In this context, the administration of cloprostenol has a marked and rapid luteolytic action (0.5–0.8 μg) in the marmoset monkey [64]. In this species, the daily use of recombinant human FSH (rehFSH; 50 IU) during the first 6 days of the follicular phase resulted in superovulation [65]. Moreover, when associated, the protocol using rehFSH 25 IU/hCG 500 UI proved to be efficient to collect oocytes [66].

Despite the difficulties in using a great number of individuals for experimentation, there is notably a need for the development of estrous synchronization protocols for South American wild mammals, especially focused on dosage and efficiency of drugs.

4. Female gametes manipulation

As with other ARTs, the main limitation for wild species is the difficulty in obtaining viable oocytes, being the recovery of preovulatory oocytes for *in vitro* culture (IVC), an available alternative [67]. The possibility to store frozen female gametes as ovary tissue fragments, isolated follicles, or either mature or immature oocytes represents an attractive alternative to cryopreservation [68]; however, the technique is not yet well-established, presenting several challenges that need to be addressed in order to be routinely used.

4.1. Recovery of female gametes

The first step in the manipulation of female gametes is to obtain viable oocytes; however, such procedures are still uncommon in South American wild mammals and are most frequently conducted in Neotropical primates. Oocyte can be collected *in vivo*, or postmortem.

The recovery of oocytes by mechanical preantral follicle (PF) isolation in ovariectomy specimens [69], the collection of ovarian biopsies by exploratory laparoscopy [70], and the puncture of antral follicles by laparotomy [71] have been described studies in capuchin monkeys (*Sapajus apella*). In marmoset monkeys, oocyte recovery was achieved through follicle aspiration after laparotomy [72], laparoscopy [65], or uni- or bilateral ovariectomy [73]. The latter was also the method described in squirrel monkeys (*Saimiri boliviensis boliviensis*) [74] for oocytes collected by ultrasound-guided puncture, after monitorization of the follicular development, in superstimulated females [75]. Finally, in *Saimiri sciureus*, follicle aspiration was conducted after laparoscopy [76] or laparotomy [77].

Regarding other South American species, in particular, wild felids, ovary collection is often performed immediately after the animal death, as it was reported for Geoffroyi and tigrina cats [78]. Also in puma and jaguar, isolation of small PFs (40-90 μm) was performed after animal death [79]. The oocyte aspiration by laparoscopy was also described for tigrina and ocelot [51], puma [80], and jaguar [55]. In vicunas (*Vicugna vicugna*), a South American wild camelid, oocytes also were recovered after follicle aspiration [81].

4.2. Characterization of ovarian follicles population

The rescue and manipulation of matured oocytes represent an important genetic source from wild species; nevertheless, it faces many obstacles, particularly the limited knowledge on the species reproductive physiology [82]. Thus, the use of immature oocytes from antral follicles or PFs allows the recovery of gametes from prepubertal, pregnant, old, or even dead animals [83, 84] The characteristics of the ovarian follicles population in various South American wild mammals are described in Table 2 and illustrated in Figure 2.

Specie	Preantral follicles population			Preantral follicles category									
				Primordial			Primary			Secondary			Authors
	Right ovary	Left ovary	Total	Total number	%	Diameter (µm)	Total number	%	Diameter (µm)	Total number	%	Diameter (µm)	
Sapajus apella	56.9 ± 21.9	49.1 ± 26.9	-	-	30.0 ± 4.3	22.1± 0.5	-	6.0 ± 1.0	27.3 ± 0.5	-	4.0 ± 0.7	61.2 ± 4.0	[85]
Saimiri sciureus	-	-	-	915.0 ± 78.8	73.3 ±1.3	20.5 ± 0.8	230.5 ± 20.8	18.6 ± 0.7	33.7 ± 2.4	115.9 ± 15.7	8.1 ± 0.9	512.5 ± 67.1	[86]
Dasyprocta leporina	4419.8 ± 532.3	5397.5 ± 574.9	-	-	86.6	18.6 ± 3.4	-	13.0	23.8 ± 5.7	-	0.4	88.6 ± 17.6	[87]
Galea spixii	220.8 ± 175.1	195.2 ± 182.8	416.0 ± 342.8	-	32.2	16.6 ± 0.3	-	63.7	28.3 ± 2.1	-	4.1	123.7 ± 18.3	[88]
Pecari tajacu	-	-	33273.5 ± 579.0 (per ovary)	30 466.6 ± 5194.8	91.6	31.8 ± 1.1	2 093.9 ± 595.0	6.3	40.9 ± 2.1	712.9 ± 95.4	2.2	196.2 ± 17.1	[89]

Table 2. Characterization of ovarian follicles population from different South American wild species

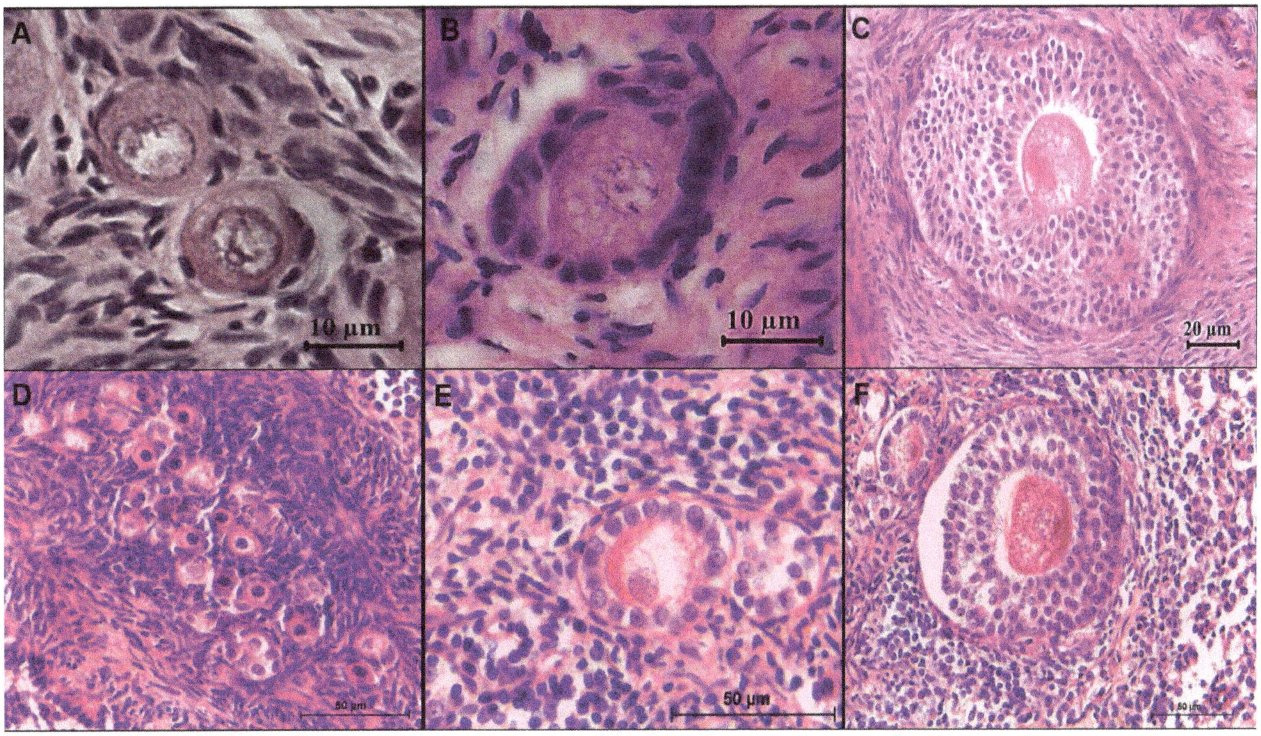

Figure 2. Photomicrograph of the ovary cortex. (A-C) Preantral follicles from *Pecari tajacu* and (D-F) *Dasyprocta leporina* Primordial (A,D), primary (B,E), and secondary (C,F) follicles.

4.3. In vitro culture and maturation of oocytes

The IVC and *in vitro* maturation (IVM) of oocytes has highlighted some of the mechanisms underlying the folliculogenesis in particular wild species, particularly among the South American wild mammals, offering new opportunities for the use of many other ARTs [85]. Oocyte IVM is a technique that provides material for the study of the final steps of oocyte meiosis. Nevertheless, the practical application of this procedure remains relatively inefficient, and the embryo production rate is much lower compared with *in vivo* matured oocytes [73]. Some of the researches that have been developed using IVC and IVM are revised below.

The competence of oocytes retrieved from antral follicles in capuchin monkeys was achieved after 36 hours of IVM, the highest maturation rate occurring in oocytes collected from dominant follicles [71]. It has also demonstrated that the IVC of ovarian cortical strips in TCM199 supplemented of β-mercaptoethanol (BME), bone morphogenetic protein 4 (BMP4), or pregnant mare serum gonadotrophin (PMSG) promoted a follicular viability similar to that of controls (89.3%) for this species, while it also increases the rate of secondary follicle formation (44.9%) [70].

The oocyte meiotic competence and the *cumulus* cell function were positively associated with the follicle size in marmoset monkeys [86], as it was demonstrated in IVM studies. It was shown that chromosome quality is crucial for cytoskeletal organization allowing a correct meiosis in IVC and IVM of the immature oocyte [73]. Moreover, abnormal spindle formation was observed in oocytes derived from small antral follicles failing to complete meiosis, contrasting almost 90% matured oocytes, surrounded by expanded *cumulus* cells at the time of isolation [66]. For this species, an alternative two-step culture system consisting in the culture of oocytes within stromal tissue fragments for 2 days has been described. Afterwards, follicles were mechanically isolated and transferred to a culture system where they grew for up to further 12 days. This process produces full-sized matured oocytes from primary and early secondary follicles [87].

An immunohistochemical assay conducted in squirrel monkey ovaries during IVC demonstrated that growth differentiation factor 9 (GDF-9) and c-Kit protein were detectable in oocyte cytoplasm from primordial to secondary follicles, whereas the Kit Ligand expression was observed in oocytes and granulosa cells from primordial to secondary follicles. On the other hand, the anti-Müllerian hormone was expressed in primary and secondary follicles, but not in the primordial ones [88]. In a different species from the same gender, the *S. boliviensis*, IVM of follicles obtained by laparoscopy requested the use of a medium containing high-energy source [89]. Moreover, oocyte IVM was also described for *S. scierus* [90].

Regarding South American wild felids, the yield of the IMV technique varies greatly among species. Immature oocytes were recovered from Geoffroyi cat, jaguar, and puma after the animal death and further matured *in vitro*. In Geoffroyi cat, from a total of 45 oocytes retrieved, 23% succeed maturation, similarly to that described for jaguar, from whose oocytes (n=21) only one-third advanced to metaphase II during IVM. In the puma, more than 100 oocytes were recovered from 8 females, from which 43.8% of the oocytes matured *in vitro* [91]. IVM has been previously described in this species after oocyte recovery using laparoscopy and transabdominal aspiration [92]. In a recent study, 42 mature oocytes were collected by laparoscopic ovum pick up from pumas, after the administration of eCG (750 UI) with hCG (500 UI). In this species, oocytes were also matured *in vitro* using TCM199 supplemented with LH, FSH, and 17β-estradiol [80].

Studies on ovarian gene expression have been conducted to better understand folliculogenesis in plains viscacha (*Lagostomus maximus*), an animal showing the highest ovulation rate among all mammals, releasing between 400 and 800 oocytes per estrous cycle [93]. The expression of germ cell–specific VASA protein, apoptotic proteins BCL2 and BAX, as well as DNA fragmentation revealed an unrestricted proliferation of germ cells in this species, without apoptosis-driven elimination, contrary to what is normally found in other mammals [94, 95].

In ungulates, IVC of collared peccaries ovarian tissue reveals that more than 50% follicles remain morphologically intact when the TCM199 medium was used whether supplemented or not with FSH. Moreover, the activation of collared peccaries PFs was stimulated by the addition of FSH to the medium during IVC, the proportion of growing PFs increased from only 31.2 ± 0.7% in the control group to more than 90% after a 7 day in TCM199+FSH [96].

4.4. Female gametes cryopreservation

The conditions required for the complete *in vitro* development of oocytes remain incompletely established in wild species; therefore, cryopreservation emerges as a promising tool for female germplasm conservation, allowing its future use in ART. It is possible to store ovarian tissue, isolated follicles, and mature or immature oocytes, for which purpose two methods are routinely used: the slow-freezing procedure and vitrification. The value of vitrification has been highlighted to avoid damages caused by conventional freezing protocols, such as the risk of ice crystal formation. It is considered a cheap method that can be performed under field conditions.

Another major factor on the success of cryopreservation of female gametes is the cryoprotectant used. In marmoset monkeys, DMSO was advantageously used as a cryoprotectant for slow freezing of ovarian tissue. It provided a higher percentage of morphologically normal primordial (26.2 ± 2.5%) and primary follicles (28.1 ± 5.4%) compared with propanediol (PROH) (12.2 ± 3.0 and 5.4 ± 2.1%, respectively) [97]. Furthermore, the development up to secondary PFs [98] and the formation of antrum after xenografting cryopreserved tissue in this species were also described [99].

Regarding South American wild felids, short-term IVC of ovarian cortex slices collected from Geoffroyi cat was performed to assess the success of the slow freezing protocol, using EG and sucrose. The results showed an increase in the number of viable PFs in frozen–thawed samples after 7 days of IVC (48%) compared with fresh cultured pieces (31%) obtained in the same sampling day. In tigrina cats, the percentage of primordial follicles after freezing and culture (44%) indicated a normal population of viable follicles [78].

Regarding Neotropical rodents, the conventional freezing was adopted to preserve ovarian fragments from agoutis. After freezing, no differences were found among groups using different cryoprotectants: DMSO (60.6 ± 3.6%), EG (64.0 ± 11.9%), or PROH (62.0 ± 6.9%).

However, only follicles cryopreserved with PROH presented a normal ultrastructure, similar to that of the control group [100]. A reduction on the morphologically normal PFs (69.5%) compared with the control group (91.2%) was also observed, after solid surface vitrification of ovarian tissue from yellow-toothed cavies. The preservation of oocytes and granulosa cell membranes and the morphological aspect of follicles were acceptable, and the transmission electronic microscopy showed that the presence of vacuoles in the oocyte and granulosa cells cytoplasm and turgid mitochondria remained the main alteration observed in some vitrified follicles [101].

In collared peccaries, the refrigeration of ovaries for up to 36 hours allowed supporting the morphological integrity and viability of PFs. Further, it was demonstrated that powdered coconut water media (66.7%) was more effective than the phosphate buffered saline (PBS, 49.4%) to preserve the morphological integrity after 36 hours storage, although without statistical differences on respect to the follicular viability [102]. Moreover, the solid surface vitrification of ovarian tissue using EG, DMSO, or dimethylformamide (DMF), at 3.0 or 6.0 M, preserved the morphological integrity of more than 70% PFs using any of these cryoprotectants [103].

5. Embryo technology

ARTs using *in vivo* embryo production by applying hormonal superovulation protocols, as well as those obtaining *in vitro* blastocyst through *in vitro* fertilization (IVF) and ICSI, allowed to establish the adequate conditions for embryonic development in different species. Furthermore, ARTs associated with embryo cryopreservation foster the preservation of genetic material, contributing to conservation of biodiversity. In general, ARTs applied to wild mammals in South America are established in accordance with procedures used in domestic mammals, despite that some wild species may have complex reproductive characteristics, compromising the rapid progress of ARTs in particular species.

5.1. In vivo embryo production

Soon after the recognition of the gametes and embryo physiology, application of ARTs in each new species begins with *in vivo* embryo production, defined as the selection of female donors, AI or natural mating, embryo collection, cryopreservation, or direct transfer into synchronized recipients. Inevitably, the number of individuals is a crucial point in research, and individual variations nurture the development of *in vitro* ARTs.

In general, superovulatory protocols consist of a single dose of eCG and/or multiple injections of FSH [104]. The status of follicular development at the moment of superovulation, as well as the stress induced by intense manipulation during ARTs, is the main factor ascribed for the variability on response [105]. In an attempt to provide efficient superovulation protocols for embryo collection, different studies have been developed in various wild mammals. An interesting example corresponds to brown brocket deer [106], in which eCG induced a good response to superovulatory protocol, promoting the formation of functional corpora lutea (7.0

± 1.8) although 66.7% (4/6) of the females showed premature luteal regression. In comparison, FSH administration resulted in a low formation of corpora lutea (2.6 ± 0.8) and lower proportion premature luteal regression (33.3%; 2/6).

There are few studies on *in vivo* embryo production protocols in wild camelids. In vicunas, two studies [107, 108] proposed ovarian superstimulatory protocols using eCG with or without CIDR® during 5 days; unfortunately, the results involved only data for follicular development, with no information on embryo recovered or transferred.

5.2. In vitro embryo production

The IVF is still the main *in vitro* technique for the reproduction of wild mammals, although this technique demands a wide quantity of viable oocytes to be fertilized. Briefly, IVF steps involve the oocyte recovery, the *in vivo* or IVM and selection of oocytes, the IVF using capacitated spermatozoa, and the preimplantation embryonic development. The aim is to develop high-quality embryos and obtain normal pregnancies, resulting in the birth of healthy offspring. In this sense, Table 3 summarizes the results of *in vitro* ARTs applied in some South American wild mammals. In general, the protocols used for domestic animals were gradually adjusted to wild species.

Although IVF embryos have been produced from capuchin monkeys [109], marmoset monkeys [66, 110], pumas [92], and tigrinas [51], modifications are needed to increase the efficiency of protocols. Currently, *in vitro* embryo production depends primarily on two major products: meiotic viable oocytes (meiotic competence) and capacitated viable spermatozoa (sperm competence). Domingues et al. [71] reported the viability of *in vitro* matured oocytes collected from nonstimulated capuchin monkeys. Lima et al. [109] established 40 hours as the optimal IVM length and showed the positive effect of FSH/LH (0.5 μg/mL/50.0 μg/mL) on *IVF* of oocytes collected also in nonstimulated females. In addition, spermatozoa obtained from semen diluted in coconut water were able to fertilize oocytes. However, the establishment of experimental conditions is still a requirement for this species.

Another interesting example of the use of IVF in wild felids [111] is that of jaguars; a stimulation protocol using FSH/LH resulted in approximately 25 follicles/female, with more than 80% of high-quality oocytes but the fertilization rates were fairly low (<25%) [55]. Contrastingly, ocelot and tigrina each treated with eCG/hCG produced an average of approximately 10 follicles with ~7–9 high-quality oocytes/female that resulted in 76 ocelot and 52 tigrina embryos [51].

Alternatively, ICSI offers a precise control of sperm competence for *in vitro* procedures. ICSI was first used in marmoset monkeys in 2007, by Grupen et al. [66], using *in vivo* matured oocytes. Still, no blastocyst was produced by ICSI, despite that 47% of blastocysts were achieved after IVF. In 2014 [110], ICSI embryos derived from *in vitro* matured marmoset oocytes developed into neonates, through the use of two strategies: the evaluation of the most suitable timing for ICSI and establishing the *in vitro* and *in vivo* embryo developmental potential. In this study, the blastocyst rate tended to be low when ICSI was performed 1–2 hours after the extrusion of the first polar body. For *in vitro* developmental competence, a

Species	Oocyte source	Maturation type	Spermatozoa		ART applied	Embryo production					Authors
			Collection/ selection	Motility (%)		Media[a]	Cleavage rate[b]	Blastocyst rate[c]	Stage reached	Embryo destination	
Sapajus apela	Without stimulation	*In vitro*	EEJ, cooled to 4°C/ Swim up	~80	IVF	SOF	20	---	4 cells	Cell staining	[114]
Callithrix jacchus	Hormonal stimulation	*In vivo*	FertiCare vibrator, artificial vagina/ Pure gradient	"/>70	IVF/ICSI	G1.2, G2.2	44.7/94.2	46.7/0.0[d]	Blastocyst	Cell staining	[66]
Callithrix jacchus	Hormonal stimulation	*In vitro*	FertiCare vibrator, artificial vagina/ Swim up	NI	IVF/ICSI	ISM1, ISM2	93.2/97.6	39.2/35.4	Blastocyst	Embryo transfer	[115]
Felis concolor	Hormonal stimulation	*In vivo*	EEJ/ Swim up	40-50	IVF	mKRB	---	---	1 cell	NI	[97]
Pecari Tajacu	Without stimulation	*In vitro*	Testis cell suspension xenografts/---	NI	ICSI	NI	37.5	---	4 cells	Xenograft assay	[42]

Abbreviations: EEJ, electroejaculation; IVF, *in vitro* fertilization; ICSI, intracytoplasmic sperm injection; ARTs, Assisted reproductive technique; NI, not informed.

[a]More information about the composition of the media see reference.

[b]Cleavage rate represents the number of cleaved oocytes in relation to the number of oocytes entering to maturation in D2.

[c]Blastocyst rate represents the number of blastocyst in relation to the number of zygotes.

[d]Blastocyst rate represents the number of blastocyst in relation to the number of cleaved embryos.

Table 3. Results of IVF and ICSI techniques applied in some South American wild mammals.

greater fertilization rate was observed in ICSI embryos (93.2%) compared with IVF embryos (82.2%). However, no differences in developmental rate (blastocyst/fertilized oocytes) were observed between ICSI (35.4%) and IVF (39.2%). In addition, for *in vivo* development of ICSI embryos, an offspring rate of 28.6% (6/21) and 2.7% (1/37) was obtained after the transference of 6 to 8 cell embryos and blastocysts, respectively.

In collared peccary [42], ICSI was used to evaluate the quality of sperm cells derived from testis cell suspension xenografts. The collared peccary presents an interesting Leydig cell cytoarchitecture, and therefore, it represents an interesting mammalian model for investigating cellular roles in male gonads. In the study, the sperm recovered from the xenografts originated 75.0% (6/8) of two-cell embryos and 37.5% (3/8) of four-cell embryos produced in 24 and 48 after ICSI, respectively.

For South American wild camelids, information on oocyte recovery and maturation are only available for vicuna: using the surgical oocyte aspiration from superstimulated females, a recovery rate of 55.4% (46/83) and an IVM rate of 41% were obtained, characterized by extrusion of the first polar body and cytoplasmic maturation [81], but no information was presented on IVF or ICSI.

5.3. Embryo cryopreservation and transfer

In recent years, advances in the embryo cryopreservation were achieved with studies in slow freezing conventional method or vitrification. Currently, embryo cryopreservation targets the establishment of a consistent vitrification protocol to be applied for the conservation of preimplantation embryos in different stages of development. Protocols developed for other species appear to be adequate for the purpose of embryo cryopreservation in wild felids. Offspring were obtained in ocelot following transfer of cryopreserved embryos [111].

In marmoset monkey, due to its small body size, the technical development of an efficient embryo transfer is being delayed. Different factors are being studied, such as nonsurgical approaches, embryo cryopreservation, use of late-stage embryos, and volume medium [112] to overcome these gaps. Using embryos in different developmental stages (10- to 16-cell, morula, and blastocysts) after vitrification in Cryotop, Ishibashi et al. [112] defended that reducing the transfer volume to 1 μL or less is essential for successful embryo transfer this species. This procedure provides pregnancy and birth rates of 80% (8/10) and 75% (9/12), while the use of larger volumes (2–3 μL) results in pregnancy and birth rates of 50% (5/10) and 27% (3/11), respectively.

6. Somatic cell processing and cloning

Somatic cell nuclear transfer (SCNT, cloning) consists of the transfer of the nucleus of donor cells into enucleated oocytes, resulting in the production of an individual genetically identical to the nucleus donor animal. In general, SCNT is performed according to the following steps: (i) preparing cytoplast receptors from oocyte recovery, selection, and maturation (cytoplast); (ii) isolation, characterization, *IVC*, and cellular cycle synchronization of the somatic cells to be used as nuclear donors (karyoplast); (iii) embryonic reconstruction by nuclear transfer, fusion of karyoplast–cytoplast complex, cellular activation, and preimplantation embryo development *in vitro;* and (iv) transfer of embryos for previously synchronized recipients, establishment of pregnancies, fetal development, and birth of offspring. In all these steps, there are determining factors for the success of the technique, including the media composition, the donor cell type and the oocyte source, and the timing of all steps down the success of the reproductive output of viable clones [113].

The use of cloning for conservation is often questionable for the risk of reducing the genetic variability. However, it is now known that SCNT not only provides the production and multiplication of a group of individual species of interest, as well as it allows additional studies in the field of reprogramming and cryopreservation of somatic tissue and cells, providing a larger scene in the reproduction of these species. Up to now, there are no studies on clone embryo production in South American wild mammals.

An important step in cloning is the selection of the cells donating a nucleus. In general, fibroblasts derived from skin biopsies are the cell model used. Numerous reasons justify its use for embryo reconstruction, such as the ease of obtaining and handling for the primary culture or subcultures, freezing patterns, and suitable transfection, in addition to obtaining inexhaustible quantities from a single explant. Moreover, the preservation of cells and tissues of the animal to be cloned allows the preservation of the genetic material of an individual indefinitely.

6.1. Generation and establishment of somatic cell primary cultures

The karyoplast preparation derived from skin biopsies involve multiple steps, harvesting of the biological material, through the isolation, establishment and characterization of cell cultures, and the cryopreservation conditions of cell populations after subculture cycle. Routinely, karyoplasts may be derived from skin biopsies of adult or fetal auricular region, according to established standards of asepsis and sterilization procedures [113]. After processing tissue samples into smaller fragments, these are distributed in Dulbecco Modified Eagles Medium (DMEM) supplemented with sodium pyruvate, sodium bicarbonate, antibiotics, amphotericin B, and fetal bovine serum. The primary cultures are monitored every 24 hours to evaluate the cell growth and total medium replacement. When 70% cell confluence is purchased on the plates, the cultures are subcultured. In general, explants are cultured under these conditions for 7–14 days [114]. To confirm the cells' characteristics, particular antibodies are used, like vimentin [113].

In native Chilean species, the establishment of donor cell was already observed in Chilean Shrew opossum (*Rhyncholestes raphanurus*) and chinchilla (*Chinchilla lanigera*) [115]. In the study, tissue fragments were subjected to culture systems and somatic cells were obtained for future cloning.

6.2. Somatic tissue conservation

The tissue cryopreservation involves some major issues, such as the conservation of cellular functionality, extended storage of the samples, and the easiness of the procedures performed outside the laboratory [115]. Tissue sample cryopreservation of wild species is an interesting step for biodiversity conservation, when there is not an appropriate culture system.

In collared peccary [114, 116], vitrification techniques (conventional or solid-surface vitrification) may be used for cryopreservation of somatic tissue, allowing the isolation of viable cells. In this study, ear fragments (9.0 mm^3) were vitrified in a solution containing DMEM plus 3.0 M DMSO, 3.0 M EG, 0.25 M sucrose, and 10% fetal calf serum. The histological analysis and IVC showed that few tissue damages were associated with solid-surface vitrification [116]. Moreover, all cells resulting in vitrified and noncryopreserved fragments showed the characteristics of suitable fibroblasts: fusiform features with oval nucleus in the center. The growth curves showed clear log and lag phases of development, with no difference between treatment and nonvitrified for population doubling time (conventional vitrification directly: 66.1 hours; solid-surface vitrification: 66.3 hours; and nonvitrified: 54.0 hours) [114].

6.3. Cloning and its perspectives

Although some domestic mammals have already been cloned, information on cloning of South American wild animals is scarce. As with other animal species, cloning allows to know the principles of nuclear reprogramming, conservation of cells and tissues for different purposes (reproductive and therapeutic), and clones can also be used as models in several studies. Thus, the efficient production of pluripotent stem cells (iPSC) allows the generation of genetic preservation by different ways, such as gamete production and embryo complementation. Dermal fibroblasts derived from jaguars were transduced using genes encoding the human transcription factors, in an experiment that highlighted the homeobox protein NANOG as crucial to the cell reprogramming in this species, and demonstrated that the technique may represent an efficient method for iPSC production from endangered felids [117].

7. Final considerations

The ARTs have applications in different productive and scientific sectors, whether or not aiming at the conservation of species. As occurred for the domestic species, the use of ART sequence followed the order of complexity and evolution of biotechnologies. Accordingly, the routine use of AI and embryo transfer in different species of mammals is evident. Moreover, *in vitro* ARTs continue to grow, using a parallel between the protocols applied to domestic species and individual characteristics of each wild ones. With the progress of *in vivo* or *in vitro* ARTs, it is expected to know, to manipulate, and to conserve the genetic material, promoting the preservation of biodiversity, especially in South American wild mammals.

Author details

Alexandre R. Silva*, Alexsandra F. Pereira, Gabriela L. Lima, Gislayne C. X. Peixoto and Ana Liza P. Souza

*Address all correspondence to: legio2000@yahoo.com

Laboratory of Animal Germplasm Conservation – UFERSA, Mossoró, RN, Brazil

References

[1] Silva AR, Morato RG, Silva LDM. The potential of gamete recovery from non-domestic canids and felids. Animal Reproduction Science. 2004;81:159–175.

[2] Garde JJ, Ortiz N, Garcia AJ, Lopez A, Gallego L. Criopreservación *postmortem* de material espermático e inceminación artificial en el ciervo iberico. [Postmortem cryo-

preservation of sperm material and artificial insemination in Iberian deer]. Archivos de Zootecnia.1998;47:351–356.

[3] Castelo TS, Silva AR. Eletroejaculação em mamíferos silvestres: principais fatores que afetam sua eficiência. [Electroejaculation in wild mammals: main factors affecting its efficiency]. Revista Brasileira de Reprodução Animal. 2014;38:208–213.

[4] Durrant BS. Semen collection, evaluation and cryopreservation in exotic animal species: maximizing reproductive potential. Ilar News Journal. 1990;32:2–10.

[5] Wildt DE, Bush M, Howard JG, Obrien SJ, Meltzer D, Van Dyk A, Ebedes H, Brand DJ. Unique seminal quality in the South African cheetah and a comparative evaluation in the domestic cat. Biology of Reproduction. 1983;29:1019–1025.

[6] Morais RN, Muciolo RG, Gomes MLF, Lacerda O, Moraes W, Moreira N, Graham LH, Swanson WF, Brown JL. Seasonal analysis of semen characteristics, serum testosterone and fecal androgens in the ocelot (*Leopardus pardalis*), margay (*L. weidii*) and tigrina (*L. tigrinus*). Theriogenology. 2002; 57:2027–2041.

[7] Morato RG, Verreschi IT, Guimarães MA, Cassaro K, Pessuti C, Barnabe RC. Seasonal variation in the endocrine-testicular function of captive jaguars (*Panthera onca*). Theriogenology. 2004;61:1273–1281.

[8] Silva HVR, Magalhães FF, Ribeiro LR, Souza ALP, Freitas CIA, Oliveira MF, Silva AR, Silva LDM. Morphometry, morphology and ultrastructure of ring-tailed coati sperm (*Nasua nasua* Linnaeus, 1766). Reproduction in Domestic Animals. 2015;50:945-951.

[9] Cunha IC, Morato RG, Santos IP, Quirino CR. Biometry of the reproductive system and the ejaculation response of maned wolf (*Chrysocyon brachyurus*) to electroejaculation procedure. In: Proceedings of the 6th ISCFR & 6th EVSSAR; 2008; Vienna. Austria; 2008, pp. 1–2.

[10] Peres MA, Benetti EJ, Milazzotto MP, Visintin JA, Miglino MA, Assumpção MEOA. Collection and evaluation of semen from the three-toed sloth (*Bradypus tridactylus*). Tissue Cells Cell. 2008;40:325–331.

[11] Mendonça MAC. Análise descritiva do perfil espermático do tamanduá-bandeira (*Myrmecophaga tridactyla* Linnaeus, 1758) de cativeiro. [Descriptive analysis of the sperm profile in captive flag anteater (*Myrmecophaga tridactyla* Linnaeus, 1758) - Master of Sciences Thesis]. São Paulo: Universidade de São Paulo; 2010.

[12] Serafim MKB, Lira RA, Costa LLM, Gadelha ICN, Freitas CIA, Silva AR. Description of semen characteristics from six-banded armadillos (*Euphractus sexcinctus*) collected by electroejaculation. Animal Reproduction Science. 2010;118:362–365.

[13] Pukazhenthi BS, Della Togna G, Padilla L, Smith D, Sanchez C, Pelican K, et al. Ejaculate traits and sperm cryopreservation in the endangered Baird's tapir (*Tapirus bairdii*). Journal of Andrology. 2011;32:260–270.

[14] Castelo TS, Souza AL, Lima GL, Peixoto GC, Campos LB, Oliveira MF, Silva AR. Interactions among different devices and electrical stimulus on the electroejaculation of captive agoutis (*Dasyprocta leporina*). Reproduction in Domestic Animals. 2015;50:492–496.

[15] Flores-Herrera H, Acuña-Hernández DG, Rivera-Rebolledo JA, González-Jiménez MA, Rodas-Martínez AZ, Swanson WF. Effect of increasing trypsin concentrations on seminal coagulum dissolution and sperm parameters in spider monkeys (*Ateles geoffroyi*). Theriogenology. 2012;78:612–619.

[16] Oliveira KG, Castro PHG, Muniz JAPC, Domingues SFS. Conservação do sêmen e liquefação do coágulo seminal de macaco-prego (Cebus apella) em água de coco em pó (ACP-118®), em diferentes temperaturas. [Spem conservation and seminal clot liquefaction in capuchin monkey (Cebus apella) using powdered coconut water (ACP-118®) at different temperatures]. Ciência Rural. 2010;40:617–621.

[17] Castelo TS, Bezerra FSB, Souza ALP, Moreira MAP, Paula VV, Oliveira MF, Silva AR. Influence of the thawing rate on the cryopreservation of semen from collared peccaries (*Tayassu tajacu*) using Tris-based extenders. Theriogenology. 2010;74:1060–1065.

[18] Chatdarong K, Thuwanut P, Morrell JM. Single-layer centrifugation through colloid selects improved quality of epididymal cat sperm. Theriogenology. 2010;73:1284–1292.

[19] Rijsselaere T, Van Soom A, Maes D, Verberckmoes S, Kruif A. Effect of blood admixture on in vitro survival of chilled and frozen-thawed canine spermatozoa. Theriogenology. 2004;61:1589–1602.

[20] Morton KM, Evans G, Maxwell WMC. Effect of glycerol concentration, Equex STM® supplementation and liquid storage prior to freezing on the motility and acrosome integrity of frozen-thawed epididymal alpaca (*Vicugna pacos*) sperm. Theriogenology. 2010;74:311–316.

[21] Ferraz MS, Menezes DJ, Pessoa GT, Cabral RM, Illera MJ, Silva AR, Carvalho MA. Collection and evaluation of epididymal sperm in captive agoutis (*Dasyprocta aguti*). Theriogenology. 2011;75:459–462.

[22] Bezerra JAB, Silva AM, Peixoto GCX, Silva MA, Oliveira MF, Silva AR. Influence of recovery method and centrifugation on epididymal sperm from collared peccaries (*Pecari tajacu* Linnaeus, 1758). Zoological Science. 2014;31:338–342.

[23] Silva AM, Bezerra JAB, Campos LB, Borges PAC, Lima GL, Castelo TS, Souza ALP, Silva AR. Características gerais de espermatozoides epididimários de préas (*Galea spixii* Wagler,1831). [General epididymal sperm characteristics of cavies (Galea spixii Wagler, 1831]. In: Proceedings of the XXII Congresso Brasileiro de Zootecnia, 2012, Cuiabá. Mato Grosso; 2012.

[24] Fickel L, Wagener A, Ludwig A. Semen cryopreservation and the conservation of endangered species. European Journal of Wildlife Research. 2007;53:81–89.

[25] Alamo D, Batista M, González F, Rodríguez N, Cruz G, Cabrera F, Gracia A. Cryopreservation of semen in the dog: use of ultra-freezers of −152 °C as a viable alternative to liquid nitrogen. Theriogenology. 2005;63:72–82.

[26] Lago AEA, Nascimento HHL, Souza ALP, Bezerra JAB, Silva AM, Castelo TS, A.R. Silva AR. *Aloe vera* use on short-term preservation of collared pecari (*Pecari tajacu*) semen. In: Proceedings of the V International Symposium on Animal Biology of Reproduction, 8-11 October 2014; Campinas. São Paulo: CBRA; 2014.

[27] Sousa PC, Santos EAA, Bezerra JAB, Lima GL, Castelo TS, Fontenele-Neto JD, Silva AR. Morphology, morphometry and ultrastructure of captive six-banded armadillo (*Euphractus sexcinctus*) sperm. Animal Reproduction Science. 2013;140:279–285.

[28] Paz RCR, Morgado TO,Viana CTR, Arruda FP, Bezerra do Nascimento DO, Guimarães LDA. Semen collection and evaluation of captive coatis (Nasua nasua). Arquivo Brasileiro de Medicina Veterinária e Zootecnia. 2012;64:318–322.

[29] Silva MA, Peixoto GCX, Lima GL, Bezerra JAB, Campos LB, Paiva ALC, Paula VV, Silva AR. Cryopreservation of collared peccaries (Tayassu tajacu) semen using a powdered coconut water (ACP-116c) based extender plus various concentrations of egg yolk and glycerol. Theriogenology. 2012;78:605–611.

[30] Castelo TS, Praxedes ECG, Souza ALP, Peixoto GCX, Lima GL, Sousa PC, Santos EAA, Campos LB, Bezerra JAB, Silva AR. Criopreservação de sêmen de cutias *(Dasyprocta aguti)* em diluente à base de água de coco em pó – ACP-109c®. [Cryopreservation of agouti (Dasyprocta aguti) semen using powdered coconut water extender – ACP-109®]. In: Proceedings of the Congresso Brasileiro de Reprodução Animal; Maio 2013; Uberlândia. Minas Gerais: CBRA; 2013.

[31] Oliveira KG, Leão DL, Almeida DV, Santos RR, Domingues SF. Seminal characteristics and cryopreservation of sperm from the squirrel monkey, *Saimiri collinsi*. Theriogenology. 2015;84:743–749.

[32] Souza ALP, Lima GL, Peixoto GCX, Castelo TS, Bezerra JAB, Silva AR. Efeito da Lipoproteína de Baixa Densidade (LDL) na congelação de sêmen de catetos (*Pecari tajacu*). Acta Veterinaria Brasilica. 2014;8:41–42.

[33] Souza ALP, Lima GL, Peixoto GCX, Bezerra JAB, Silva AM, Oliveira MF, Silva AR. Innovative use of *Aloe vera* as a cryoprotectant for the collared peccaries (*Pecari tajacu*) semen freezing. In: Proceedings of the V International Symposium on Animal Biology of Reproduction, 8-11 October 2014; Campinas. São Paulo: CBRA; 2014.

[34] Bezerra JAB, Praxedes ECG, Silva AM, Souza ALP, Castelo TS, Campos LB, Silva AR. Objective assessment of the effect of different extenders on frozen-thawed epididymal sperm from collared peccaries (*Pecari tajacu*). In: Proceedings of the V Interna-

tional Symposium on Animal Biology of Reproduction, 8-11 October 2014; Campinas. São Paulo: CBRA; 2014.

[35] Silva AM, Apolinario CAC, Almeida LM, Souza ALP, Castelo TS, Lago AEA, Silva AR. Criopreservação de espermatozoides epididimários de preá (*Galea spixii* Wagler, 1831). [Cryopreservation of cavies (*Galea spixii* Wagler, 1831) epididimal sperm]. In: Proceedings of the XXI Congresso Brasileiro de Reprodução Animal; 2015; Belo Horizonte. Minas Gerais: CBRA; 2015.

[36] Silva MA, Peixoto GCX, Santos EAA, Castelo TS, Oliveira MF, Silva AR. Recovery and cryopreservation of epididymal sperm from agouti (*Dasyprocta aguti*) using powdered coconut water (ACP-109c) and Tris extenders. Theriogenology. 2011;76:1084–1089.

[37] Leão KM. Técnicas de inseminação artificial [Techniques for artificial insemination – Graduate monography]. Botucatu: Faculdade de Medicina Veterinária e Zootecnia da Universidade Estadual Paulista; 2003.

[38] Swanson WF, Horohov DW, Godke RA. Production of exogenous gonadotropin-neutralizing immunoglobulins in cats following repeated eCG/hCG treatment and relevance for assisted reproduction in felids. Journal of Reproduction and Fertility. 1995;105:35–41.

[39] Morrell JM, Nubbemeyer R, Heistermann M, Rosenbusch J, Küderling I, Holt W, Hodges JK. Artificial insemination in *Callithrix jacchus* using fresh or cryopreserved sperm. Animal Reproduction Science. 1998;52:165–174.

[40] Woelders H, Chaveiro A. Theoretical prediction of optimal freezing programmes. Cryobiology. 2004;49:258–271.

[41] Abrishami M, Anzar M, Yang Y, Honaramooz A. Cryopreservation of immature porcine testis tissue to maintain its developmental potential after xenografting into recipient mice. Theriogenology. 2010;73:86–96.

[42] Campos-Junior PHA, Costa GMJ, Avelar GF, Lacerda SMSN, Costa NN, Orashi OM, Miranda MS, Barcelos LS, Jorge EC, Guimarães DA, França LR. Derivation of sperm from xenografted testis cells and tissues of the peccary (*Tayassu tajacu*). Reproduction. 2014;147:291–299.

[43] Borges PAC, Bezerra JAB, Praxedes ECG, Lima GL, Souza ALP, Peixoto GCX, Silva AR. Cryopreservation of testicular tissue: a potential tool for the conservation of male genetic material from collared peccaries (*Pecari tajacu*). In: Proceedings of the V International Symposium on Animal Biology of Reproduction, 8-11 October 2014; Campinas. São Paulo: CBRA; 2014.

[44] Durrant BS. The importance and potential of artificial insemination in CANDES (companion animals, non-domestic, endangered species). Theriogenology. 2009;71:113–122.

[45] Micheletti T, Brown JL, Walker SL, Cubas ZS, Furtado PV, Putman SB, Moraes W, Liveira MJ, Oliveira CA, Moreira N. The use of altrenogest to ovoid hyperestrogenism after eCG-hCG ovulation induction in southern tigrina (*Leopardus guttulus*). Theriogenology. 2015;84:575–582.

[46] Pfeifer LFM, Leonardi CEP, Castro NA, Viana JHM, Siqueira LGB, Castilho EM, Singh J, Krusser RH, Rubin MIB. The use of PGF2a as ovulatory stimulus for timed artificial insemination in cattle. Theriogenology. 2014;81:689–695.

[47] Kriegsfeld LJ, Ubuka T, Bentley GE, Tsutsui K. Seasonal control of gonadotropin-inhibitory hormone (GnIH) in birds and mammals. Frontiers in Neuroendocrinology. 2015;37:65–75.

[48] Ahmed HMM, Yeh JY, Lin WJ, Forsberg NE, Cheng WTK, Ou BR. Validation of a luciferase bioassay to detect the progestative activity in gilts whose estrus was induced by an uterotonic herb (*Ligusticum chuanxiong*). Livestock Science. 2014;163:159–164.

[49] Howard J, Wildt DE. Approaches and efficacy of artificial insemination in felids and mustelids. Theriogenology. 2009;71:130–148.

[50] Pelican KM, Wildt DE, Pukazhenthi B, Howard J. Ovarian control for assisted reproduction in the domestic cat and wild felids. Theriogenology. 2006;66:37–48.

[51] Swanson WF, Paz RCR, Morais RN, Gomes MLF, Moraes W, Adania CH. Influence of species and diet on efficiency of *in vitro* fertilization in two endangered Brazilian felids—the ocelot (*Leopardus pardalis*) and tigrina (*Leopardus tigrinus*). Theriogenology. 2002;57:593.

[52] Silva ME, Colazo MG, Ratto MH. GnRH dose reduction decreases pituitary LH release and ovulatory response but does not affect corpus luteum (CL) development and function in llamas. Theriogenology. 2012;77:1802–1810.

[53] Crichton EG, Bedows E, Miller-Lindholm AK, Baldwin DM, Armstrong DL, Graham LH, et al. Efficacy of porcine gonadotropins for repeated stimulation of ovarian activity for oocyte retrieval and in vitro embryo production and cryopreservation in Siberian tigers (*Panthera tigris altaica*). Biology of Reproduction. 2003;68:105–113.

[54] Phillips LG, Simmons LG, Bush M, Howard JG, Wildt DE. Gonadotropin regimen for inducing ovarian activity in captive wild felids. JAVMA. 1982;181:1246–1250.

[55] Morato RG, Crichton EG, Paz RCR, Zuge RM, Moura CA, Nunos AVL, Teixeira RH, Porto L, Priscila M, Guimaraes MABV, Correa SHR, Barnabe RC, Armstrong DL, Loskutoff NM. Ovarian stimulation and successful *in vitro* fertilization in the jaguar (*Panthera onca*). Theriogenology. 2000;53:339.

[56] Morato RG, Bueno MG, Malmheister P, Verreschi ITN, Barnabe RC. Changes in the fecal concentrations of cortisol and androgen metabolites in captive male jaguars (Panthera onca) in response to stress. Brazilian Journal of Medical and Biological Research. 2004;37:1903–1907.

[57] Swanson WF. Application of assisted reproduction for population management in felids: the potential and reality for conservation of small cats. Theriogenology. 2006;66:49–58.

[58] Brown JL. Female reproductive cycle of wild female felids. Animal Reproduction Science. 2011;124:155–162.

[59] Johnson AEM, Freeman EW, Colgin M, McDonough C, Songsasen N. Induction of ovarian activity and ovulation in an induced ovulator, the maned wolf (*Chrysocyon brachyurus*), using GnRH agonist and recombinant LH. Theriogenology. 2014;82:71–79.

[60] Chittick E, Rotstein D, Brown T, Wolfe B. Pyometra and uterine adenocarcinoma in a melengestrol acetate-implanted captive coati (*Nasua nasua*). Journal of Zoo and Wildlife Medicine. 2001;32:245–251.

[61] Zanetti ES, Polegato BF, Duarte JMB. Comparison of two methods of synchronization of estrus in brown brocket deer (*Mazama gouazoubira*). Animal Reproduction Science. 2010;117:266–274.

[62] Maia KM, Peixoto GC, Campos LB, Silva AM, Castelo TS, Ricarte ARF, Silva AR. Estrous synchronization in captive collared peccaries (*Pecari tajacu*) using a prostaglandin F2α analog. Zoological Society. 2014;31:836–839.

[63] Michael AE, Abayasekara DRE, Webley GE. The luteotrophic actions of prostaglandins E2 and F2α on dispersed marmoset luteal cells are differentially mediated via cyclic AMP and protein kinase C. Journal of Endocrinology. 1993;138:291–298.

[64] Webley GE, Michael AE, Abayasekara DRE. The relationship between the production and the anti-gonadotrophic action of prostaglandin F2a in luteal cells from the marmoset monkey (*Callithrix jacchus*) in the early and mid-luteal phase. General and Comparative Endocrinology. 2010;166:436–442.

[65] Marshall VS, Browne MA, Knowles L, Golos TG, Thomson JA. Ovarian stimulation of marmoset monkeys (*Callithrix jacchus*) using recombinant human follicle stimulating hormone. Journal of Medical Primatology. 2003;32;57–66.

[66] Grupen CG, Gilchrist RB, Nayudu PL, Barry MF, Schulz SJ, Ritter LJ, Armstrong DT. Effects of ovarian stimulation, with and without human chorionic gonadotrophin, on oocyte meiotic and developmental competence in the marmoset monkey (*Callithrix jacchus*). Theriogenology. 2007;68:861–872.

[67] Pukazhenthi B, Comizzoli P, Travis AJ, Wildt DE. Applications of emerging technologies to the study and conservation of threatened and endangered species. Reproduction, Fertility and Development. 2006;18:77–90.

[68] Lermen D, Blömeke B, Browne R, et al. Cryobanking of viable biomaterials: implementation of new strategies for conservation purposes. Molecular Ecology. 2009;18:1030–1033.

[69] Domingues SFS, Ferreira HS, Muniz JAPC, Lima AKF, Ohashi OM, Figueiredo JR, Silva LDM. Mechanical isolation of capuchin monkey (*Cebus apella*) preantral ovarian follicles. Arquivo Brasileiro de Medicina Veterinária e Zootecnia. 2003;55:301–308.

[70] Brito AB, Santos RR, van den Hurk R, Lima JS, Miranda MS, Ohashi OM, Domingues SF. Short-term culture of ovarian cortical strips from capuchin monkeys (*Sapajus apella*): a morphological, viability, and molecular study of preantral follicular development *in vitro*. Reproductive Sciences. 2013;20:990–997.

[71] Domingues SF, Caldas-Bussiere MC, Petretski MD, Ohashi OM, Lima JS, Santos RR, Cordeiro MS, Gomes de Castro PH. Effects of follicular phase and oocyte–*cumulus* complexes quality on the protein profile and *in vitro* oocyte meiosis competence in *Cebus apella*. Fertility and Sterility. 2010;93:1662–1667.

[72] Marshall VS, Browne MA, Knowles L, Golos TG, Thomson JA. Ovarian stimulation of marmoset monkeys (*Callithrix jacchus*) using recombinant human follicle stimulating hormone. Journal of Medical Primatology. 2003;32:57–66.

[73] Delimitreva S, Tkachenko OY, Berenson A, Nayudu PL. Variations of chromatin, tubulin and actin structures in primate oocytes arrested during *in vitro* maturation and fertilization—what is this telling us about the relationships between cytoskeletal and chromatin meiotic defects? Theriogenology. 2012;77:1297–1311.

[74] Yeoman RR, Helvacioglu A, Williams LE, Aksel S, Abee CR. Restoration of oocyte maturational competency during the nonbreeding season with follicle-stimulating hormone stimulation in squirrel monkeys (*Saimiri boliviensis boliviensis*). Biology of Reproduction. 1994;50:329–335.

[75] Schuler AM, Westberry JM, Scammell JG, Abee CR, Kuehl TJ, Gordon JW. Ovarian stimulation of squirrel monkeys (*Saimiri boliviensis boliviensis*) by using pregnant mare serum gonadotropin. Comparative Medicine. 2006;56:13–17.

[76] Dukelow WR, Jarosz SJ, Jewett DA, Harrison RM. Laparoscopic examination of the ovaries in goats and primates. Journal of Laboratory Animal Science. 1971;21:594–597.

[77] Kuehl TJ, Dukelow WR. Maturation and *in vitro* fertilization of follicular oocytes of the squirrel monkey (*Saimiri sciureus*). Biology of Reproduction. 1979;21:545–556.

[78] Wiedemann C, Zahmel J, Jewgenow. Short-term culture of ovarian cortex pieces to assess the cryopreservation outcome in wild felids for genome conservation. BMC Veterinary Research. 2013;9:37.

[79] Jewgenow K, Stolte M. Isolation of preantral follicles from nondomestic cats–viability and ultrastructural investigations. Animal Reproduction Science. 1996;44:183–193.

[80] Baldassare H, Carelli JB, Requena LA, Rodrigues MG, Ferreira S, Salomão J, Jorge Neto PN. Efficient recovery of oocytes from onça parda (*Puma concolor*) by laparo-

scopic ovum pick-up of gonadotropin-stimulated females. Animal Reproduction. 2015;12:717.

[81] Chaves MG, Miragaya MH, Capdevielle EF, Rutter B, Giuliano SM, Agüero A. *In vitro* maturation of vicuna oocytes recovered by surgical aspiration of follicles from superstimulated ovaries. Biocell. 2004;28:545.

[82] Luvoni GC. Cryosurvival of *ex situ* and *in situ* feline oocytes. Reproduction in Domestic Animals. 2012;47:266–268.

[83] Tingen C, Kim A, Woodruff TK. The primordial pool of follicles and nest breakdown in mammalian ovaries. Molecular Human Reproduction. 2009;15:795–803.

[84] Comizzoli P, Songsasen N, Wildt DE. Protecting and extending fertility for females of wild and endangered mammals. Cancer Treatment and Research. 2010;156:87–100.

[85] Matos MH, Lima-Verde IB, Luque MC, Maia JE Jr, Silva JR, Celestino JJ, Martins FS, Báo SN, Lucci CM, Figueiredo JR. Essential role of follicle stimulating hormone in the maintenance of caprine preantral follicle viability *in vitro*. Zygote. 2007;15:173–182.

[86] Gilchrist RB, Nayudu PL, Hodges JK. Maturation, fertilization, and development of marmoset monkey oocytes *in vitro*. Biology of Reproduction. 1997;56:238–246.

[87] Nayudu PL, Wu J, Michelmann HW. In vitro development of marmoset monkey oocytes by pre-antral follicle culture. Reproduction in Domestic Animals. 2003;38:90–96.

[88] Scalercio SRRA, Brito AB, Domingues SFS, Santos RR, Amorim CA. Immunolocalization of growth, inhibitory, and proliferative factors involved in initial ovarian folliculogenesis from adult common squirrel monkey (*Saimiri collinsi*). Reproductive Sciences. 2015;22:68–74.

[89] Yeoman RR, Williams LE, Abee CR. Low oxygen inhibits but complex high-glucose medium facilitates *in vitro* maturation of squirrel monkey oocyte-granulosa cell complexes. Journal of Assisted Reproduction and Genetics. 1999;16:102–107.

[90] Asakawa T, Chan PJ, Dukelow WR. Time sequence of *in vitro* maturation and chromosomal normality in metaphase I and metaphase II of the squirrel monkey (*Saimiri sciureus*) oocyte. Biology of Reproduction. 1982;27:118–124.

[91] Johnston LA, Donoghue AM, Obrien SJ, Wildt DE. Rescue and maturation *in vitro* of follicular oocytes collected from nondomestic Felid Species. Biology of Reproduction. 1991;45:898–906.

[92] Miller AM, Roelke ME, Goodrowe KL, Howard JG, Wildt DE. Oocyte recovery, maturation and fertilization *in vitro* in the puma (*Felis concolor*). Journal of Reproduction and Fertility. 1990;88:249–258.

[93] Weir BJ. The reproductive physiology of the plains viscacha, *Lagostomus maximus*. Journal of Reproduction and Fertility. 1971;25:355–363.

[94] Jensen FC, Willis MA, Leopardo NP, Espinosa MB, Vitullo AD. The ovary of the gestating South American plain vizcacha (*Lagostomus maximus*): suppressed apoptosis and corpora lutea persistence. Biology of Reproduction. 2008;79:240–246.

[95] Leopardo NP, Jensen F, Willis MA, Espinosa MB, Vitullo AD. The developing ovary of the South American plains vizcacha, *Lagostomus maximus* (Mammalia, Rodentia): massive proliferation with no sign of apoptosis-mediated germ cell attrition. Reproduction. 2011;141:633–641.

[96] Silva AR, Lima GL, Lima LF, Rocha RMP, Oliveira MF, Rodrigues APR, Figueiredo JR. Effect of FSH supplementation in the TCM 199 medium for *in vitro* culture of ovarian preantral follicles from collared peccaries (*Pecari tajacu*). In: Proceedings of the Systems Biology of Reproduction Fertility A global challenge; 2014; Grand Rapids. EUA: SSR; 2014. p. 442.

[97] Schönfeldt V, Chandolia R, Kiesel L, Nieschlag E, Schlatt S, Sonntag B. Assessment of follicular development in cryopreserved primate ovarian tissue by xenografting: prepubertal tissues are less sensitive to the choice of cryoprotectant. Reproduction. 2011;141:481–490.

[98] Schönfeldt V, Chandolia R, Kiesel L, Nieschlag E, Schlatt S, Sonntag B. Advanced follicle development in xenografted prepubertal ovarian tissue: the common marmoset as a nonhuman primate model for ovarian tissue transplantation. Fertility and Sterility. 2011;95:1428–1434.

[99] Candy CJ, Wood MJ, Whittingham DG. Ovary and ovulation: follicular development in cryopreserved marmoset ovarian tissue after transplantation. Human Reproduction. 1995;10:2334–2338.

[100] Wanderley LS, Luz HK, Faustino LR, Lima IM, Lopes CA, Silva AR, Báo SN, Campello CC, Rodrigues AP, Figueiredo JR. Ultrastructural features of agouti (*Dasyprocta aguti*) preantral follicles cryopreserved using dimethyl sulfoxide, ethylene glycol and propanediol. Theriogenology. 2012;77:260–267.

[101] Praxedes ECG, Lima GL, Silva AM, Bezerra JAB, Souza ALP, Campos LB, Silva AR. Vitrificação de tecido ovariano de preá (*Galea spixi*). [Vitrification of cavies (*Galea spixi*) ovarian tissue]. In: Proceedings of the XXI Congresso Brasileiro de Reprodução Animal, 2015, Belo Horizonte. Minas Gerais: CBRA; 2015.

[102] Lima GL, Santos EAA, Lima LF, Luz VB, Rodrigues APR, Silva AR. Short-term preservation of Pecari tajacu ovarian preantral follicles using phosphate buffered saline (PBS) or powdered coconut water (ACP®) media. Arquivo Brasileiro de Medicina Veterinária e Zootecnia. 2014;66:1623–1630.

[103] Lima GL, Luz VB, Alves AMCV, Lunardi FO, Souza ALP, Peixoto GCX, Rodrigues APR, Oliveira MF, Silva AR. Vitrification of collared peccaries (*Tayassu tajacu*) ovarian tissue using various cryoprotectants preliminary results. In: Proceedings of the IV

International Symposium on Animal Biology of Reproduction. 2012; Campinas. São Paulo: ISABR; 2012.

[104] Asher GW, Jabbour HN, Thompson JG, Tervit HR, Morrow CJ. Superovulation of farmed red deer (*Cervus elaphus*) and fallow deer (*Dama dama*): incidence of ovulation and changes in plasma hormone concentrations during the pré-ovulatory period in relation to ova recovery and fertilization. Animal Reproduction Science. 1995;38:137–154.

[105] González R, Berlinguer F, Espeso G, Ariu F, del Olmo A, Garde JJ, Gomendio M, Ledda S, Rolda ERS. Use of a neuroleptic in assisted reproduction of the critically endangered Mohor gazelle (*Gazella dama mhorr*). Theriogenology. 2008;70:909–922.

[106] Zanetti ES, Munerato MS, Cursino MS, Duarte JMB. Comparing two different superovulation protocols on ovarian activity and fecal glucocorticoid levels in the brown brocket deer (*Mazama gouazoubira*). Reproductive Biology and Endocrinology. 2014;12:24.

[107] Aller JF, Rebuffi GE, Cancino AK. Superovulation response to progesterone-eCG treatment in vicuna (*Vicugna vicugna*) in semicaptive conditions. Theriogenology. 2002;57:576.

[108] Aba MA, Miragaya MH, Chaves MG, Capdevielle EF, Rutter B, Agüero A. Effect of exogenous progesterone and eCG treatment on ovarian follicular dynamics in vicunas (*Vicugna vicugna*). Animal Reproduction Science. 2005;86:153–161.

[109] Lima JS, Leão DL, Sampaio RV, Brito AB, Santos RR, Miranda MS, Orashi OM, Domingues SFS. Embryo production by parthenogenetic activation and fertilization on *in vitro* matures oocytes from *Cebus apella*. Zygote. 2012;21:162–166.

[110] Takahashi T, Hanazawa K, Inoue T, Sato K, Sedohara A, Okahara J, Suemizu H, Yagihashi C, Yamamoto M, Eto T, Konno Y, Okano H, Suematsu M, Sasaki E. Birth of healthy offspring following ICSI in *in vitro*-matured common marmoset (*Callithrix jacchus*) oocytes. Plos One. 2014;9:e95560.

[111] Swanson WF, Brown JL. International training programs in reproductive sciences for conservation of Latin American felids. Animal Reproduction Science. 2004;82–83:21–34.

[112] Ishibashi H, Motohashi HH, Kumon M, Yamamoto K, Okada H, Okada T, Seki Kazuhiko. Efficient embryo transfer in the common marmoset monkey (*Callithrix jacchus*) with a reduced transfer volume: a non-surgical approach with cryopreserved late-stage embryos. Biology of Reproduction. 2013;88:1–5.

[113] Pereira AF, Santos MLT, Borges AA, Queiroz Neta LB, Santos MVO, Feitosa AKN. Isolation and characterization of skin-derived donor cells for nuclear transfer. Acta Veterinaria Brasilica. 2014;8:311–316.

[114] Borges AA, Queiroz Neta LB, Lima GL, Santos MVO, Santos MLT, Oliveira MF, Silva AR, Pereira AF. Atividade proliferativa de células provenientes de tecido auricular

de catetos (*Pecari tajacu*) após distintas técnicas de vitrificação. [Proliferative activity of cells derived from the ear tissue of collared peccaries (Pecari tajacu) after diferente vitrification techniques]. In: Proceeding of the XXV Congresso Brasileiro de Zootecnia; Fortaleza, Ceará; 2015.

[115] Tovar H, Navarrete F, Rodríguez L, Skewes O, Castro FO. Cold storage of biopsies from wild endangered native Chilean species in field conditions and subsequent isolation of primary culture cell lines. In Vitro Cellular & Developmental Biology – Animal. 2008;44:309–320.

[116] Borges AA, Lima GL, Santos MLT, Queiroz Neta LB, Praxedes ECG, Santos MVO, Oliveira MF, Silva AR, Pereira AF. Histological evaluation of ear tissue of collared peccary (*Pecari tajacu*) after different vitrification techniques. Animal Reproduction. 2015;12:230.

[117] Verma R, Liu J, Holland MK, Temple-Smith P, Williamson M, Verma PJ. Nanog is an essential factor for induction of pluripotency in somatic cells from endangered felids. BioResearch Open Access. 2013;2:72–76.

5

Cryopreservation of Sheep Produced Embryos – Current and Future Perspectives

Ricardo Romão, Carla C. Marques, Elisa Bettencourt and Rosa M.L.N. Pereira

Abstract

Due to economical and scientific limitations, sheep embryo reproductive technologies are less commercially applied than in other animal species. However, it is very clear that, in the near future, those techniques are expected to have a central role in animal production as a consequence of genetic and reproductive demands. One drawback is that results obtained after sheep embryo cryopreservation are unattractive for commercial purposes. It is expected that a successful cryopreservation of sheep embryos can push forward all other reproductive biotechnologies in this species, such as multiple ovulation and embryo transfer (MOET), artificial insemination, or *in vitro* production of embryos. This paper tries to discuss the current and future perspectives of cryopreservation of *in vivo*- and *in vitro*-produced sheep embryos concerning advantages and limitations for its practical use and possible solutions for improving methods to allow a higher survival rate of cryopreserved embryos.

Keywords: Sheep, *in vitro*-produced embryos, *in vivo*-produced embryos, cryopreservation

1. Introduction

In the last decade, the production of sheep embryos did not get the improvement researchers would like to announce, and its application is still lower compared to other species [1,2]. It is very clear, however, that these techniques will have a central role in animal production in the next few years as a consequence of genetic and reproductive challenges. In the agricultural context, their use will also follow the needs of a growing demand for agricultural products from the world emerging economies [3].

One of the constraints in this area is sheep embryo cryopreservation. Cryopreservation is of primordial importance to preserve embryos so they can be maintained as genetic reserves or for field application. The use of cryopreserved embryos can facilitate work planning, eliminate distance limitations, and especially reduce sanitary risks [4,5]. Although attempts have been made to understand the effects of freezing on sheep embryos since 1976 [6,7], the results are less than desirable [1,8]. A successful cryopreservation of sheep embryos could push forward the other reproductive biotechnologies in this species, such as MOET, artificial insemination (AI), or *in vitro* production (IVP) of embryos. These latter associated techniques are still expensive, consequently limiting their widespread. However, the improvement in the number of produced embryos as well as the number of cryopreserved embryos resulting in lambs after transfer could increase their use by technicians and farmers.

In sheep, as in other species, embryos may be *in vivo* or *in vitro* produced (IVP). *In vivo* embryo production consists of collecting them directly from the uterus either after natural service or after AI, whereas IVP involves oocyte collection from sheep ovaries followed by laboratorial maturation and fertilization. The advantages and utilization of each technique have been previously discussed [9,10]. Transfer of fresh *in vivo*-derived embryos achieves the highest lambing rates, with reported results of 46.4–66.7% [4], 67.8% [11], 75% [12], and 81.2% [13], whereas fresh IVP embryos have lower results, ranging from 32.8% [12] or 37.5% [11] to 40% [13]. However, after cryopreservation, these rates are even lower. The reported lambing rates after transfer of cryopreserved IVP embryos are not higher than 21.7% [13], 19.4–23.8% [14], or 23–26.6% [11], whereas the results for *in vivo*-derived cryopreserved embryos are 32–36% [4], 50% [15], 60% [16], or 60.1–75.1% [14].

The above results highlight low cryotolerance as the main obstacle on sheep embryo cryopreservation, especially for IVP embryos. In the last 15 years, researchers have tried to understand the reason for the low cryotolerance of IVP embryos and keep on searching for new methods to enhance the results in this species. Both *in vivo* and IVP sheep cryopreserved embryos present ultrastructural damage [17,18] corresponding to the expression of damage to cells during the process. Cryopreserved embryos show less microvilli and lower mitochondria activity compared to fresh ones [17–19]. The lower cryotolerance of IVP blastocysts can be also related to the excessive accumulation of lipids [13,20–23], which can be favored by their culture in serum-containing media [22,24,25]. Likewise, the nature and/or concentration of cryoprotectant(s) and the freezing protocol used are also determinants for embryo cryosurvival [19]. According to Tveden-Nyborg et al. [26], IVP sheep embryos are characterized by a slower developmental rate with detrimental effects in placentation, increasing fetal loss, and lowering lambing rates. Moreover, vitrified/warmed IVP sheep embryos presented decreased concentrations of individual amino acids in the amniotic fluid during the second trimester of gestation possibly due to an impaired placental vasculogenesis and/or to a reduced placental transport [27]. In summary, there are a vast number of variables that could potentially influence the outcome of cryopreservation and more studies are urgently needed to improve its success.

The present paper briefly reviews the current status of sheep embryo cryopreservation, discussing the main applications of this technique, the limitations of its use, and the progress and future perspectives worldwide.

2. Main applications of sheep cryopreserved embryos

One of the best ways to preserve the genetic resources of endangered populations is through the establishment of germplasm banks in which a long storage of biological material can be achieved. The Food and Agriculture Organization (FAO) has established the minimum number of individuals in each population so they can have a classification of endangered [28]. This classification is the base for choosing the target breeds or species needing to be urgently preserved. Following Rio de Janeiro convention, in 1992, all countries must have a plan of conservation of autochthone genetic resources, also recognizing their qualities concerning adaptation to local conditions as well as their potential use in agricultural production niches [29]. At actual knowledge, cryopreservation is a reliable way of long-term conservation of genetic resources [30]. One of the current alternative strategies for maintenance of some breeds or strains is through embryo cryopreservation, as is now routinely used in mouse strains. For instance, for a long-term maintenance of a colony, cryopreserved embryos have the advantages of saving costs, readiness in distant transport, and health guarantee and also of preventing genetic drift occurrence [31,32]. Germplasm exchanges are nowadays crucial [33] and embryo transfer can have a major role in reducing or eliminating some transmissible diseases in germplasm livestock changes [23,34,35]. For commercial proposes, embryo transfer should be chosen for those reasons in which other cheaper techniques are not as advantageous. For example, Cognié and Baril [35] estimated a 10 times higher cost when comparing it (using either *in vivo* or *in vitro* embryos) with AI, and this represents an elevated cost attending the economic value of the animal species.

On behalf of developing new and improved reproductive-assisted techniques, mainly by research teams, there has always been an interest for research in sheep embryo production. The birth of the first IVP lambs in 1991 [36,37] and the first cloned animal (a lamb) by nuclear transfer in 1997 [38] are only two examples. Based on the International Embryo Transfer Society (IETS) data, a decade ago, Thibier [39] reported 6674 fresh and 2907 frozen sheep embryos transferred in the world in 2003, whereas, for the year 2012, Perry [40] referred 8124 fresh and 4120 frozen embryos, mainly represented by Australia, South America, and South Africa. These numbers, although considered underestimated because of the difficulty in retrieving data, show an evolution on the use of this reproductive biotechnology and, in our opinion, reflect the demands of an emerging global market in this area.

In sheep, as in other species, several methods can be used for *ex situ* conservation of genetic resources, and attention in reproductive cells has been focused in oocyte, spermatozoa, zygote, or embryo cryopreservation [41,42]. In parallel, in the last years, research has been conducted in other methods such as ovary, testicle [42–45], and somatic cell cryopreservation [46,47] as well as in new methods in embryo technology such as somatic cell nucleus transfer and transgenesis [46,48].

Taking into account the background of other species, we can mention that, for example, in the United States, 79% of all bovine embryos collected in 2011 were cryopreserved [49]. In sheep, cryopreservation of embryos is a crucial developing technique for commercial application, and in our opinion, when good results for cryopreservation of sheep embryos have been attended, it will be the dominant technique of sheep embryo production.

3. Embryo production

3.1. *In vivo* embryo production

Despite recent development at *in vitro* embryo production in small ruminants [8,11,50], *in vivo*-produced embryos still represent the greatest source used for the implementation of breeding and conservation programs of endangered livestock [34,40]. To increase the number of embryos that can be recovered and transferred from each donor female, *in vivo* embryo production is generally associated with multiple ovulation and embryo transfer (MOET) programs. Despite the increased efficiency of *in vivo* embryo production in the last years, there are still some problems to be solved. These problems are mainly related to the great variations observed in the ovarian response, fertilization rates, and embryo quality of hormonally stimulated ewes [9,35,51,52].

3.1.1. Estrus synchronization and superovulation programs

To overcome the reproductive seasonality in this species and to obtain embryos from several females simultaneously while improving superovulatory response, MOET is usually associated with estrus synchronization programs [34]. Estrus synchronization can be performed using an exogenous progestagen administered through the application of an intravaginal sponge or using a controlled internal drug release (CIDR) device [35,53]. During the breeding season, in cycling females, estrus synchronization can also be achieved injecting a luteolytic agent, PGF2α [54]. The application of an intravaginal sponge with synthetic progestogens such as fluorogestone acetate (FGA) or medroxiprogesterone (MAP) for 12 to 14 days is frequently used for estrus synchronization. Although allowing good estrus synchronization, the fertility is usually lower compared to natural estrus [55]. Several reasons have been raised to explain this lower fertility, including (a) the asynchrony between the beginning of estrus and the ovulation time [56], (b) a negative influence on the transport of spermatozoa [57], (c) the reduced expression of estradiol and progesterone receptors in oviduct and endometrial cells, (d) a reduced progesterone secretion by the corpus luteum (CL) [58], (e) an inadequate process of maturation and acquisition of competence by the oocyte [59,60], and (f) the development and lack of ovulation of large persistent follicles [53,59–61] due to altered levels of progesterone after the first 6 days of treatment [62].

Several alternatives have been implemented to overcome the negative effects of long-term treatment with progesterone, including the application of superovulatory treatment starting after natural estrus detection without the use of sponges [63]. The insertion of a second progesterone device to maintain constant levels of progesterone treatment has also been performed [51].

Multiple ovulation is mainly acquired using FSH, although some programs still use equine chorionic gonadotrophin (eCG) alone or associated with FSH [52,64–66]. Despite the increase in the ovulation rate, the number of embryos obtained with eCG is often lower than expected. The long half-life of eCG allows it to be applied in a single administration, which naturally simplifies the process and reduces the associated costs [65]. However, when given in high doses, eCG has some undesirable effects, changing the hormonal profile and therefore conditioning the ovulation and fertilization rates as well as the embryo collection rate and viability [67,68]. Actually, commercial FSH is used in most of the MOET programs. The short half-life of FSH implies, however, a 12-hour interval of repeated administration for a period of 4 days [69]. FSH may be of porcine (pFSH) [66,70] or ovine (oFSH) source [52,66,71]. Usually, the treatment begins 72 hours before the withdrawal of exogenous progesterone. The response to this treatment is dependent on the number of applied doses [72,73] and used protocol (constant or decreasing doses) [51,71,73]. There are also intrinsic factors that may affect the superovulatory response in small ruminants, namely, breed, age, nutrition, and reproductive status, the latter being directly conditioned by the reproductive seasonality in this species [51,66]. In addition, individual differences are of crucial importance in the embryo development process [51].

The genetic influence on the variability of superovulatory response was identified since 1986 [74], and in general, the most prolific breeds have a better response to superovulation treatments [75]. To deal with this variability, a simple protocol for identification of higher responders to superovulation allowing a preselection of ewe donors was recently suggested [76]. Additionally, the influence of seasonality in response to superovulation treatment is obviously more noticeable in breeds located in higher latitudes, mainly affecting the viability of obtained embryos [77,78]. Conversely, in temperate regions, a less pronounced seasonality has been identified without differences in the superovulatory response between animals treated inside or outside the breeding season [66,79–81]. However, embryos collected after the administration of the treatment within the breeding season had higher quality [66,78].

The ovulation rate is directly related to the number of small follicles (2–3 mm) at the onset of treatment [82], but in sheep the number of viable embryos depends on a more restricted population of follicles of 3 mm in diameter [83]. The influence of reproductive seasonality in the efficacy of superovulatory response may also be related to the stage of follicular development during the treatment implementation [78]. According to González-Bulnes et al. [58,84], the presence of a CL at the beginning of FSH treatment may improve the number of viable embryos. Moreover, before the application of the multiple ovulation treatment and to improve the superovulatory response, GnRH agonists or antagonists can also be administered [59]. These latter strategies intend to inhibit the final stage of follicular development [75], suppressing the existence of follicles more than 3 mm [85] and simultaneously implementing the development of small follicles [86].

3.1.2. Oocyte fertilization and embryo recovery

Even controlling the limiting factors associated with superovulation treatments, differences still exist when comparing the total embryo production/total embryos able to be cryopreserved to the number of embryos previewed from the follicles observed in the superovulated ovary. In fact, several failures were described, being the most evident ones associated with failures in the ovulation and fertilization processes [51,87], embryo degeneration [88], and inefficiency of embryo recover methods [9,34].

Anovulation has been described as one of the main reasons for fertilization failures, especially due to the associated higher estradiol concentrations [87], influencing the uterine environment and affecting oocytes [89] and spermatozoa [90] transport. However, other authors described no effect of these anovulatory follicles [91]. Another reason for fertilization failure may be related to asynchrony among ovulations. Although a good synchrony concerning the preovulatory wave of LH has been described, the interval among different ovulations in each female can extend from 6 to 12 hours, with an average of 6 hours [9], impairing the fertilization process. A better synchrony among ovulations can be achieved through the application of GnRH 30 to 36 hours after progestagen removal [87]. Conversely, to overcome the deficient spermatozoa transport, intrauterine insemination at 48 hours after progestagen removal is frequently applied [92,93]. Protocols performing both intrauterine insemination and natural mating were associated with the best fertilization rates compared to the individual use of one of these methods [72,92].

Embryo degeneration can result from a deficient oocyte maturation process or from extrinsic factors that may affect the spermatozoon or the embryo itself. This degeneration is mainly related to changes in the endocrine environment of the oviduct or uterus and to anomalies of luteal function [88,94]. In fact, the number and quality of embryos can be affected by changes in the uterus, namely, modifications in endometrial development and its nutritional competence, due to the superovulatory treatment [88]. It is clearly stated that a normal luteal function assuring high progesterone levels is essential for early embryonic development. Premature luteal regression has been described in superovulated sheep [71,81,95] being more frequent after estrus synchronization with a double treatment of PGF2α [95] and during seasonal anestrus [96].

In sheep, embryo recover is performed at day 6 or 7 (estrus day = day 0), most of the times by laparotomy [35]. This technique allows recovery rates of approximately 80%, but adherences after the second surgery may compromise following recovery rates [35,92]. The use of heparinized saline solution can minimize the occurrence of adherences allowing no changes in recovery rates after two [97] and three [71] successive surgeries. Less invasive techniques, such as laparoscopy, minimize the risk of surgical adherences, offering advantages in animal welfare, but the embryo recovery rates they offer are lower (65% and 70% vs. 80%) [92]. The application of a modified laparoscopic technique allowed repeated recoveries in the same ewe, keeping a recovery rate lower than the one obtained with laparotomy, hence offering no commercial advantage [98]. In a conservation program, the total number of embryos recovered and frozen after a MOET program is approximately 7 and 6 per superovulated ewe, respectively, associated with an embryo recovery rate between 60 and 70% [66].

3.2. *In vitro* embryo production

In vitro embryo production simulates the natural formation of an embryo and comprises the stages of oocyte maturation (IVM), *in vitro* fertilization (IVF), and *in vitro* embryo culture (IVC). In these steps, the last one is critical for determining the blastocyst yield [99]. As stated also in other species, *in vitro* techniques reduce the rate of produced embryos and, as previously discussed, lead to inferior embryo survival after transfer compared to *in vivo*-derived embryos [23,35,99,100]. Concerning sheep embryo production from abattoir-derived oocytes, Cocero et al. [101] obtained 34.6% and 40.4% blastocyst yield at days 7 and 8, respectively, Romão et al. [50] obtained 42.6–46.9% at day 6, and Cognié et al. [9] obtained 25% at day 8.

In terms of lambing rate, IVP embryo survival is 25% lower compared to *in vivo*-produced embryos mainly due to an increased embryo loss at days 30 to 40 [10]. Other authors [102] have discussed the principal mechanisms underlying embryonic losses, suggesting that the preimplantation period of development is the most sensitive to cellular damage. Specifically, the importance of permanent embryo demise that occurs in the first week of development was also pointed out [103]. In fact, the extent and regulation of cell death during preimplantation development is likely to be critical for later development of the concepts. Therefore, it is of primordial importance to implement tools and culture conditions that can promote better IVP results.

In vitro embryo production is not viewed as a controlled and totally defined technique and in sheep has been associated with the large offspring syndrome (LOS) [20,104], also reported in other species such as bovine [105]. This syndrome is characterized by high birth weights, increased gestation length, frequent dystocia, elevated abortion, perinatal mortality rates, and various morphological deviations [20,106]. These abnormalities seem to be promoted very early in developmental stages, in IVM or IVF [3,104], and apparently can be justified by the use of serum supplementation and coculture with somatic cells [3]. Moreover, these problems could be extended after birth, as higher growing rates were found until weaning age in lambs born from IVP [107]. It is known that IVP embryos have altered morphology compared with *in vivo* derived [18,108]. On the contrary, an increase in the lipid content of IVP was observed compared to *in vivo* bovine morulae and blastocysts regardless of the composition of the medium of culture, although differences also depend on the culture medium used [22]. These changes were further correlated to the lower cryotolerance of IVP embryos as stated before.

Several studies have tried to establish the most successful laboratory method of IVP, including oocyte IVM, IVF, and IVC of embryos, as all these steps are crucial to obtain good results.

3.2.1. Collection of oocytes

The anatomical constraints to access the sheep uterus, mainly by the peculiar anatomy of the cervix, have reduced the chances of an easy collection of oocytes and embryos, although future interest in new techniques [109] could improve welfare issues and results. There are several techniques of oocyte collection, namely, follicular aspiration or ovary slicing in abattoir-derived ovaries [110,111] and ovum pick-up systems applied in living animals. Ovary slicing technique provides cumulus-oocyte complexes (COCs) of better quality and the number of harvested oocytes collected is higher compared to follicle aspiration [110,112]. Sheep ovum pick-up systems are applied in living animals using minimal invasive procedures such as laparoscopic ovum pick-up (LOPU) technique guided by laparoscopy [113–115]. As referred before, these procedures when used *in vivo* are usually associated with follicular stimulation treatments [116] to raise the number of collected COCs [117], allowing better maturation, fertilization, and *in vitro* development capacities [101,118]. Moreover, the development of these *in vivo* techniques of oocyte collection is important, as abattoir-derived oocytes usually cannot be used in genetic programs because of their unknown sanitary and even genetic status, being applied mainly for research proposes [115].

Oocyte quality seems to be the key factor to achieve high embryo production ratios [9,119]. This quality depends on the competence of the oocyte to complete meiosis, which is affected by the follicle size [35]. It is also clearly established that prepubertal oocytes do not have the developmental potential of those collected from adult ewes [118], although, in the future, the results obtained with juvenile oocyte collection could have potential interest for decreasing half the generation interval in genetic programs [112].

3.2.2. *In vitro Maturation (IVM)*

Collected oocytes to be fertilized must undergo nuclear and cytoplasmic maturation. During IVM, oocytes must reach nuclear metaphase II stage. Besides nuclear maturation, several modifications in cytoplasmic organelles and compounds need to be accomplished. Oocyte cytoplasmic maturation and metabolism are mandatory, as ATP seems to have an important role in determining oocyte quality by supporting energy requirements for further development [120,121].

IVM is usually performed at 38–39°C in a humidified atmosphere in air and 5% CO_2 [8,18,50, 112]. Tissue Culture Medium (TCM199) supplemented with bicarbonate, pyruvate, LH, FSH, and estradiol (E_2) and with 10% fetal calf serum [10,101] or 10% estrous sheep serum [50,122] is the widest used medium for IVM. Nevertheless, other media and supplementation can be also used. For example, Shabankareh et al. [123] obtained good results using human menopausal serum and Birler et al. [124] showed that, in sheep, SOF medium improves the rate of cleavage compared to TCM199 medium. Also, Guler et al. [125] obtained better results adding follicular fluid to maturation medium or using epidermal growth factor (EGF) conjugated with FSH and E_2 [50]. The positive effect of EGF or other EGF-like ligands in COCs has been shown during IVM [126,127], whereas other researchers [101,128] preferred EGF and cysteamine. Cumulus cells are important to IVM success and subsequent fertilization. These cells and their physical contact with the oocyte are of primordial importance not only for the resumption and progression of nuclear maturation but also for cytoplasmic maturation [129–131].

3.2.3. *In vitro Fertilization (IVF)*

Fertilization of matured oocytes can be performed with frozen-thawed or fresh semen. Frozen semen is easily available for routine use, and some authors argued that early embryo development is not influenced by the method of semen preservation [132], whereas others demonstrated that fresh semen could improve embryo production rates and quality [50]. Eventually, freezing and thawing disrupt the stability of ram sperm chromatin, reducing its fertilization efficiency [133,134]. On the contrary, cryopreserved semen is also advantageous, as it can be stored in seasonal propitious period. However, when a ram of proven fertility is available, its can be used either fresh or frozen-thawed.

Fertilization is performed with previously capacitated spermatozoa using media containing heparin or sheep serum [10,122] that causes Ca^{2+} influx into the sperm acrosome and consequent capacitation [135]. The choice of the best semen fraction is usually achieved by Percoll gradient centrifugation [10] or swim-up [8,136].

3.2.4. *In vitro Culture (IVC)*

The goal for this phase is to simulate the events that would occur in the oviduct and uterus, so that putative zygotes obtained after fertilization would undergo optimal development into blastocysts. For this reason, culture can be performed *in vivo* using surrogate oviducts [137]. IVC, on the contrary, is achieved in a controlled atmosphere at 38.5–39°C with 5% O_2, 5% CO_2, and 90% N_2. This medium is usually composed of synthetic oviductal fluid supplemented with amino acids, bovine serum albumin (BSA), and/or serum [112,122], although the use of some media with oviduct cells or fetal fibroblasts was also reported [138]. In this stage, it is important to keep an oxygen low concentration to prevent oxidation [139] and the deleterious effects of reactive oxygen species (ROS). Some low molecular weight thiol compounds are also added to IVM and IVC media, such as cysteamine, β-mercaptoethanol, and cysteines, to increase the synthesis of glutathione (GSH), which in turn reduces the oxidative stress [140]. The embryos spend 6 to 7 days in culture before being transferred in fresh or after cryopreservation.

4. Cryopreservation of sheep embryos

The science of cryobiology aims to preserve cells at low temperatures (-196°C), intending to avoid the negative effects of the process, which could preclude the main objective. For cryopreservation, cells should follow the freezing and warming processes. Cryopreservation of sheep embryos is a difficult task, and practical results are not as encouraging as in cattle mainly due to the reduced embryo cryotolerance in this species. Several reasons have been indicated to justify this fact, mainly cellular [21,22], metabolic [141,142], and biochemical [143] changes. In an ultrastructural evaluation, IVP blastocysts exhibited less microvilli and a less extensive network of intercellular junctions, specifically an apparent lack of desmosomal junctions, a higher number of lipid droplets, and incidence of cellular debris compared to their *in vivo* counterparts [18]. After cryopreservation, both *in vivo* and IVP embryos exhibited ultrastructural damages that were much more severe in the latter [17,18]. These differences could certainly explain the observed lower cryotolerance of IVP embryos.

Although it seems that very low temperatures will disrupt cell components, the most complicated freezing window is between 15°C and -60°C and especially the step of intracellular ice formation between -5°C and -15°C [144]. The major risks for fracture of the *zona pellucida* or cytoplasm range from -50°C to -150°C [145]. Below -150°C, the risks for embryo damage are lower [146]. In embryo cells of species with high lipid content, as in sheep, the lipid droplets combine with the cytoskeleton, membranes, organelles, and other structures of the cytoplasm. Therefore, when the embryos are cooled below 15°C until -5°C, irreversible and fatal structural damages occur, mostly in the traditional freezing methods [146,147].

Differences in freezing methods were extensively identified depending on the speed of freezing particularly between slow and rapid cooling [23,144,148]. Three major methods have been used to preserve embryos: slow freezing, conventional vitrification or an adaptation of the last one, vitrification in open pulled straws (OPS) [149] or in other specialized embryo vitrification devices (copper or gold electron microscope grids [150] and pipette tips and cryotop [151,152], among others) referred as an ultrarapid vitrification. The major difference among them is the acceleration of the speed of freezing. Whereas slow freezing tries to differentiate the cooling rate in sequential steps of freezing, vitrification increases the freezing speed by reducing volume, thus minimizing ice formation. These techniques were compared using *in vivo*-derived or IVP embryos in sheep [14,52,153] or bovine [154]. Bettencourt et al. [52] described no differences in lambing or embryo survival rates among slow freezing, conventional vitrification, and vitrification in OPS in *in vivo*-derived embryos. Although the achieved results by each method can be relevant, the cost and application in routine field use is also of primordial importance [14,153].

4.1. Slow freezing

The method of slow freezing tries to control the descending cooling event in several steps, regulating extracellular and intracellular water exchange, with a balance between ice crystal formation and structural damage, cryoprotectant toxicity, and osmotic damage [146,155]. The method is possible using programmable freezers with a rapid cooling until -6°C to -7°C, when ice formation is induced (seeding), and then a lower rate of cooling (0.3–1°C/min) causing freezing of extracellular ice and increasing the concentration of extracellular solution and cellular dehydration; at -30°C, straws are then plunged in liquid nitrogen [23,31].

Despite being used by many teams in the world, it is argued that slow freezing has a limited future in embryology, and according to Vajta and Nagy [146], "the rate of advancement in oocyte and embryo cryopreservation will depend on the rate by which embryologists and decision-makers adopt the new approaches".

4.2. Vitrification

Vitrification was found to be interesting in embryo cryopreservation because it reduces cryoinjuries caused by ice formation [156]. In fact, vitrification is based on embryo manipulation into different carrier tools applied to minimize the volume and to submerge the sample quickly into liquid nitrogen, allowing an ultrafast freezing speed, which avoids ice crystal formation, thus eliminating its deleterious effects in the cell [157]. This technique combines the use of small volumes with high concentration of two or more cryoprotectants [158] and it can be more adapted to IVP embryos [149,159,160].

According to several authors, the slow freezing techniques are time consuming and laborious [154], whereas vitrification is simple, rapid, and inexpensive [149,161,162]. Moreover, no publications have demonstrated that the results obtained by vitrification were significantly worse than those obtained by slow freezing [146]. Concerns in safety of vitrification related to transmissible diseases are partially justified, but new methods have been developed to avoid these constraints [163].

Several authors described good results with this cryopreservation technique in sheep. For example, Baril et al. [153] and Bettencourt et al. [52] achieved good field results (50–60% lambing rate) with vitrification of *in vivo*-derived sheep embryos even with direct transfer of cryopreserved embryos, and Folch et al. [4] achieved lambing rates of 32–36%, also in *in vivo*-derived embryos. In IVP embryos, Ptak et al. [164] found no differences in pregnancy rates between fresh and vitrified transferred embryos (47% vs. 42%). However, they achieved significant differences in lambing rates (41% vs. 23%), showing that vitrification of IVP embryos is not yet an optimized technique.

The success of vitrification depends on the stage at which embryos are cryopreserved apart from the used method. In fact, there are different levels of sensitivity to low temperatures depending on the stage of embryo development [165]. Garcia-Garcia et al. [166,167] and Shirazi et al. [168] showed that cryotolerance of IVP embryos to conventional slow freezing or vitrification increased as the developmental stage of embryos progressed perhaps due to the higher cryotolerance of blastocysts compared with early developmental embryos. In fact, Garcia-Garcia et al. [167], using slow freezing, described similar rates of viability between fresh and frozen-thawed embryos (92.5% vs. 83.7%) frozen at the blastocyst stage.

4.2.1. Vitrification in OPS and other specialized embryo devices

To achieve a higher rate of temperature reduction, the OPS method uses thinner, superfine straws, with their ends having half the diameter of the conventional ones. This characteristic enables filling the OPS by capillary action with a volume of approximately 1 μL, different from the conventional straw that contains 5 μL [146,149]. Therefore, this can allow freezing rates of 20,000°C/min [159] (i.e., 10 times higher than in 0.25 mL straws) and thereby overcome some problems of vitrification as toxicity of cryoprotectants or difficulties in the permeability of membranes that can lead to intracellular ice formation and osmotic overswelling [160]. Allowing higher speed of freezing, this type of vitrification is helpful in preventing embryo and *zona pellucida* fracture that occurs at low temperatures, especially with appropriate adjustments of warming parameters [152]. The method was introduced by Vajta et al. [149], and it has been successfully used ever since in several animal species. It is a robust and feasible method for animal embryo vitrification more than other new, but delicate, techniques [146].

Vitrification by OPS is so effective that results in lambing rate of *in vivo*-derived sheep embryos vitrified by this method are persuasive. Green et al. [169] reported higher pregnancy rates when using vitrification by OPS and direct transfer, thus enhancing the field application of this technique. Nevertheless, these good results with OPS vitrification cannot avoid the reduced cryotolerance of IVP embryos. Dattena et al. [16] achieved lambing rates of 60% for *in vivo*-produced embryos vitrified in OPS but only 24% for IVP.

4.3. Cryoprotectants

The role of these substances is to reduce damage to cryopreserved embryos, minimizing ice formation. They usually are classified as permeable and nonpermeable cryoprotectants, being also important in osmotic dehydration [23,146].

The negative effects of cryoprotectants are related to their osmotic and toxic effects and closely dependent on the time of exposure and the concentration used [147,148,159]. The intent to reduce cryoprotectants to a minimum is a goal that has been tried in both slow freezing and vitrification to minimize their deleterious effects. According to Liebermann et al. [158], a balance between the maximization of cooling rate and the minimization of cryoprotectant concentration is the key point for a successful cryopreservation. Mainly in vitrification, where it is necessary to deal with higher concentrations of cryoprotectants, one of the strategies is to use more than one. This strategy reduces individual toxicity of cryoprotectants and also permits adding them in a stepwise equilibration (two or three steps), with increasing concentration or after cooling from 4°C to subzero temperatures when their toxicity is lower [31,146,159].

Ethylene glycol is a frequent choice as a permeable cryoprotectant due to its high penetration rate and low toxicity besides protecting the membranes and cytoplasmic structures of embryo from cryoinjury [19]. However, other cryoprotectants such as acetamide, glycerol, raffinose, and dimethylsulfoxide (DMSO) can be used in several combinations. Ethylene glycol and glycerol [146,170] or ethylene glycol and DMSO [8,18,50,171] are often employed. Varago et al. [172] used dimethylformamide and ethylene glycol to vitrify ovine embryos but concluded that the conventional freezing with ethylene glycol was the most efficient method to cryopreserve ovine embryos.

Mono- and disaccharides, including sucrose, trehalose, glucose, and galactose, can also be added to vitrification media as nonpermeable cryoprotectants, with sucrose being the most commonly used [23,146]. Other substitute substances such as polymers and proteins have been tried to replace the former indicated.

4.4. Embryo warming/thawing

Embryo thawing/warming is also an essential step of the cryopreservation process and a specific procedure should be performed to avoid embryo damage. It comprises the rise of temperature, passing through the above-mentioned critical windows [144–146], and the cryoprotectant removal while maintaining embryo posterior viability. The thawing/warming rate and the protocol used for cryoprotectant removal need to be adjusted to the cooling procedure and embryo characteristics [23]. Thus, in embryos cryopreserved by slow freezing or vitrification, the best method depends on the specific protocol and cryoprotectants used (time and temperature and cryoprotectant characteristics and concentrations) but also on the origin of the embryo (*in vivo* or IVP derived). For instance, in slow freezing, if the freezing method stops at temperatures of -30°C to -40°C, before plunging in nitrogen, a moderately rapid warming (200–350°C/min) is required to maximize the survival rate but, if freezing reached -60°C or less, warming rate should be slow in the range of 25°C/min [161], although injuries were reported [173].

Warming of vitrified embryos is usually performed directly into a solution at body temperature, although it can be advisable to wait 1–3 seconds in air to avoid fracture damage [146]. Vitrified embryos need to be warmed ultrarapidly in the presence of nonpermeating cryoprotectants (usually sucrose) to dilute and remove the very high levels of intracellular permeating cryoprotectants [161]; this can lead to changes in dynamics of water in the embryo cells that can cause damage due to osmotic overshrinkage [31,160].

Regardless of the method used for embryo cryopreservation, some researchers [1,14,169] obtained promising results after direct transfer of vitrified sheep embryos for field use. These authors reported that the interval between thawing/warming and transfer has a great influence on embryo subsequent viability and valorize the use of direct transfer to spread this technique in field conditions.

4.5. Evaluation of cryopreserved embryos

The aim of embryo production and storage is to give practitioners the possibility of using them as an important resource for improving sheep production worldwide. Nevertheless, the effective use of these techniques depends on the guarantee that these embryos have the necessary quality enabling the reliability of its use. The need for predicting embryo ability to produce an offspring is essential in this context.

Embryo evaluation can use invasive and noninvasive techniques [174]. The last group, in assessing embryos without damage, has been widely used based on morphological changes that occur as the result of the sequential cleavage and this classification can be important in preimplantation embryos and especially in embryos that are being transferred [175]. Ushijima et al. [176] proposed an 11-stage classification for bovine embryos based on the number of cleavages. Embryo stage and quality related to morphology are usually based on the descriptions published by the IETS [177]. This ranking classifies embryos in grades 1 (excellent and good), 2 (fair), and 3 (poor) and become the standard reference worldwide. In cryopreserved embryos, stereomicroscopic evaluation of morphology after thawing has been used for transfer selection [178] to avoid transferring nonviable embryos. Leoni et al. [179] stated that reexpansion of blastocoelic cavity within 8 hours after vitrification/warming can be considered a reliable marker of its quality and developmental potential both *in vitro* and *in vivo*. However, some authors argue that morphological evaluation of thawed embryos is not accurate and the use of direct transfer of embryos can result in an improvement of 7–8% in offspring born [9]. Also, Green et al. [169] found that direct transfer improves the viability of transferred vitrified sheep embryos maybe because the time elongation from warming to transfer has a detrimental effect on subsequent embryo viability. This fact could suggest that embryo evaluation after embryo warming is dispensable as predictive of success. In fact, it is necessary to find other associated evaluation methods besides those based on embryo morphological evaluation to provide more accurate information. For example, Hernandez-Ledezma et al. [180] showed that good-quality hatched blastocysts produced more trophoblast protein, thus being an indicator of embryo quality before transfer.

Selection of embryos by the commonly used morphological criteria has, as expected, some individual variations because it can be a subjective evaluation. Indeed, studies in *in vivo*-produced ovine morulae and blastocysts have shown that certain abnormalities remain undetected by stereomicroscopy [19].

Invasive methods of embryo evaluation are used mainly for research proposes, validating the above-described techniques with greater accuracy [175]. Embryo postthawing viability and blastocoele reexpansion in different time periods of embryo development have been widely used to determine its feasibility [13,14,22,181]. Although several methods can be performed to predict the success of sheep embryos survival, the ultimate method, the outcome of the *in vivo* transfer to ewes, as the final goal of all the practical applications, cannot be tested after these invasive techniques. Staining methods as propidium iodide and TUNEL assay [182,183] are used for evaluating the number of viable cells. Hosseini et al. [184] established an easy approach for *in vivo*-derived and IVP embryos by evaluating the competence of cell membrane that is maintained in viable cells and can differentiate them from those necrotic or apoptotic, even without other morphological signs of cell death [102].

Ultrastructural evaluation of embryos can predict the damage caused in cell structures by the cryopreservation process and the consequences in their functions [185]. Although this is an expensive and invasive technique, the obtained outputs due to detailed information are of the utmost importance. However, according to Vajta et al. [186], in IVP bovine embryos, relevant ultrastructural changes observed immediately after thawing were progressively restored over a 24-hour period. Thus, it may be important to perform the ultrastructural evaluation of thawed embryos at different time periods.

Research on sheep embryo ultrastructure is limited compared to bovine species. However, recent publications have provided new and enough information allowing its discussion. Bettencourt et al. [187] described the ultrastructure of *in vivo*-produced sheep embryos, and previously, there were also some insights about the effects of cryopreservation in this source of embryos [17,19,188,189]. Concerning IVP sheep embryos, Dalcin et al. [2] revealed its ultrastructure as well as subtle changes in mitochondrial activity or cytoskeletal integrity caused by the cryopreservation process. Later on, Romão et al. [18] performed the ultrastructural characterization of fresh and vitrified *in vivo* and IVP sheep embryos. However, some details about IVP sheep embryos still need to be investigated.

Fresh IVP sheep embryos lacking desmosomal junctions have a reduction in the microvilli, an increase of debris in the periviteline space, and a high amount of lipid drops [17]. According to Romão et al. [18], the ultrastructural observation of these embryos also showed a reduced number of mitochondria and a lower number of mature mitochondria, having, on the contrary, a larger number of vesicles with light and dense content compared to their *in vivo* counterpart. These lower mitochondria content can be associated with deficient metabolism in IVP embryos.

After thawing/warming, the cryopreserved sheep embryos both *in vivo* and *in vitro* showed signs of ultrastructural injury represented by differences in mitochondria, with presence of more immature ones and lower total number [17,19]. Mitochondria are important for embryo development and metabolism [108,178,190] and thus can be a predictor of the embryo development potential [17,187]. However, in cryopreserved IVP sheep embryos, their function is reduced or absent [2]. Cryopreserved embryos also show cytoskeleton disturbance with actin microfilament disorganization [2,19] that can be caused by cryoprotectants [190]. The ultrastructural evaluation of cryopreserved IVP blastocysts confirmed the presence of more severe cellular damage in these embryos compared to those produced *in vivo* [18], pointing out the need of research to improve IVP sheep embryo quality and cryotolerance.

Having in mind that the evaluation of sheep embryos is important as a prognostic tool for the success rate after embryo transfer, the search for more accurate and practical methods is necessary, especially when dealing with stored cryopreserved embryos.

4.6. Methods of improving the cryopreservation of sheep embryos

Different strategies to increase the cryotolerance of embryos in sheep as in other species were developed mainly by improving the cryopreservation procedures [148], changing the composition of IVC media or other IVP techniques [8,50,192] and/or through the decrease of their lipid content [8,121]. As referred, methods of ultrarapid vitrification have been developed to further increase the freezing speed. However, the ideal device and combination of cryoprotectants have not yet been achieved. Therefore, this technique continues to be a challenge in embryo cryopreservation [148,155]. The number of created devices, such as glass micropipette, solid surface vitrification, cryoloop, microdrop, cryotop, cryotip, electron microscopy grids, or nylon mesh [15,152,155,161,193–195], clearly shows the lack of consensus. On the contrary, recent research pointed out the possibility of a further decrease in the temperature for embryo cryopreservation, from -196°C to -207°C, through the use of slush nitrogen [196], improving embryo cryosurvival at least in the mouse. According to Arav [148], the next evolutionary step in embryo cryopreservation will be to preserve them in the dry state at room temperature, allowing home storage for future use. In this alternative method, vitrification is followed by drying, and embryo storage is done in a dry state, avoiding their maintenance in liquid nitrogen containers and reducing the cost of the entire process.

Attempts were also experienced in IVC conditions, as it was argued that the effect of changing culture conditions could only be seen after embryo cryopreservation and warming. For example, Dattena et al. [192] confirmed higher sheep embryo cryosurvival results when including BSA and hyaluronan during IVC, and Gad et al. [197] realized that altered culture conditions at different time points of the preimplantation period lead to adaptations of the embryos, changing their gene expression and developmental ability. As referred, the initial stages of development are those with higher cryosensitivity, from oocytes to late embryo, with the acquisition of cryotolerance in an intermediate stage of development. Accordingly, Garcia-

Garcia et al. [167] and Lin et al. [198] found differences in the cryotolerance of IVP embryos vitrified by OPS, depending on its developmental stage, with better results in morulae/ blastocyst, defining the eight-cell stage as the critical point for acquiring cryotolerance. Therefore, it is not surprising that, in the last years of research, cryopreservation techniques have focused mainly in embryos at morulae and blastocyst stages and less in oocytes or zygotes. In addition, embryos that develop early have better survival rates after cryopreservation, and usually, these are male embryos [199].

Differences in species' cryosensitivity of embryos are responsible for different approaches in their cryopreservation. As indicated before, one of the major concerns in embryo cryopreservation is its lipid content that can hamper the process [13,20–23,200,201], as it happens in species such as sheep or pigs. This phenomenon is more pronounced in IVP embryos [18,202, 203] and in those produced in serum-containing media [20,22,24,204]. Serum is useful in oocyte and embryo culture as a source of albumin that balances the osmolality, acting as a free radical scavenger, and also with an additional important nutritive role [205]. However, the fatty acids and lipoproteins of the serum seem to be the source of the cytoplasmic lipids of embryos, hampering embryo quality [20,204,206], albeit the perturbations induced by the presence of serum in sheep embryo culture are higher before rather than after compaction [207].

Conversely, embryo lipid content effect on chilling sensitivity is not totally elucidated at the moment. However, it seems that lipid droplets interact directly with the intermediate filaments of the cytoskeleton, and changes within these organelles during the cryopreservation process may lead to irreversible damages [208,209]. Also, it has been observed that cryopreserved embryos have ultrastructural changes that are visible as degenerated cells, disruption of cell membranes, and mitochondrial injuries, mainly in poor-quality embryos [17–19]. Mitochondrial changes, namely, a reduced total number and an increase in the proportion of immature mitochondria [18,187], have been observed in poor-quality and IVP embryos and were associated with culture in serum-containing medium [210]. These changes have been also associated with inefficient lipid metabolism, with the presence of cytoplasmic vesicles and lipid drops and fewer lysosome-like vesicles [18], compromising ATP production [108,197,204].

Due to the negative effect of embryo lipid content and composition in their chilling sensitivity, attempts have been made to reduce its amount in IVP embryos either by chemical or by physical approaches. Nevertheless, lipids are important in the embryo cell metabolism as a source of energy and are also essential for membrane formation and as intracellular messengers [200,211]. Thus, these strategies should be carefully evaluated. For instance, in cattle, lipolytic agents or chemical delipidators have been successfully applied, increasing the cryotolerance of vitrified embryos (phenazine ethosulfate [212], *trans*-10 *cis*-12 conjugated linoleic acid (CLA) [22,208], and forskolin [213]). These chemicals that regulate metabolism were used to reduce embryo lipid content, inducing smaller lipid droplets and fat indexes, thus improving embryonic cryosurvival [22,208,214]. The addition of CLA during culture of sheep IVP embryos also increased their cryotolerance [8]. Likewise, Nagashima et al. [215] demonstrated that the high lipid content of pig embryos was responsible for their higher chilling sensitivity and also that embryos delipidated by centrifugation and microaspiration of polarized lipids became more tolerant to chilling. The lipid content was also pointed out as responsible for the chilling and freezing sensitivity of IVP cattle embryos, and when lipid droplets were displaced by ultracentrifugation (mechanical delipidation), their cryosurvival was improved [216,217]. On the contrary, Romão et al. [8] found a reduced viability in IVP sheep embryos submitted to an ultracentrifugation process mostly due to *zona pellucida* and membrane fracture. This problem could be overcome by using cytochalasin D during the ultracentrifugation process.

Cytoskeleton relaxant/stabilizers such as cytochalasin B or D were another approach investigated to improve embryo cryosurvival. Initially, they were used for cattle and pig embryo vitrification to prevent cellular disruption, specifically to the embryonic cytoskeleton during and after cryopreservation [218–220] with either no benefits [221] or with positive results [220,222]. These molecules induce microfilament depolymerization before and during vitrification [219], acting as microfilaments inhibitors, thus preventing actin polymerization and making the plasmatic membrane and cytoskeleton more elastic. Therefore, embryo microfilaments and cytoskeletal architecture were not only destroyed during micromanipulation but also become more resistant to the osmotic stress induced by the exposure and removal of cryoprotectants during the vitrification/warming process [218,220]. Dobrinsky et al. [223] obtained an 82% birth rate in pig vitrified embryos using cytochalasin B. Moreover, the association of two strategies, cytoskeleton relaxants and ultracentrifugation, were successfully attempted by different authors in IVP sheep, cattle, and pig embryos [8,222,224].

In spite of the above-described methods for improving sheep embryo cryosurvival, several limitations persist, impairing their spread as a routine technique. Therefore, possible solutions for improving methods to allow a higher survival rate of cryopreserved embryos should be further investigated.

5. Conclusions and future perspectives

Although sheep embryo technology is less applied than in other animal species (e.g., bovine) due to economical and scientific limitations, it is expected that in the next few years, these techniques will have a central role in sheep production. Cryopreservation is of primordial importance to preserve embryos so that they can be maintained as genetic reserves or for application simultaneously to other reproductive technologies. Vitrification and other ultrarapid vitrification protocols provide nowadays an alternative to slow freezing of sheep embryos. The effective use of these techniques depends on the guarantee that these embryos have enough quality, enabling the reliability of their use. New strategies, allied to emergent technologies, could be implemented to preserve the embryos such as to maintain them in a dry state at room temperature or to warm them with laser pulses. Manipulation of the lipid content of the embryo, one key factor hampering the efficiency of cryopreservation in this species, is a promising strategy. At the moment, *in vivo* and *in vitro* embryo production and cryopreservation remains a challenging technique that is growing worldwide in all species. Continued research to establish the optimal protocol(s) must persist to use embryo production and cryopreservation as an essential tool in reproductive medicine, guaranteeing superior results that will enable the expansion of clinical/biotechnological services.

Acknowledgements

This work is funded by FEDER funds through the Operational Programme for Competitiveness Factors-COMPETE and national funds through FCT-Foundation for Science and Technology under the Strategic Project PEst-C/AGR/UI0115/2011 and the project PTDC/CVT-REP/2863/2012 and supported by Instituto de Ciências Agrárias e Ambientais Mediterrânicas (ICAAM), Universidade de Évora-Núcleo da Mitra.

Author details

Ricardo Romão[1], Carla C. Marques[2], Elisa Bettencourt[1] and Rosa M.L.N. Pereira[2,3*]

*Address all correspondence to: rosalnp@gmail.com

1 Escola de Ciências e Tecnologia, Instituto de Ciências Agrárias e Ambientais Mediterrânicas (ICAAM), Universidade de Évora, Polo da Mitra, Évora, Portugal

2 Unidade de Biotecnologia e Recursos Genéticos, INIAV-Santarém, Quinta da Fonte Boa, Portugal

3 CIISA, Faculdade de Medicina Veterinária da Universidade de Lisboa, Avenida da Universidade Técnica, Lisboa, Portugal

References

[1] Isachenko V, Alabart JL, Dattena A, Nawroth F, Cappai P, Isachenko E, Cocero, MJ, Olivera, J, Roche A, Accardo C, Kriovokharchneko A, Folch J. New technology for vitrification and field (microscope-free) warming and transfer of small ruminant embryos Theriogenology. 2003;59:1209–1218.

[2] Dalcin L, Silva RC, Paulini F, Silva BD, Neves JP, Lucci CM. Cytoskeleton structure, pattern of mitochondrial activity and ultrastructure of frozen or vitrified sheep embryos. Cryobiology. 2013;67:137–45.

[3] Galli C, Lazzari G. The manipulation of gametes and embryos in farm animals. Reproduction in Domestic Animals. 2008;43:1–7.

[4] Folch J, Olivera J, Aguilar B, Alabart JL, Sanchez P, Echegoyen E, Cocero MJ. Resultados obtenidos em la transferência de embriones dentro del programa genético de la U.P.R.A. carnes Oviaragon. In: Resumenes de las XXV Jornadas SEOC; 2000 p. 559–561.

[5] Guérin B, Le Guiennea B, Thibier M. A secure health status associated with the production and trade of in vitro derived cattle embryos. Livestock Production Science. 2000;62:271–285.

[6] Willadsen SM. Factors affecting the survival of sheep embryos during deep-freezing and thawing. In: The Freezing of Mammalian Embryos. Ciba Foundation, Symposium, 52 (New Series). Elsevier Excerpta Medica/North-Holland, Amsterdam; 1977 p. 175–201 [abstract].

[7] Willadsen SM, Polge C, Rowson LEA, Moor RM. Deep freezing of sheep embryos. Journal of Reproduction and Fertility. 1976;46:151–154.

[8] Romão R, Marques CC, Baptista MC, Barbas JP, Horta AEM, Bettencourt E, Carolino N, Pereira RMLN. Cryopreservation of in vitro produced sheep embryos: effect of different protocols of lipid reduction. Theriogenology. 2015;84:118–126.

[9] Cognié Y, Baril G, Poulin N, Mermillod P. Current status of embryo technologies in sheep and goat. Theriogenology. 2003:59:171–188.

[10] Cognié Y, Poulin N, Locatelli Y, Mermillod P. State-of-the-art production, conservation and transfer of *in-vitro*-produced embryos in small ruminants. Reproduction, Fertility and Development. 2004;16:437–445.

[11] Martínez AG, Valcácel A, Furnus CC, de Matos DG, Iorio, G, de las Heras MA. Cryopreservation of *in vitro*-produced ovine embryos. Small Ruminant Research. 2006;63:288–296.

[12] Papadopoulos S., Rizos D, Duffy P, Wade M, Quinn K, Boland MP, Lonergan P. Embryo survival and recipient pregnancy rates after transfer of fresh or vitrified, *in vivo* or *in vitro* produced ovine embryos. Animal Reproduction Science. 2002;74:35–44.

[13] Dattena M, Ptak G, Loi P, Cappai P. Survival and viability of vitrified in vitro and in vivo produced ovine blastocysts. Theriogenology. 2000;53:1511–1519.

[14] Dattena M, Accardo C, Pilchi S, Isachenko V, Mara L, Chessa B, Cappai P. Comparison of different vitrification protocols on viability after transfer of ovine blastocysts in vitro produced and *in vivo* derived. Theriogenology. 2004;62:481–493.

[15] Gibbons A, Cueto MI, Pereyra Bonnet F. A simple vitrification technique for sheep and goat embryo cryopreservation. Small Ruminant Research. 2011;95:61–64.

[16] Dattena M, Isachenko V, Alabart JL, Folch J, Accardo C, Cappai P. Comparison between two embryo transfer methods of vitrified sheep blastocysts. In: Proceedings of the 17th Scientific Meeting AETE, Lyon; 2001, p. 113.

[17] Bettencourt EMV, Bettencourt CM, Silva JNCE, Ferreira P, de Matos CP, Oliveira E, Romão RJ, Rocha A, Sousa M. Ultrastructural characterization of fresh and cryopreserved *in vivo* produced ovine embryos. Theriogenology. 2009;71:947–958.

[18] Romão R, Bettencourt E, Pereira RMLN, Marques CC, Baptista MC, Barbas JP, Oliveira E, Bettencourt C, Sousa M. Ultrastructural characterization of fresh and vitrified *in vitro*- and *in vivo*-produced sheep embryos. Anatomia, Histologia, Embryologia. 2015 [in press]. doi: 10.1111/ahe.12191.

[19] Cocero MJ, Moreno Díaz de la Espina S, Aguilar B. Ultrastructural characteristics of fresh and frozen-thawed ovine embryos using two cryoprotectants. Biology of Reproduction. 2002;66:1244–1258.

[20] Thompson JG, Gardner DK, Pugh PA, McMillan WH, Tervit HR. Lamb birth weight is affected by culture system utilized during in vitro pre-elongation development of ovine embryos. Biology of Reproduction. 1995;53:1385–1391.

[21] Rizos D, Ward F, Duffy P, Boland MP, Lonergan P. Consequences of bovine oocyte maturation, fertilization or early embryo development *in vitro* versus *in vivo*: implications for blastocyst yield and blastocyst quality. Molecular Reproduction and Development. 2002;61:234–248.

[22] Pereira RM, Baptista MC, Vasques MI, Horta AEM, Portugal PV, Bessa RJB, Chagas e Silva J, Silva Pereira M, Marques CC. Cryo-survival of bovine blastocysts is enhanced by culture with trans-10 cis-12 conjugated linoleic acid (10t, 12c CLA). Animal Reproduction Science. 2007;98:293–301.

[23] Pereira RM, Marques CC. Animal oocyte and embryo cryopreservation. Cell Tissue Bank. 2008;9:267–277.

[24] Cho S-R, Cho S-K, Lee S-L, Lee H-J, Choe S-Y, Rho G-J. Enhanced cryosurvival of bovine blastocysts produced in vitro in serum-free medium. Journal of Assisted Reproduction and Genetics. 2002;19:487.

[25] Abe H, Hoshi H. Evaluation of bovine embryos produced in high performance serum-free media. Journal of Reproduction and Development. 2003;49:193–202.

[26] Tveden-Nyborg P, Peura TT, Hartwich KM, Walker SK, Maddox-Hyttel P. Morphological characterization of pre- and peri-implantation in vitro cultured, somatic cell nuclear transfer and in vivo derived ovine embryos. Reproduction. 2005;130:681–694.

[27] Nieddu SM, Mossa F, Strina A, Ariu F, Pau S, Ledda M, Sotgia S, Carru C, Ledda S. Differences in amniotic amino acid concentrations between pregnancies obtained with transfer of vitrified thawed in vitro–produced embryos and with natural mating in sheep. Theriogenology. 2015;83:687–692. doi: 10.1016/j.theriogenology.2014.11.002.

[28] Henson EL. *In situ* conservation of livestock and poultry. FAO Animal Production health paper; 1992. http://www.fao.org/docrep/004/T0559E/T0559E00.htm#TOC [Accessed 2013-11-20].

[29] Båge R. Collection and preservation of female gametes and embryos. Farm animal reproduction: conserving local genetic resources. Proceedings from a minisymposium at Lithuanian Veterinary Academy, Kaunas, Lithuania, 2003 p. 22.

[30] Yao Y-C, Qi M-Y, Lu M-H, Wang S-M, Li W, Han H-B. Long-term cryopreservation had no adverse effect on viability of embryos and their offspring in sheep. Animal Reproduction Science. 2012;136:42–46.

[31] Woods EJ, Benson JD, Agca Y, Critser JK. Fundamental cryobiology of reproductive cells and tissues. Cryobiology. 2004;48:146–156.

[32] Mochida K, Hasegawa A, Li M-W, Fray MD, Kito S, Vallelunga JM, Lloyd KCK, Yoshik Ai, Obata Y, Ogura A. High osmolality vitrification: a new method for the simple and temperature-permissive cryopreservation of mouse embryos. PLoS ONE. 2013;8:e49316.

[33] Thibier M. Embryo transfer: a comparative biosecurity advantage in international movements of germplasm. Revue scientifique et technique. 2011;30:177–188.

[34] Thibier M, Guérin B. Embryo transfer in small ruminants: the method of choice for health control in germplasm exchanges. Livestock Production Science. 2000;62:253–270.

[35] Cognié Y, Baril G. Le point sur la production et le transfert d'embryons obtenus in vivo et in vitro chez la brebis et la chèvre. INRA Productions Animales. 2002;15:199–207.

[36] Czlonkowska M, Eysymont U, Guszkiewicz A, Kossakowski M, Dziak J. Birth of lamb after in vitro maturation, fertilisation, and co-culture with oviductal cells. Molecular Reproduction and Development 1991;30:34–38.

[37] Pugh PA, Fukui Y, Tervit HR, Thompson JG. Developmental ability of *in vitro* matured sheep oocytes collected during the nonbreeding season and fertilized *in vitro* with frozen ram semen. Theriogenology 1991;36:771–778.

[38] Wilmut I, Schnieke AE, McWhir J, Kind AJ, Campbell KH. Viable offspring derived from fetal and adult mammalian cells. Nature. 1997;385:810–813.

[39] Thibier M. Stabilization of numbers of in vivo collected embryos in cattle but significant increases of in vitro bovine produced embryos in some parts of the world. Data Retrieval Committee Annual Report. IETS Newsletter 2004;22:12–19.

[40] Perry G. Statistics of embryo collection and transfer in domestic farm animals. 22nd Annual Report of Data Retrieval Committee. IETS; 2013.

[41] Mullen SF, Fahy GM. A chronologic review of mature oocyte vitrification research in cattle, pigs, and sheep. Theriogenology. 2012;78:1709–1719.

[42] Mara L, Casub S, Cartac A, Dattena M. Cryobanking of farm animal gametes and embryos as a means of conserving livestock genetics. Animal Reproduction Science. 2013;138:25–38.

[43] Arav A, Revel A, Nathan Y, Bor A, Gacitua H, Yavin S, Gavish Z, Uri M, Elami A. Oocyte recovery, embryo development and ovarian function after cryopreservation and transplantation of whole sheep ovary. Human Reproduction. 2005;20:3554–3559.

[44] Faustino LR, Santos RR, Silva CMG, Pinto LC, Celestino JJH, Campello CC, Figueiredo JR, Rodrigues APR. Goat and sheep ovarian tissue cryopreservation: effects on the morphology and development of primordial follicles and density of stromal cell. Animal Reproduction Science. 2010;122:90–97.

[45] Merdassi G, Mazoyer C, Guerin JF, Saad A, Salle B, Lornage J. Examination of viability and quality of ovarian tissue after cryopreservation using simple laboratory methods in ewe. Reproductive Biology and Endocrinology. 2011;9:1–7.

[46] McCreath K, Howcroft J, Campbell KH, Colman A, Schnieke A, Kind A. Production of gene-targeted sheep by nuclear transfer from cultured somatic cells. Nature. 2000;405:1066–1069.

[47] Liu J, Westhusin M, Long C, Johnson G, Burghardt R, Kraemer D. Embryo production and possible species preservation by nuclear transfer of somatic cells isolated from bovine semen. Theriogenology. 2010;74:1629–1635.

[48] Betteridge K. Farm animal embryo technologies: achievements and perspectives. Theriogenology. 2006;65:905–913.

[49] Hasler J. Bovine embryo transfer: are efficiencies improving? Proceedings, Applied Reproductive Strategies in Beef Cattle, Sioux Falls, USA; 2012. p. 319–338.

[50] Romão R, Marques CC, Baptista MC, Vasques MI, Barbas JP, Horta AEM, Carolino N, Bettencourt E, Plancha C, Rodrigues P, Pereira RM. Evaluation of two methods of in vitro production of ovine embryos using fresh or cryopreserved semen. Small Ruminant Research. 2013;110:36–41.

[51] González-Bulnes A, Baird DT, Campbell BK, Cocero MJ, Gracía-Garcia RM, Inskeep EK, López-Sebastián A, McNeilly AS, Santiago-Moreno J, Souza CJH, Veiga-López A. Multiple factors affecting the efficiency of multiple ovulation and embryo transfer in sheep and goats. Reproduction Fertility and Development. 2004;16:421–435.

[52] Bettencourt EM, Bettencourt CM, Chagas e Silva J, Ferreira P, Matos CP, Romão RJ, Rocha A. Fertility rates following the transfer of ovine embryos cryopreserved using three protocols. Small Ruminant Research. 2009:82:112–116.

[53] Viñoles C, Forsberg M, Banchero G, Rubianes E. Effect of long-term and short-term progestogen treatment on follicular development and pregnancy rate in cyclic ewes. Theriogenology. 2001;55:993–1004.

[54] Baril G, Brebion P, Chesné P. Practical training handbook on sheep and goat embryo transfer. FAO. 1993;115:1014–1099.

[55] Robinson TJ, Moore NW, Lindsay DR, Fletcher IC, Salamon S. Fertility following synchronization of oestrus in the sheep with intravaginal sponges. I. Effects of vagi-nal douche, supplementary steroids, time of insemination, and numbers and dilution of spermatozoa. Australian Journal of Agricultural Research. 1970;21:767–781.

[56] Scaramuzzi RJ, Downing JA, Carflpbell BK, Cognié Y. Control of fertility and fecundity of sheep by means of hormonal manipulation. Australian Journal of Biological Sciences. 1988;41:37–45.

[57] Pearce DT, Robinson TJ. Plasma progesterone concentrations, ovarian and endocrinological responses and sperm transport in ewes with synchronized oestrus. Journal of Reproduction and Fertility. 1985;75:49–62.

[58] González-Bulnes A, Berlinguer F, Cocero MJ, Garcia-Garcia RM, Leoni G, Naitana S, Rosati I, Succu S, Veiga-López A. Induction of the presence of corpus luteum during superovulatory treatments enhances *in vivo* and *in vitro* blastocysts output in sheep. Theriogenology. 2005;64:1392–1403.

[59] Oussaid B, Mariana J, Poulin N, Fontaine J, Lonergan P, Beckers JF, Cognié Y. Reduction in the developmental competence of sheep oocytes by inhibition of LH pulses during the follicular phase with a GnRH antagonist. Journal of Reproduction and Fertility. 1999;117:71–77.

[60] Berlinguer F, Gonzaléz-Bulnes A, Succu S, Leoni GG, Veiga-López A, Mossa F, Garcia-Garcia RM, Bebbere D, Galioto M, Cocero MJ, Naitana S. GnRH antagonist enhance follicular growth in FSH-treated sheep but affect developmental competence of oocytes collected by ovum pick-up. Theriogenology. 2006;65:1099–1109.

[61] Ungerfeld R, Rubianes E. Effectiveness of short-term progestogen primings for the induction of fertile oestrus with eCG in ewes during late seasonal anoestrus. Animal Science. 1999;68:349–353.

[62] Greyling JPC, Kotze WF, Taylor GJ, Hagendijk WJ, Cloete F. Synchronization of oestrus in sheep: use of different doses of progestagen outside the normal breeding season. South African Journal of Animal Science. 1994;24:33–37.

[63] Mayorga I, Mara L, Sanna D, Stelletta C, Morgante M, Casu S, Dattena M. Good quality sheep embryos produced by superovulation treatment without the use of progesterone devices. Theriogenology. 2011;75:1661–1668. doi: 10.1016/j.theriogenology.2010.12.029.

[64] Cognié Y, Baril G, Poulin N, Mermillod P. Current status of embryo technologies in sheep and goats. Theriogenology. 2003;59:171–188.

[65] Simonetti L, Forcada F, Rivera OE, Carou N, Alberio RH, Abecia JÁ, Palacin I. Simplified superovulatory treatments in Corriedale ewes. Animal Reproduction Science. 2008;104:227–237.

[66] Bettencourt EM, Bettencourt CM, Chagas e Silva N, Ferreira P, Manito CI, Matos CM, Romão RJ, Rocha A. Effect of season and gonadotrophin preparation on superovulatory response and embryo quality of Portuguese Black Merino ewes. Small Ruminant Research. 2008;74:134–139.

[67] Bindon BM, Piper LR. Physiological basis of ovarian response to PMSG in sheep and cattle. In: Shelton JN, Trounson AO, Moore NW, James JW, editors. Embryo Transfer in Cattle, Sheep and Goats. Australian Society for Reproductive Biology, Canberra; 1982. p. 1–5.

[68] Armstrong DT, Evans G. Factors influencing success of embryo transfer in sheep and goats. Theriogenology. 1983;19:31–42.

[69] Cognié Y. State of art in sheep and goat embryo transfer. Theriogenology. 1999;51:105–116.

[70] D'Alessandro A, Martemucci G, Toteda F, Gambacorta M, Manchisi A. Superovulation and embryo production in ewes using a commercial p-FSH. Small Ruminant Research. 1996;19:255–261.

[71] Forcada F, Lozano JM, Abecia JÁ, Zuniga O. Repeated superovulation of high prolificacy Rasa Aragonesa ewes before culling as an inexpensive way to obtain high quality embryos. Livestock Production Science. 2000;66:263–269.

[72] D'Alessandro AG, Martemucci G. Efficiency of superovulatory treatment with FSH-p for in vivo embryo production in dairy ewes: multiple versus single dose regimen. Journal of Animal and Veterinary Advances. 2004;3:388–393.

[73] D'Alessandro AG, Martemucci G, Taibi L. How the FSH/LH ratio and dose numbers in p-FSH administration treatment regimen and insemination schedule affect superovulatory response in ewes Theriogenology. 2005;63:1764–1774.

[74] Bindon BM, Piper LR, Cahill LP, Driancourt MA, O'Shea T. Genetic and hormonal factors affecting superovulation. Theriogenology. 1986;25:53–70.

[75] Dufour JJ, Cognié Y, Mermillod P, Mariana JC, Romain RF. Effects of the Booroola Fec gene on ovarian follicular populations in superovulated Romanov ewes pretreated with a GnRH antagonist. Journal of Reproduction and Fertility. 2000;118:85–94.

[76] Bruno-Galarraga M, Cueto M, Gibbons A, Pereyra-Bonnet F, Subiabre M, González-Bulnes A. Preselection of high and low ovulatory responders in sheep multiple ovulation and embryo transfer programs. Theriogenology. 2015;84:784–790.

[77] Mitchell LM, Reis A, Staines ME, Rooke JA, Dingwall WS, McEvoy TG. Laparoscopic oocyte aspiration from ewes feed diets with or without 3% fish oil at contrasting stages of their breeding season. Theriogenology. 2002;57:614 [abstract].

[78] Mitchell LM, Dingwall WS, Mylne MJA, Hunton J, Matthews K, Gebbie FE, McCallum GJ, McEvoy TG. Season affects characteristics of the pre-ovulatory LH surge and embryo viability in superovulated ewes. Animal Reproduction Science. 2002;74:163–174.

[79] López Sebastián A, González de Bulnes A, Santiago Moreno J, Gómez Brunet A, Townsend EC, Inskeep EK. Effects of follicular status at treatment on follicular development and ovulation in response to FSH in Spanish merino ewes. Theriogenology. 1999;52:505–514.

[80] Chagas e Silva J, Lopes da Costa L, Cidadão R, Robalo Silva J. Plasma progesterone profiles, ovulation rate, donor embryo yield and recipient embryo survival in native Saloia sheep in the fall and spring breeding seasons. Theriogenology. 2003;60:521–532.

[81] Forcada F, Abecia JA, Cebrián-Pérez JA, Muiño-Blanco MT, Valares JA, Palacínn I, Casao A. The effect of melatonin implants during the seasonal anestrus on embryo production after superovulation in aged high-prolificacy Rasa Aragonesa ewes. Theriogenology. 2006;65:356–365.

[82] González-Bulnes A, Santiago-Moreno J, Cocero MJ, López-Sebastián A. Effects of FSH commercial preparation and follicular status on follicular growth and superovulatory response in Spanish Merino ewes. Theriogenology. 2000;54:1055–1064.

[83] Veiga-López A, Gonzaléz-Bulnes A, Garcia-Garcia RM, Domínguez V, Cocero MJ. Effects of previous ovarian status in ovulation rate and early embryo development in response to superovulatory FSH treatments in sheep. Theriogenology. 2005;63:1973–1983.

[84] González-Bulnes A, Garcia-Garcia RM, Santiago-Moreno J, López-Sebastián A, Cocero. Effects of follicular status on superovulatory response in ewes is influenced by presence of corpus luteum at first FSH dosage. Theriogenology. 2002;58:1607–1614.

[85] Campbell BK, Dobson H, Scaramuzzi RJ. Ovarian function in ewes made hypogonadal with GnRH antagonist and stimulated with FSH in the presence or absence of low amplitude LH pulses. Journal of Endocrinology. 1998;156:213–222.

[86] González-Bulnes A, Souza CJH, Scaramuzzi RJ, Campbell BK, Baird DT. Long-term suppression of reproductive function by a single dose of gonadotropin-releasing hormone antagonists in a sheep model. Fertility and Sterility. 2006;86:1121–1128.

[87] Jabbour HN, Evans G. Ovarian and endocrine response of Merino ewes to treatment with PMSG and/or FSH-p. Animal Reproduction Science. 1991;26:93–106.

[88] González-Bulnes A, Garcia-Garcia RM, Castellanos V, Santiago-Moreno J, Ariznavarreta C, Domínguez V, López-Sebastián A, Tresguerres JAF, Cocero MJ. Influence of

maternal environment on the number of transferable embryos obtained in response to superovulatory FSH treatments in ewes. Reproduction Nutrition Development. 2003;43:17–28.

[89] Murray JF; Downing JA, Scaramuzzi RJ, Evans G. Heterogeneity in ovarian steroid secretion response to treatment with PMSG in ewes during the breeding season and anestrus. Theriogenology. 1994;42:1337–1347.

[90] Evans G, Armstrong DT. Reduction of sperm transport in ewes by superovulation treatments. Journal of Reproduction and Fertility. 1984;70:47–53.

[91] Veiga-López A, Gonzaléz-Bulnes A, Tresguerres JAF, Dominguez V, Ariznavarreta C, Cocero MJ. Causes, characteristics and consequences of anovulatory follicles in superovulated sheep. Domestic Animal Endocrinology. 2006;30:76–87.

[92] Bari F, Khalid M, Haresign W, Murray A, Merrell B. Effect of mating system, flushing procedure, progesterone dose and donor ewe age on the yield and quality of embryos within a MOET programme in sheep. Theriogenology. 2000;53:727–742.

[93] Brebion P, Baril G, Cognié Y, Vallet JC. Transfert d'embryons chez les ovins et les caprins. Annales de Zootechnia. 1992;41:331–339.

[94] Barnes FL. The effects of early uterine environment on the subsequent development of embryo and fetus. Theriogenology. 2000;53:649–658.

[95] Schiewe MC, Fitz TA, Brown JL, Stuart LD, Wildt DE. Relationship of oestrus synchronization method, circulating hormones, luteinizing hormone and prostaglandin F-2 alpha receptors and luteal progesterone concentration to premature luteal regression in superovulated sheep. Journal of Reproduction and Fertility. 1991;93:19–30.

[96] Ryan JP, Hunton JR, Maxwell MC. Increased production of sheep embryos following superovulation of Merino ewes with a combination of pregnant mare serum gonadotrophin and follicle stimulating hormone. Reproduction, Fertility and Development. 1991;3:551–560.

[97] Cordeiro MF, Lima-Verde JB, Lopes-Júnior ES, Teixeira DIA, Farias LN, Salles HO, Simplício AA, Rondina D, Freitas VJF. Embryo recovery rate in Santa Inês ewes subjected to successive superovulatory treatments with pFSH. Small Ruminant Research. 2003;49:19–23.

[98] Bari F, Khalid M, Wolf B, Haresign W, Murray A, Merrell B. The repeatability of superovulatory response and embryo recovery in sheep. Theriogenology. 2001;56:147–155.

[99] Lonergan P, Fair T. The art of studying early embryo development: progress and challenges in ruminant embryo culture. Theriogenology. 2014;81:49–55.

[100] Tominaga K. Cryopreservation and sexing of in vivo- and in vitro-produced bovine embryos for their practical use. Journal of Reproduction and Development. 2004;50:29–38.

[101] Cocero MJ, Alabart JL, Hammami S, Martí JI, Lahoz B, Sánchez P, Echegoyen E, Beckers JF, Folch J. The efficiency of *in vitro* ovine embryo production using an undefined or a defined maturation medium is determined by the source of the oocyte. Reproduction in Domestic Animals. 2011;46:463–470.

[102] Betts DH, King WA. Genetic regulation of embryo death and senescence. Theriogenology. 2001;55:171–191.

[103] Betts DH, Madan P. Permanent embryo arrest: molecular and cellular concepts. Molecular Human Reproduction. 2008;14:445–453.

[104] Holm P, Walker SK, Seamark RF. Embryo viability, duration of gestation and birth weight in sheep after transfer of *in vitro* matured and in vitro fertilized zygotes cultured *in vitro* or *in vivo*. Journal of Reproduction and Fertility. 1996;107:175–181.

[105] Lazzari G, Wrenzycki C, Herrmann Duchi R, Kruip T, Niemann H, Galli C. Cellular and molecular deviations in bovine *in vitro*-produced embryos are related to the large offspring syndrome. Biology of Reproduction. 2002;67:767–775.

[106] Shirazi A, Ahmadi E, Jadidi M, Shams-Esfandabadi N, Heidari B. Acephalous lamb from an *in vitro*-produced sheep embryo. The Canadian Veterinary Journal. 2009;50:501–505.

[107] Çörekçi Ş, Yilmaz A. A comparative study on the growth and survival characteristics of lambs produced by the transfer of *in vitro* produced (IVP) embryos. Turkish Journal of Veterinary and Animal Sciences. 2004;28:831–835.

[108] Crosier AE, Farin PW, Dykstra MJ, Alexander JE, Farin CE. Ultrastructural morphometry of bovine compact morulae produced in vivo or in vitro. Biology of Reproduction. 2000;62:1459–1465.

[109] Gusmão AL. State-of-the-art in the transcervical embryo collection in goats and sheep. Acta Scientiae Veterinariae. 2011;39:37–42.

[110] Wani NA, Wani GM, Khan MZ, Salahudin S. Effect of oocyte harvesting techniques on *in vitro* maturation and *in vitro* fertilization in sheep. Small Ruminant Research. 2000;36:63–67.

[111] Zeinoaldini S, Jafari Z, Sarmast F, Torbati E, Davachi ND. Different harvesting techniques used in ovine in vitro embryo production. International Journal of Medical Sciences (Scimetr). 2013;1:1–5.

[112] Amiridis GS, Csehb S. Assisted reproductive technologies in the reproductive management of small ruminants. Animal Reproduction Science. 2012;130:152–161.

[113] Rodríguez C, Anel L, Alvarez M, Anel E, Boixo JC, Chamorro CA, de Paz P. Ovum pick-up in sheep: a comparison between different aspiration devices for optimal oocyte retrieval. Reproduction in Domestic Animals. 2006;41:106–113.

[114] Cox JF, Alfaro V. *In vitro* fertilization and development of OPU derived goat and sheep oocytes. Reproduction in Domestic Animals. 2007;42:83–87.

[115] Gibbons A, Bonnet FP, Cueto MI, Salamone D, Catala M. Colheita de oócitos guiada por laparoscopia em caprinos e ovinos. Acta Scientiae Veterinariae. 2008;36:223–230.

[116] Gibbons A, Bonnet FP, Cueto MI, Catala M, Salamone DF, Gonzalez-Bulnes A. Procedure for maximizing oocyte harvest for *in vitro* embryo production in small ruminants. Reproduction in Domestic Animals. 2007;42:423–426.

[117] Crocomo LF, Marques Filho WC, Landim Alvarenga FC, Bicudo SD. Peculiaridades da coleta de oócitos para produção *in vitro* de embriões ovinos. Revista Brasileira de Reprodução Animal. 2012;36:25–31.

[118] O'Brien JK, Catt SL, Ireland KA, Maxwell WM, Evans G. *In vitro* and *in vivo* developmental capacity of oocytes from prepubertal and adult sheep. Theriogenology. 1997;47:1433–1443.

[119] Dieleman SJ, Hendriksen PJM, Viuff D, Thomsen PD, Hyttel P, Knijn HM, Wrenzycki C, Kruip TAM, Niemann H, Gadella BM, Bevers MM, Vos PLAM. Effect of *in vivo* prematuration and *in vivo* final maturation on developmental capacity and quality of pre-implantation embryos. Theriogenology. 2002;57:5–20.

[120] Dunning KR, Robker RL. Promoting lipid utilization with L-carnitine to improve oocyte quality. Animal Reproduction Science. 2012;134:69–75.

[121] Prates EG, Nunes JT, Pereira RMLN. A role of lipid metabolism during cumulus-oocyte complex maturation: impact of lipid modulators to improve embryo production. Mediators of Inflammation. 2014, Article ID 692067, p. 11.

[122] Pereira RM, Mesquita P, Batista M, Baptista MC, Barbas JP, Pimenta J, Santos IC, Marques MR, Vasques MI, Silva Pereira M, Santos Silva F, Oliveira Sousa MC, Fontes CM, Horta AEM, Prates JÁ, Marques CC. Doppel gene polymorphisms in Portuguese sheep breeds: insights on ram fertility. Animal. Reproduction Science. 2009;114:157–166.

[123] Shabankareh HK, Sarsaifi K, Mehrannia T. *In vitro* maturation of ovine oocytes using different maturation media: effect of human menopausal serum. Journal of Assisted Reproduction and Genetics. 2011;28:531–537.

[124] Birler S, Pabuccuoglu S, Alkan S, Ak K, Evecen M, Ileri IK. Effects of different maturation and culture media on IVF of sheep oocytes. Pakistan Veterinary Journal. 2001;21:100–102.

[125] Guler A, Poulin N, Mermillod P, Terqui M, Cognié Y. Effect of growth factors, EGF and IGF-1, and estradiol on *in vitro* maturation of sheep oocytes. Theriogenology. 2000;54:209–218.

[126] Cotterill M, Catt SL, Picton HM. Characterisation of the cellular and molecular responses of ovine oocytes and their supporting somatic cells to pre-ovulatory levels of LH and FSH during *in vitro* maturation. Reproduction. 2012;144;195–207.

[127] Zhou P, Liu DJ, Canga M, Maa YZ, Yang DS, Li HJ, Wang LM, Boua S, Feng HL. TGFα and EGFR in ovine preimplantation embryos and effects on development. Animal Reproduction Science. 2008;104:370–381.

[128] Wani AR, Khan MZ, Sofi KA, Lone FA, Malik AA, Bhat FA. Effect of cysteamine and epidermal growth factor (EGF) supplementation in maturation medium on *in vitro* maturation, fertilization and culturing of embryos in sheep. Small Ruminant Research. 2012;106:160–164.

[129] Van Soom A, Tanghe S, De Pauw I, Maes D, de Kruif A. Function of the cumulus oophorus before and during mammalian fertilization. Reproduction in Domestic Animals. 2002;37:144–151.

[130] Shirazi A, Shams-Esfandabadi N, Hosseini SM, Karimi I. The presence of cumulus cells on nuclear maturation of sheep oocytes during in vitro maturation. Small Ruminant Research. 2007;68:291–295.

[131] Bogliolo L, Ariu F, Fois S, Rosati I, Zedda MT, Leoni G, Succu S, Pau S, Ledda S. Morphological and biochemical analysis of immature ovine oocytes vitrified with or without cumulus cells. Theriogenology. 2007;68:1138–1149.

[132] Hollinshead FK, Evans G, Evans KM, Catt SL, Maxwell WMC, O'Brien JK. Birth of lambs of a pre-determined sex after in vitro production of embryos using frozen-thawed sex-sorted and re-frozen-thawed ram spermatozoa. Reproduction. 2004;127:557–568.

[133] Peris SI, Morrier A, Dufour M, Bailey JL. Cryopreservation of ram semen facilitates sperm DNA damage: relationship between sperm andrological parameters and the sperm chromatin structure assay. Journal of Andrology. 2004;25:224–233.

[134] Khalifa T, Lymberopoulos A. Changeability of sperm chromatin structure during liquid storage of ovine semen in milk-egg yolk and soybean lecithin based extenders and their relationships to field-fertility. Cell Tissue Bank. 2013;14:687–698.

[135] Freitas VJF, Andrade MLL, Cajazeiras JB, Luz JV. *In vitro* embryo production in small ruminants raised in northeastern Brazil. Acta Scientiae Veterinariae. 2007;35:781–786.

[136] Suttiyotin P, Thwaites CJ. Evaluation of ram semen motility by a swim-up technique. Journal of Reproduction and Fertility. 1993;97:339–345.

[137] Lazzari G, Colleoni S, Lagutina I, Crotti G, Turini P, Tessaro I, Brunetti D, Duchi R, Galli C. Short-term and long-term effects of embryo culture in the surrogate sheep

oviduct versus in vitro culture for different domestic species. Theriogenology. 2010;73:748–757.

[138] Bernardi ML. Produção *in vitro* de embriões ovinos. Acta Scientiae Veterinariae. 2005;33:1–16.

[139] Harvey AJ. The role of oxygen in ruminant preimplantation embryo development and metabolism. Animal Reproduction Science. 2007:98:113–128.

[140] De AK, Malakar D, Akshey YS, Jena MK, Garg S, Dutta R, Sahu S. *In vitro* development of goat (*Capra hircus*) embryos following cysteamine supplementation of the in vitro maturation and *in vitro* culture media. Small Ruminant Research. 2011;96:185–190.

[141] Khurana NK, Niemann H. Effects of cryopreservation on glucose metabolism and survival of bovine morulae and blastocyst derived *in vitro* or *in vivo*. Theriogenology. 2000;54:313–326.

[142] Farin PW, Crosier AE, Farin CE. Influence of *in vitro* systems on embryo survival and fetal development in cattle. Theriogenology. 2001;55:151–170.

[143] Massip A, Mermillod P, Dinnyes A. Morphology and biochemistry of *in vitro* produced bovine embryos: implications for their cryopreservation. Human Reproduction. 1995;10:3004–3011.

[144] Gao D, Critser JK. Mechanisms of cryoinjury in living cells. ILAR J. 2000;41:187–196.

[145] Rall WF, Meyer TK. Zona fracture damage and its avoidance during the cryopreservation of mammalian embryos. Theriogenology. 1989;31:683–692.

[146] Vajta G, Nagy ZP. Are programmable freezers still needed in the embryo laboratory? Review on vitrification. Reproductive BioMedicine Online. 2006;12:779–796.

[147] Dinnyes A, Meng Q, Polgar Z, Boonkuso D, Somfai T. Cryopreservation of mammalian embryos. Acta Scientiae Veterinariae. 2006;34:171–190.

[148] Arav A. Cryopreservation of oocytes and embryos. Theriogenology. 2014;81:96–102.

[149] Vajta G, Holm P, Kuwayama M, Booth PJ, Jacobsen H, Greve T, Callesen H. Open pulled straw (OPS) vitrification: a new way to reduce cryoinjuries of bovine ova and embryos. Molecular Reproduction and Development. 1998;51:53–58.

[150] Chen SU, Lien YR, Cheng YY, Chen HF, Ho HN, Yang YS. Vitrification of mouse oocytes using closed pulled straws (CPS) achieves a high survival and preserves good patterns of meiotic spindles, compared with conventional straws, open pulled straws (OPS) and grids. Human Reproduction. 2001;16:2350–2356.

[151] Hochi S, Terao T, Kamei M, Hirao M, Hirabayashi M. Vitriication of pronuclear stage rabbit zygotes by different ultra-rapid cooling procedures Theriogenology. 2003;59:303.

[152] Kuwayama M. Highly efficient vitrification for cryopreservation of human oocytes and embryos: the cryotop method. Theriogenology. 2007;67:73–80.

[153] Baril G, Traldi AL, Cognié Y, Leboeuf B, Beckers JF, Mermillod P. Successful direct transfer of vitrified sheep embryos. Theriogenology. 2001;56:299–305.

[154] Assumpção MEOA, Milazzotto MP, Simões R, Nicacio AC, Mendes CM, Mello MRB, Visintin JA. *In vitro* survival of *in v itro*-produced bovine embryos cryopreserved by slow freezing, fast freezing and vitrification. Animal Reproduction. 2008;3/4:116–120.

[155] Gajda B, Smorąg V. Oocyte and embryo cryopreservation—state of art and recent developments in domestic animals. Journal of Animal and Feed Sciences. 2009;18:371–387.

[156] Rall WF, Fahy GM. Ice-free cryopreservation of mouse embryos at -196°C by vitrification. Nature. 1985;313:573–575 [abstract].

[157] Dalcin L, Lucci CM. Criopreservação de embriões de animais de produção: princípios criobiológicos e estado atual. Revista Brasileira de Reprodução Animal Belo Horizonte. 2010:149–159.

[158] Liebermann J, Nawroth F, Isachenko V, Isachenko E, Rahimi G, Tucker MJ. Potential importance of vitrification in reproductive medicine. Biology of Reproduction. 2002;67:1671–1680.

[159] Massip A. Cryopreservation of embryos of farm animals. Reproduction in Domestic Animals. 2001;36:49–55.

[160] Kasai M, Mukaida T. Cryopreservation of animal and human embryos by vitrification. Reproductive BioMedicine Online. 2004;9:164–170.

[161] Gupta MK, Lee H-T. Cryopreservation of oocytes and embryos by vitrification. Korean Journal of Reproductive Medicine. 2010;37:267–291.

[162] Saragusty J, Arav A. Current progress in oocyte and embryo cryopreservation by slow freezing and vitrification. Reproduction. 2011;141:1–19.

[163] Yu XL, Deng W, Liu FJ, Li YH, Li XX, Zhang YL, Zan LS. Closed pulled straw vitrification of in vitro-produced and in vivo-produced bovine embryos. Theriogenology. 2010;73:474–479.

[164] Ptak G, Dattena M, Loi P, Tischner M, Cappai P. Ovum pick-up in sheep: efficiency of *in vitro* embryo production, vitrification and birth of offspring. Theriogenology. 1999;52:1105–1114.

[165] Balasubramanian S, Rho G-J. Effect of chilling on the development of in vitro produced bovine embryos at various cleavage stages. Journal of Assisted Reproduction and Genetics. 2006;23:55–61.

[166] Garcia-Garcia RM, Gonzalez-Bulnes A, Domínguez V, Veiga-Lopez A, Cocero MJ. Culture of early stage ovine embryos to blastocyst enhances survival rate after cryopreservation. Theriogenology. 2005;63:2233–2242.

[167] Garcia-Garcia RM, Gonzalez-Bulnes A, Domínguez V, Veiga-Lopez A, Cocero MJ. Survival of frozen-thawed sheep embryos cryopreserved at cleavage stages. Cryobiology. 2006;52:108–113.

[168] Shirazi A, Soleimani M, Karimi M, Nazari H, Ahmadi E, Heidari B. Vitrification of *in vitro* produced ovine embryos at various developmental stages using two methods. Cryobiology. 2010;60:204–210.

[169] Green RE, Santos BFS, Sicherle CC, Landim-Alvarenga FC, Bicudo SD. Viability of OPS vitrified sheep embryos after direct transfer. Reproduction in Domestic Animals. 2009;44:406–410.

[170] Guignot F, Bouttier A, Baril G, Salvetti P, Pignon P, Beckers JF, Touzé JL, Cognié J, Traldi AS, Cognié Y, Mermillod P. Improved vitrification method allowing direct transfer of goat embryos. Theriogenology. 2006;66:1004–1011.

[171] Kartberg A-J, Hambiliki F, Arvidsson T, Stavreus-Evers A, Svalander P. Vitrification with DMSO protects embryo membrane integrity better than solutions without DMSO. Reproductive BioMedicine Online. 2008;17:378–384.

[172] Varago FC, Moutacas VS, Carvalho BC, Serapião RV, Vieira F, Chiarini-Garcia H, Brandão FZ, Camargo LS, Henry M, Lagares MA. Comparison of conventional freezing and vitrification with dimethylformamide and ethylene glycol for cryopreservation of ovine embryos. Reproduction in Domestic Animals. 2014;49:839–844.

[173] Hochi S, Semple E, Leibo SP. Effect of cooling and warming rates during cryopreservation on survival of in vitro-produced bovine embryos. Theriogenology. 1996;46:837–847.

[174] Bączkowski T, Kurzawa R, Głąbowski W. Methods of embryo scoring in *in vitro* fertilization. Reproductive Biology. 2004;4:5–22.

[175] Van Soom A, Vanroose G, de Kruif A. Blastocyst evaluation by means of differential staining: a practical approach. Reproduction in Domestic Animals. 2001;36:29–35.

[176] Ushijima H, Akiyama K, Tajima T. Classification of morphological changes based on the number of cleavage divisions in bovine embryos. Journal of Reproduction and Development. 2009;55:83–87.

[177] Stringfellow DA, Seidel SM, editors. Manual of the International Embryo Transfer Society, Savoy, IL, USA; 1998.

[178] Abe H, Matsuzaki S, Hoshi H. Ultrastructural differences in bovine morulae classified as high and low qualities by morphological evaluation. Theriogenology. 2002;57:1273–1283.

[179] Leoni GG, Berlinguer F, Succu S, Bebbere D, Mossa F, Madeddu M, Ledda S, Bogliolo L, Naitana S. A new selection criterion to assess good quality ovine blastocysts after vitrification and to predict their transfer into recipients. Molecular Reproduction and Development. 2008;75:373–382.

[180] Hernandez-Ledezma, JJ, Mathialagan N, Villanueva C, Sikes JD, Roberts RM. Expression of bovine trophoblast interferons by *in vitro*-derived blastocysts is correlated with their morphological quality and stage of development. Molecular Reproduction and Development. 1993;36:1–6.

[181] Rizos D, Gutierrez-Adan A, Moreira P, O'Meara C, Fair T, Evans ACO, Boland MP, Lonergan P. Species-related differences in blastocyst quality are associated with differences in relative mRNA transcription. Molecular Reproduction and Development. 2004;6:381–386.

[182] Knijn HM, Gjørret JO, Vos PLAM, Hendriksen PJM, van der Weijden BC, Maddox-Hyttel P, Dieleman SJ. Consequences of *in vivo* development and subsequent culture on apoptosis, cell number, and blastocyst formation in bovine embryos. Biology of Reproduction 2003;69:1371–1378.

[183] Fabian D, Gjørret JO, Berthelot F, Martinat-Botte F, Maddox-Hyttel P. Ultrastructure and cell death of in vivo derived and vitrified porcine blastocysts. Molecular Reproduction and Development. 2005;70:155–165.

[184] Hosseini SM, Hajian M, Asgari V, Forozanfar M, Abedi P, Nasr-Esfahani MH. Novel approach of differential staining to detect necrotic cells in preimplantation embryos. Iranian Journal of Fertility and Sterility. 2007;1:103–106.

[185] Smith GD, Silva e Silva CA. Developmental consequences of cryopreservation of mammalian oocytes and embryos. Reproductive BioMedicine Online. 2004;9:171–178.

[186] Vajta G, Hyttel P, Callesen H. Morphological changes of *in-vitro*-produced bovine blastocysts after vitrification, in-straw direct rehydration, and culture. Molecular Reproduction and Development. 1997;48:9–17.

[187] Bettencourt EM, Bettencourt CM, Chagas e Silva JN, Ferreira P, Oliveira E, Romão R, Rocha A, Sousa M. Ultrastructural characterization of *in vivo*-produced ovine morulae and blastocysts. Zygote. 2015;23:583–593.

[188] Ferrer F, Garcia C, Villar J, Arias M. Ultrastructural study of the early development of the sheep embryo. Anatomia Histologia Embryologia. 1995;24:191–196.

[189] Rizos D, Fair T, Papadopoulos S, Boland M, Lonergan P. Developmental, qualitative and Ultraestructural differences between ovine and bovine embryos produced *in vivo* and *in vitro*. Molecular Reproduction and Development. 2002;62:320–327.

[190] Crosier AE, Farin PW, Dykstra MJ, Alexander JE, Farin CE. Ultrastructural morphometry of bovine blastocysts produced in vivo or in vitro. Biology of Reproduction. 2001;64:1375–1385.

[191] Dobrinsky JR. Cellular approach to cryopreservation of embryos. Theriogenology. 1996;45:17–26.

[192] Dattena M, Mara L, Bin T AA, Cappai P. Lambing rate using vitrified blastocysts is improved by culture with BSA and hyaluronan. Molecular Reproduction and Development. 2007;74:42–47.

[193] Kong IK, Lee SI, Cho SG. Comparison of open pulled straw (ops) vs glass micropipette (GMP) vitrification in mouse blastocysts. Theriogenology. 2000;53:1817–1826.

[194] Kelly J, Kleemann D, Kuwayama M, Walker S. Pregnancy rates for *in vitro* and *in vivo* produced ovine embryos vitrified using the minimum volume cooling cryotop method. Reproduction, Fertility and Development. 2005;17:194–194.

[195] Kuwayama M. Evidence-based embryo cryopreservation. Journal of Mammalian Ova Research. 2005;22:28–32.

[196] Lee HJ, Elmoazzen H, Wright D, Biggers J, Rueda BR, Heo YS, Toner M, Toth TL. Ultra-rapid vitrification of mouse oocytes in low cryoprotectant concentrations. Reproductive Biomedicine Online. 2010;20:201–208.

[197] Gad A, Hoelker M, Besenfelder U, Havlicek V, Cinar U, Rings F, Held E, Dufort I, Sirard M-A, Schellander K, Tesfaye D. Molecular mechanisms and pathways involved in bovine embryonic genome activation and their regulation by alternative *in vivo* and *in vitro* culture conditions. Biology of Reproduction. 2012;87:1–13.

[198] Lin TA, Chen CH, Sung LY, Carter MG, Chen YE, Du F, Ju JC, Xu J. Open-pulled straw vitrification differentiates cryotolerance of *in vitro* cultured rabbit embryos at the eight-cell stage. Theriogenology. 2011;75:760–768.

[199] Nedambale TL, Dinnyés A, Yang X, Tian C. Bovine blastocyst development *in vitro*: timing, sex, and viability following vitrification. Biology of Reproduction. 2004;71:1671–1676.

[200] Tsujii H, Khandoker Y, Hamano K. lipid in mammalian embryo development. Journal of Mammalian Ova Research. 2001;18:73–80.

[201] Seidel Jr GE. Modifying oocytes and embryos to improve their cryopreservation. Theriogenology. 2006;65:228–235.

[202] Leibo SP, Pollard JW, Martino A. Chilling and freezing sensitivity of "reassembled" in vitro-derived bovine embryos. Theriogenology. 1995;43:265.

[203] Leibo SP, Martino A, Kobayashi S, Pollard JW. Stage-dependent sensitivity of oocytes and embryos to low temperatures. Animal Reproduction Science. 1996;42:45–53.

[204] Abe H, Shiku H, Aoyagi S, Hoshi H. *In vitro* culture and evaluation of embryos for production of high quality bovine embryos. Journal of Mammalian Ova Research. 2004;21:22–30.

[205] Thompson JG. *In vitro* culture and embryo metabolism of cattle and sheep embryos—a decade of achievement. Animal Reproduction Science. 2000;30:273–280.

[206] Lapa M, Marques CC, Baptista MC. Vasques MI, Horta AEM, Portugal PV, Bessa RJB, Pereira RM. Fatty acid profile of oocytes and blastocysts during in vitro bovine embryo production. Reproduction in Domestic Animals. 2005;40:365.

[207] Rooke JA, McEvoy TG, Ashworth CJ, Robinson JJ, Wilmut I, Young LE, Sinclair KD. Ovine fetal development is more sensitive to perturbation by the presence of serum in embryo culture before rather than after compaction. Theriogenology. 2007;67:639–647.

[208] Pereira RM, Carvalhais I, Pimenta J, Baptista MC, Vasques MI, Horta AE, Santos IC, Marques MR, Reis A, Pereira MS, Marques CC. Biopsied and vitrified bovine embryos viability is improved by *trans*10, *cis*12 conjugated linoleic acid supplementation during in vitro embryo culture. Animal Reproduction Science. 2008;106:322–332.

[209] Zehmer JK, Huang Y, Peng G, Pu J, Anderson RG, Wand Liu P. A role for lipid droplets in inter-membrane lipid traffic. Proteomics. 2009;9:914–921.

[210] Abe H, Yamashita S, Itoh T, Satoh T, Hoshi H. Ultrastructure of bovine embryos developed from *in vitro*-matured and -fertilized oocytes: comparative morphological evaluation of embryos cultured either in serum-free medium or in serum-supplemented medium. Molecular Reproduction and Development. 1999;53:325–335.

[211] McEvoy TG, Coull GD, Broadbent PJ, Hutchinson JS, Speake BK. Fatty acid composition of lipids in immature cattle, pig and sheep oocytes with intact *zona pellucida*. Journal of Reproduction and Fertility. 2000;118:163–170.

[212] Barceló-Fimbres M, Seidel Jr GE. Effects of fetal calf serum, phenazine ethosulfate and either glucose or fructose during *in vitro* culture of bovine embryos on embryonic development after cryopreservation. Molecular Reproduction and Development. 2007;74:1395–1405.

[213] Sanches BV, Marinho LSR, Filho BDO, Pontes JHF, Basso AC, Meirinhos MLG, Silva-Santos KC, Ferreira CR, Seneda MM. Cryosurvival and pregnancy rates after exposure of IVF-derived *Bos indicus* embryos to forskolin before vitrification. Theriogenology. 2013;80:372–377.

[214] Accorsi MF. Assessment of bovine embryos grown *in vitro* in the presence of fatty acids and their survival after cryopreservation. Dissertation. Faculdade de Zootecnia e Engenharia de Alimentos, Universidade de São Paulo, Pirassununga, Brasil; 2008.

[215] Nagashima H, Kashiwazaki N, Ashman RJ, Grupen CG, Seamark RF, Nottle MB. Removal of cytoplasmic lipid enhances the tolerance of porcine embryos to chilling. Biology of Reproduction. 1994;51:618–622.

[216] Ushijima H, Yamakawa H, Nagashima H. Cryopreservation of bovine pre-morula-stage *in vitro* matured/*in vitro* fertilized embryos after delipidation and before use in nucleus transfer. Biology of Reproduction. 1999;60:534–539.

[217] Diez C, Heyman Y, Le Bourhis D, Guyader-Joly C, Degrouard J, Renard JP. Delipidating *in vitro*-produced bovine zygotes: effect on further development and consequences for freezability. Theriogenology. 2001;55:923–936.

[218] Dobrinsky JR, Pursel VG, Long CR, Johnson LA. Birth of piglets after transfer of embryos cryopreserved by cytoskeletal stabilization and vitrification. Biology of Reproduction. 2000;62:564–570.

[219] Dobrinsky JR. Advancements in cryopreservation of domestic animal embryos. Theriogenology. 2002;57:285–302.

[220] Franco M, Hansen PJ. Effects of hyaluronic acid in culture and cytochalasin B treatment before freezing on survival of cryopreserved bovine embryos produced *in vitro*. In Vitro Cellular and Developmental Biology Animal. 2006;42:40–44.

[221] Mezzalira A, Vieira AD, Barbieri DP, Machado MF, Thaler Neto A, Bernardi ML, Silva AM, Rubin MIB. Vitrificação de oócitos e embriões bovinos produzidos *in vitro* e expostos à citocalasina B. Brazilian Journal of Veterinary Research and Animal Science. 2002;39:260–265.

[222] Tominaga K, Shimizu M, Ooyama S, Isaike Y. Effect of lipid polarization by centrifugation at different developmental stages on post-thaw survival of bovine in vitro produced 16-cell embryos. Theriogenology. 2000;53:1669–1680.

[223] Dobrinsky JR, Nagashima H, Purse VG, Schreier LL, Johnson LA. Cryopreservation of morula and early blastocyst stage swine embryos: birth of litters after embryo transfers. Theriogenology. 2001;55:303 [abstract].

[224] Li R, Murphy CN, Spate L, Wax D, Isom C, Rieke A, Walters EM, Samuel M, Prather RS. Production of piglets after cryopreservation of embryos using a centrifugation-based method for delipidation without micromanipulation. Biology of Reproduction. 2009;80:563–565.

6

The Use of Reproductive Technologies to Produce Transgenic Goats

Vicente J. F. Freitas, Luciana M. Melo, Dárcio I.A. Teixeira, Maajid H. Bhat, Irina A. Serova, Lyudmila E. Andreeva and Oleg L. Serov

Abstract

Recombinant DNA technology has revolutionized the production of therapeutic proteins. Thus, genes of a great number of human proteins have already been identified and cloned. The use of farm animals as bioreactors may be the better choice to produce recombinant therapeutic proteins. For this activity, the term "pharming" was created, referring to the use of genetic engineering to obtain a transgenic or genetically modified animal. Considering the advantages and disadvantages of livestock species, goats appear as a very good model. In addition, the first human commercially approved biological drug (antithrombin (AT)) was produced from the milk of transgenic goats. The aim of this chapter is to present various reproductive technologies used to obtain transgenic goats secreting recombinant proteins in milk. Initially, this chapter presents the methods for embryo production (in vivo and in vitro) to realize the DNA microinjection in pronuclear embryos. Thus, the techniques of superovulation of donors (in vivo embryo production) and ovarian stimulation for oocyte recovery (in vitro embryo production) are described. Also, the methods for DNA microinjection and embryo transfer are detailed in this chapter. Finally, this chapter describes the reproductive procedures used for obtaining transgenic goats by cloning.

Keywords: Transgenesis, DNA microinjection, Cloning, Assisted reproductive technologies, Embryo

1. Introduction

The transgenesis technique consists of introducing a DNA construction into the genome of a pluricellular organism, which then appears in most cells and, posteriorly, is transmitted to the offspring. Typically, the term "transgenesis" is used to plants and animals, whereas other organisms (yeast, bacteria, and cultured cells) harboring a foreign DNA fragment are named as "recombinant". The first gene transfer into mouse using isolated DNA showed that it was possible to generate animals stably harboring foreign DNA and to modify phenotypic properties [1, 2]. Subsequently, the use of gene microinjection was extended to pigs, sheep, and rabbits [3]. After 4 years, the gene replacement was achieved by homologous recombination [4].

Initially, in the genetic engineering area, the recombinant proteins were expressed in bacteria and yeast. Subsequently, it was proved that several human recombinant proteins might not be efficiently produced in those systems because human proteins do not undergo post-translational modifications in bacteria. Additionally, in yeast cells, the modifications those proteins suffer are different from those occurring in human cells. For these reasons, the great challenge was to develop expression systems ensuring correct post-translational modifications in recombinant proteins. According to the studies performed to date, transgenic animals offer attractive advantages to prepare recombinant proteins, such as low cost production and high protein quality (for review see [5]).

Application of genetic modified animals (GMA) technology to domestic species has been limited due to the high cost of this kind of research. Thus, the selection of species to be used as bioreactors depends on several factors (Table 1); however, the quantity of protein required and the timescale for production are key factors. Additionally, feasibility and the costs of keeping and breeding the animals should also be considered [6].

Species	Pros	Cons	Reference*
Rabbit	Short generation interval	Very low milk yield	[3]
Pig	Short pregnancy length Production of multiple offspring	Low milk yield Difficult DNA microinjection	[3]
Sheep	Short pregnancy length Production of multiple offspring	Low milk yield Difficult DNA microinjection	[3]
Goat	Short pregnancy length Production of multiple offspring Good milk yield	Difficult DNA microinjection	[9]
Cattle	Very good milk yield	Log generation interval High maintenance cost Difficult DNA microinjection	[10]

Table 1. The pros and cons for many animal species producing recombinant proteins in milk.

Considering the advantages and disadvantages of each livestock species, goats appear as an excellent model for their use as bioreactors. In addition, it was from the milk of transgenic goats that was produced and approved the commercialization of the first human biological drug (antithrombin (AT)). This approval first occurred in Europe, by The European Agency for the Evaluation of Medicinal Products [7], and after in the United States, by the Food and Drug Administration [8].

Transgenic goats are traditionally produced through the microinjection of a DNA construct into the pronuclei of early embryos (Figure 1). Since the first report on transgenic goats using this method to produce the human tissue plasminogen activator [9], several other human proteins have been similarly produced such as hepatitis B surface antigen [11], lysozyme [12], lactoferrin [13], and granulocyte colony stimulating factor (hG-CSF) [14]. Following the report of the birth of the first cloned sheep [15], new improvements in production of transgenic goats were obtained by the use of somatic cell nuclear transfer (SCNT), a method allowing to incorporate a DNA construct into target cells in culture using lipid-mediated transfection, followed by the selection of properly integrated donor cells for transfer. The efficiency obtained from the two methods is lower in nuclear transfer reconstructed embryos than in pronuclear microinjection, although all of the animals born are transgenic and most of them produce the recombinant protein in milk during induced lactation [16].

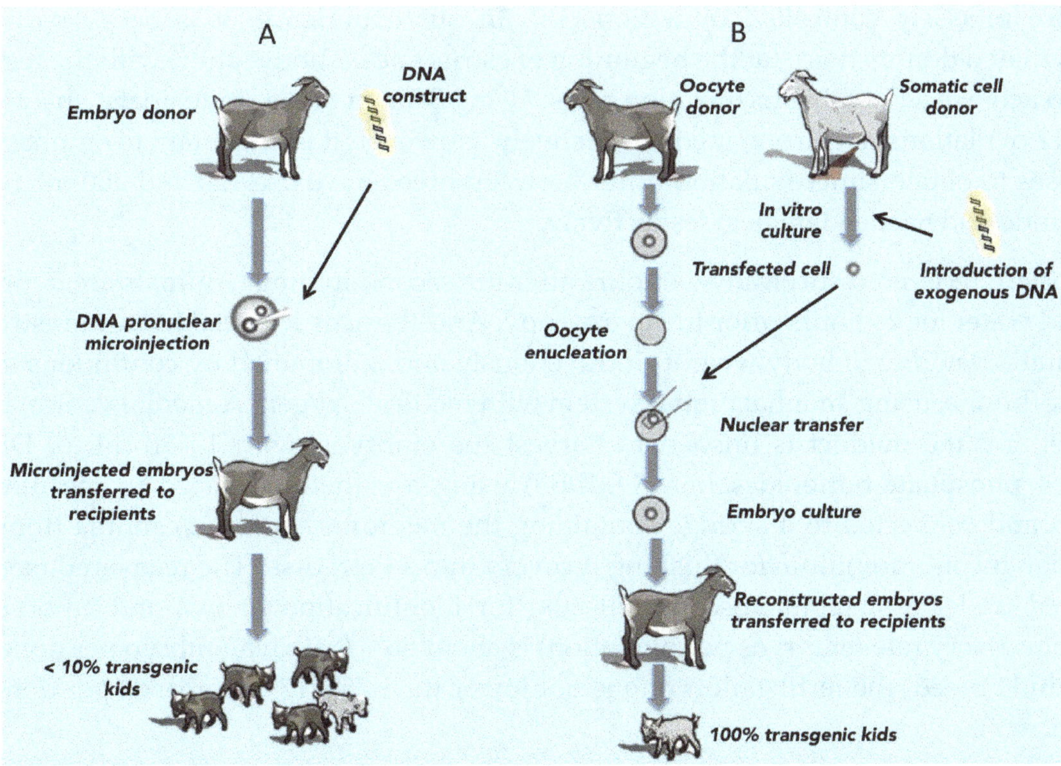

Figure 1. Traditional methods used to obtain transgenic goats. A: DNA pronuclear microinjection. B: Somatic cell nuclear transfer (SCNT).

2. Pronuclear microinjection

The intention of the production of the first genetically manipulated goat embryos was to obtain transgenic animals able to secrete pharmaceuticals in their milk, in particular the human tissue plasminogen activator [9]. Upon its success, several other human proteins have been produced in goats using pronuclear microinjection. However, the technique presents a poor overall efficiency, especially when compared to that obtained in mice. In goats, up to 1% of the injected zygotes give birth to a transgenic kid [17]. The method to produce transgenic goats using this method can be summarized in the following steps: obtaining pronuclear embryos (in vivo or in vitro), embryo microinjection with a DNA construct, transfer of microinjected embryos into recipient goats, and early pregnancy diagnosis by ultrasound. The methods presented here on microinjection technology to generate transgenic goats are based on our extensive experience in goat reproductive technology and are complemented by information from the literature.

2.1. In vivo embryo production

A closely timed protocol to recover pronuclear embryos was developed in our laboratory (Table 2); groups of four donors are used for reasons of efficiency. This protocol involves estrus synchronization with the use of vaginal sponges, superovulation with porcine follicle stimulating hormone (pFSH; Folltropin-V, Vetrepharm, Belleville, Canada), and the administration of gonadotrophin-releasing hormone (GnRH; Fertagyl; Intervet, Boxmeer, The Netherlands) to ensure precisely controlled ovulation [18]. In our experience, it is necessary to mate superovulated donors twice (at the beginning of estrus and 12 hours after) with the respective bucks to achieve acceptable fertilization rates. Using Saanen or Canindé goats, an average of 24 and 12 ovulations can be obtained, respectively. However, it is important to emphasize that FSH doses to obtain superovulation differ between breeds: we use 120 and 200 mg of pFSH for Canindé and Saanen females, respectively.

Embryos are recovered surgically 24 hours after the second mating. Animals are deprived of food and water for 24 hours prior to laparoscopy. Anesthesia is induced by administration of thiopental (20 mg/kg of body weight) intravenously and maintained by continuous infusion of 3% isoflurane, using an inhalational system with medical oxygen. A medial ventral incision is made, and the oviduct is flushed to harvest the embryos, with 15–20 mL of Dulbecco modified phosphate-buffered saline (DMPBS) using a catheter inserted in the uterotubal junction and connected to a syringe containing the medium. A plastic cannula (tom cat) is inserted in the infundibulum for flushing recovery into a Petri dish. The recovered medium is examined under a stereomicroscope (70–80×) for identification of ova and embryos. The average recovery rate (embryos per ovulation) is about 80% (>10 ova/embryo per donor) and, for Canindé breed, the fertilization rate is not lower than 75% (7–8 embryos per donor) [19, 20] (Figure 2).

Thereafter, the ova found are observed with an inverted microscope (300×) for a qualitative evaluation. Presumable zygotes are centrifuged (12000× *g* for 5 min) to allow a better visualization of the pronuclei, while ova are placed in drops of M16 medium (Sigma–Aldrich) supplemented with 10% fetal bovine serum (FBS). Then, embryos presenting visible pronuclei are immediately microinjected, whereas the others are placed for in vitro culture at 38.5°C in a humidified atmosphere with 5% CO_2 for a short period. Goat zygotes show nontransparent cytoplasm and approximately 125–130 μm in diameter (Figure 3A and B). Nontransparent cytoplasm is due to the presence of a large amount of lipid granules that hinder the visualization of pronuclei. In addition, pronuclei seem to be visible sometimes, but this impression can, however, appear false after attentive examination with an inverted microscope equipped with interferential contrast optics (Nomarsky). The presence of the second polar body is a rather marked indication that the egg has been fertilized. The presence of the second polar body is typical for all zygote stages. It can be located in contact with the first polar body or not far away from it. The first polar body most often stays at the degradation stage; sometimes zygotes with three polar bodies occur if by that moment the division of the first polar body has already occurred [19].

Day	Time	Event in	
		Donors	Recipients
−10	Morning	Insertion of vaginal sponge (60mg medroxyprogesterone acetate)	
−2	8:00 a.m. 8:00 p.m.	50 μg cloprostenol + 50 mg pFSH 50 mg pFSH	50 μg cloprostenol + 300 iu eCG
−1	8:00 a.m. 8:00 p.m.	30 mg pFSH 30 mg pFSH	
0	8:00 a.m. 8:00 p.m.	20 mg pFSH + sponge removal 20 mg pFSH	Sponge removal
1	8:00 p.m.	100 μg GnRH + first natural mating	Estrus detection
2	8:00 a.m.	Second natural mating	Estrus detection
3	8:00 – 10:00 a.m. 2:00 – 5:00 p.m.	Surgical embryo recovery	Transfer of microinjected embryos

GnRH, gonadotrophin-releasing hormone; pFSH, porcine follicle- stimulating hormone.

Table 2. Time schedule to produce pronuclear goat embryos for subsequent DNA microinjection.

The pronuclei of goat zygotes are visible without centrifugation in only 30% of times (Figure 3C). To facilitate visualization of pronuclei, all the fertilized eggs are subjected to centrifugation, which contributes to precipitation of the lipid granules. The pronuclei of late zygotes are located closely to one another, usually in the center of the cytoplasm though closer to the pole with dark granules; one of the pronuclei (male) is somewhat larger than the other (female) (Figure 3D and F). Pronuclei in goat, unlike those in mouse, rabbit, or swine, are not visualized; morphologically, they rather resembled sheep and cow pronuclei. The two pronuclei may not be always simultaneously observed, and one of them can be located in the lipid granules. However, even after very careful examination of the centrifuged zygotes in a microscope with

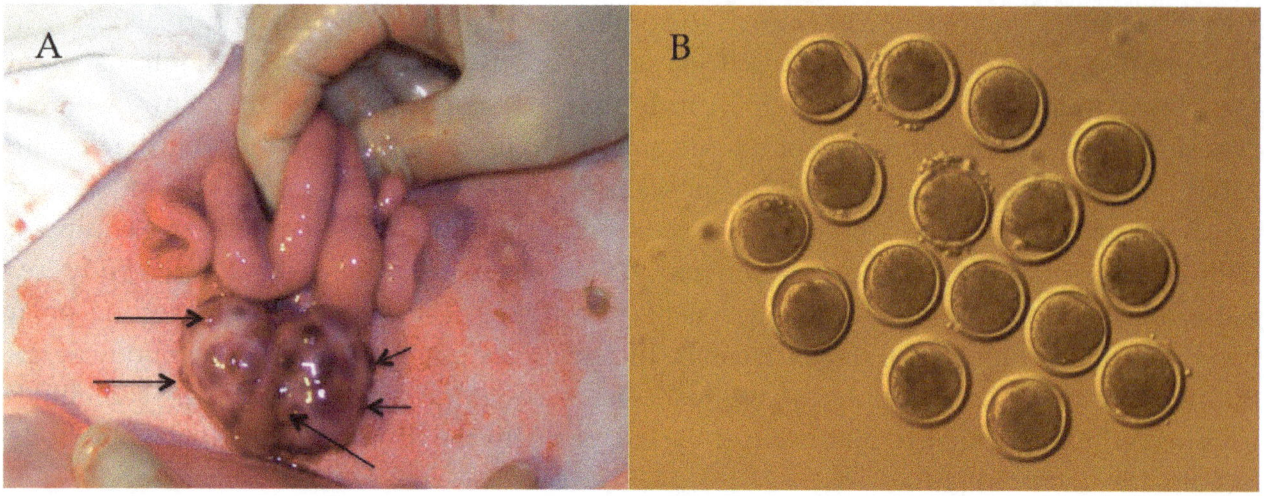

Figure 2. Embryo recovery in goats. A: Evaluation of ovarian response (arrows show some ovulation points) for further oviduct flushing. B: Presumable embryos observed under stereomicroscope.

Nomarsky optics, the pronuclei are not always clearly observed. Zygotes of different goat breeds differ in the degree of visualization of pronuclei. For instance, in Canindé goats, the pronuclei visualization is possible in almost 100% of examined zygotes, whereas in Saanen this rate was only slightly higher than 70% [20].

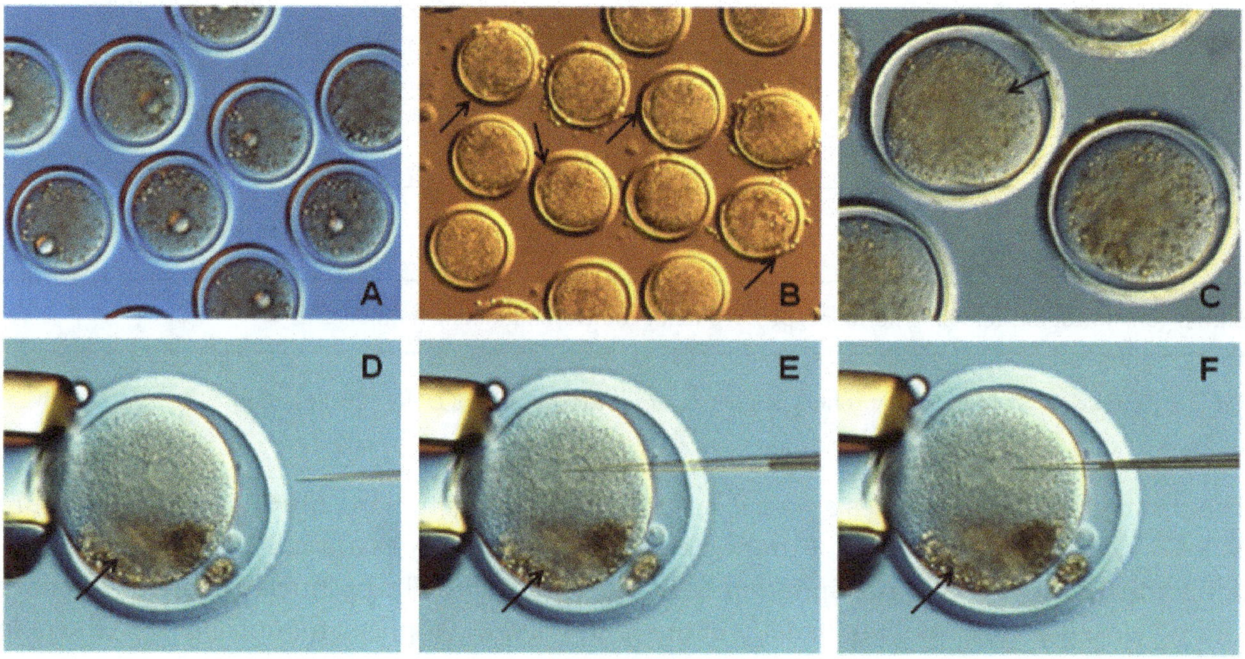

Figure 3. Morphology of goat ova and embryos with microinjection details. A: Unfertilized oocytes; the secondary polar body is absent (×200). B: Zygotes; arrows indicate the secondary polar bodies but pronuclei are not visible (×200). C: Zygotes; arrow indicates pronuclei visible without centrifugation (×300). D, E, and F: Microinjection into centrifuged pronuclei of zygotes; arrow indicates the pole with lipid granules.

2.2. In vitro embryo production

Although the in vivo production method results in zygotes of high developmental capacity, the procedure is characterized by a great deal of variability in the number of pronuclear embryos. In addition, this technique causes adhesion formation following laparotomy and, consequently, limits the repeat use of the donors. Laparoscopic ovum pick-up (LOPU) followed by in vitro embryo production appears as an efficient method for the exponential dissemination of high genetic value sheep and goats [21]. Similarly described for bovines, the use of in vitro-matured oocytes is also advantageous in goats [22]. A year-round provision of immature oocytes can be obtained from hormone-primed goats by LOPU, contrasting to the seasonal fluctuations observed in the quantity of oocytes obtained from slaughterhouse-derived ovaries. Thus, the laparoscopic approach provides a minimally invasive, efficient means of obtaining immature oocytes for subsequent embryo production and microinjection.

The estrous cycle of goats should be manipulated with gonadotropins. Different hormonal protocols have been tested to improve the oocyte production in quantity and quality obtained from goats submitted to LOPU [22, 23]. These protocols enable the synchronous recruitment of a large population of follicles. Progestagen impregnated vaginal sponges (45 mg MAP, Progespon; Syntex, Buenos Aires, Argentina) are employed for 10 days to synchronize the female cycle; they should be left in the vagina until the moment for oocyte aspiration to avoid ovulations. Indeed, a dose of prostaglandin $F_{2\alpha}$ analog (50 μg cloprostenol, Ciosin; Coopers, São Paulo, Brazil) is used at the seventh day of progestagen treatment to promote luteolysis. In a breed-dependent total dose (120–200 mg), pFSH is administered in five injections (decreasing doses), 12 hours apart, beginning at prostaglandin injection. A 24-hour interval between the last pFSH administration and LOPU is used. Goats can be treated seven or more times using this protocol without a decrease in the ovarian response (Table 3).

Session	Punctured follicles	Recovered COCs	Recovery rate (%)
1	21.3 ± 7.1	12.9 ± 5.1	60.8 ± 21.6
2	21.8 ± 8.3	15.6 ± 5.9	71.6 ± 12.7
3	20.0 ± 5.7	15.2 ± 5.1	75.8 ± 11.7
4	16.5 ± 7.9	12.9 ± 6.8	78.8 ± 14.4
5	16.3 v 4.8	11.9 ± 3.8	73.3 ± 9.6
6	19.0 ± 5.9	13.9 ± 3.4	73.2 ± 16.0
7	14.9 ± 5.8	10.6 ± 3.4	70.9 ± 16.4

Table 3. Punctured follicles, recovered cumulus–oocyte complexes (COCs), and recovery rate in goats hormonally treated for seven times per session [25]. The values are mean (±SD) and percentages.

LOPU starts in the morning of the 10th day of progestagen treatment, after sponge removal. Briefly, goats are deprived of food and water for 36 hours and 24 hours prior to LOPU, respectively. Using a similar protocol for anesthesia to that previously described for embryo recovery, the female is placed on a cradle at a 45° angle to avoid accidents at trocars punction (Figure 4A). One trocar is used to pass the endoscope into the abdominal cavity, cranial to the udder, and to the left of the midline. Once the abdominal cavity is expanded with air, a second trocar inserted in the right side is used to introduce atraumatic grasping forceps. Using gentle manipulation, the uterine horns are pulled to allow visualization of each stimulated ovary. A third trocar, inserted in the midline, is used to pass the oocyte aspiration needle (Figure 4B). The system comprises a 5-mm laparoscope associated to a 22-G needle and a vacuum pump regulated to 35 mmHg. Follicles of 2–7 mm are punctured from the side, the needle paralleling the base of the follicle (Figure 4C). The collection medium is composed of HEPES buffered TCM 199 supplemented with 20 IU/mL heparin and 40 mg/mL gentamicin sulfate. To avoid adherences, after LOPU, each ovary is flushed with 100 mL of heparinized saline.

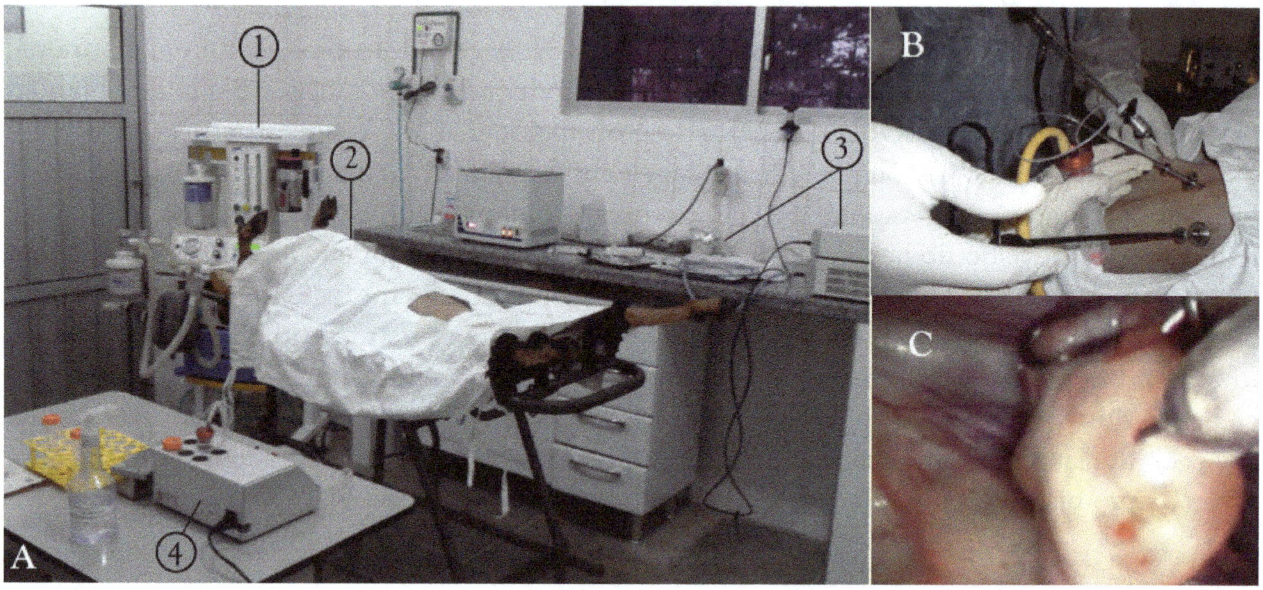

Figure 4. Structure to perform the laparoscopic ovum pick-up (LOPU) in goats. A: Room for oocyte recovery showing the inhalation anesthesia equipment ①, goat on a cradle ②, video-laparoscopy equipment ③, and vacuum pump ④. B: Points for insertion of the endoscope, follicular puncture needle, and the atraumatic grasping forceps. C: Moment of the follicular puncture with needle in a vacuum system.

As shown in Figure 5, collected COCs are visualized under a stereo zoom microscope and graded according to the classification used in our laboratory [24]. Only oocytes graded as 1 and 2 are considered acceptable and used for in vitro maturation (IVM). COCs are then washed four times and transferred into maturation medium consisting of TCM 199 supplemented with 10 ng/mL epidermal growth factor (EGF) and 100 μM cysteamine in four-well Petri dishes, each well containing 45–50 oocytes in 500 μL of maturation medium. COCs are incubated for 24 hours at 38.5°C and 5% CO_2.

Oocyte maturation includes meiotic resumption and progression to the fertilizable stage of metaphase II after emission of the first polar body and related events in oocyte cytoplasm and surrounding cumulus cells. Therefore, the result of IVM depends on the intrinsic quality of immature oocytes, but the maturation conditions can widely modulate the final competence

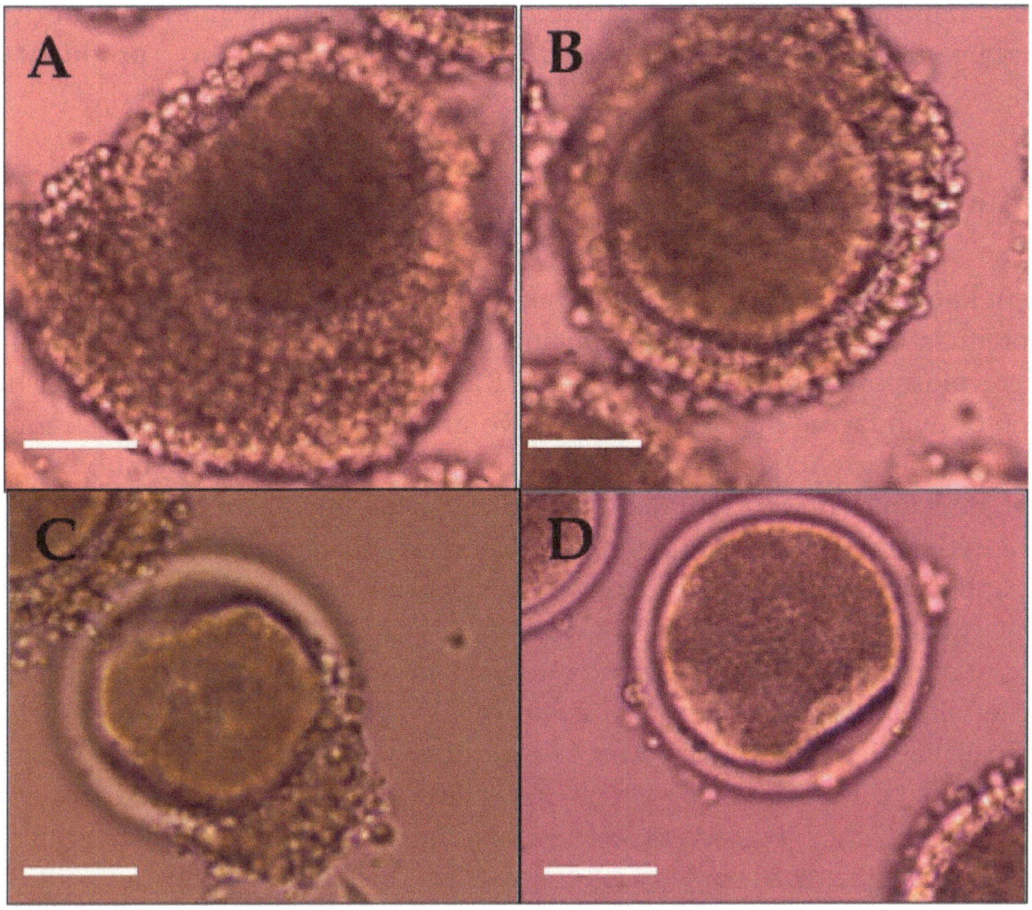

Figure 5. Criteria for grading *cumulus*–oocyte complexes (COCs) recovered from goats. A: Grade 1 COCs show multilayered compact *cumulus* and finely granulated oocyte cytoplasm. B: Grade 2 COCs show one to three layers of *cumulus* cells and finely granulated oocyte cytoplasm. C: Grade 3 COCs show incomplete or no cellular investment or heterogeneous oocyte cytoplasm. D: Grade 4 oocyte evidences an abnormal shape and heterogeneous oocyte cytoplasm, whereas apoptotic oocytes are surrounded by jelly-like *cumulus*-corona cells investment. Scale bar represents 50 μm.

of IVM oocytes. The IVM procedure is commonly performed using TCM 199 enriched with amino acids and glucose, supplemented with hormones and heat-inactivated serum. To make IVM simpler, more safe, and repeatable, our laboratory proposed a maturation medium using just defined compounds—TCM 199 supplemented with 10 ng/mL EGF and 100 mM cysteamine—and obtained good results in embryo development of adult goat oocytes [26].

Motile sperm from frozen/thawed semen are separated by centrifugation (700× *g* for 15 min) on 2 mL of Percoll discontinuous density gradient (45%/90%). Then, dilution of viable sperm in fertilization medium to achieve a final concentration of 2.0×10^6 sperm/mL is performed. The matured COCs are transferred into plates containing washing medium. The washing and fertilization medium are synthetic oviduct fluid (SOF) medium containing 10% of heat inactivated estrus goat serum, 5 mg/mL heparin, and 4 mg/mL gentamicin. Groups of 45–50 oocytes are transferred into four-well Petri dishes with 450 mL of fertilization medium and 50 mL of sperm suspension is added to each well. Sperm and oocytes are co-incubated for 18 hours at 38.5°C in a humidified atmosphere of 5% CO_2 in air [27, 28]. Oocytes are then stripped off cumulus cells and centrifuged to facilitate pronuclear visualization. All procedures for DNA microinjection are similar to those made during the in vivo embryo production.

3. Somatic cell nuclear transfer

The current knowledge on SCNT results from the hard work and curiosity of scientists who began more than 100 years ago. Hans Spemann was the first to demonstrate that salamander nuclei were pluripotent up to 16-cell stage [28]. The fascinating study demonstrated the production of twin larvae by cutting the embryo in half using his son's hair strand. Further research from other groups confirmed that nuclei from early amphibian embryos transferred to appropriate cytoplasm were totipotent. The further experiments tried to transfer nuclei from more advanced species; disappointingly, the results were not encouraging. Following the lead, Illmensee and Hoppe [29] reported successful nuclear transfer of embryonic nuclei in mice by directly injecting inner cell mass into enucleated zygotes. Willadsen [30, 31] reported for the first time the nuclear transfer using enucleated, metaphase-II sheep oocytes. After 10 years, Campbell et al. [32] reported the production of first offspring using transfer of cultured embryonic cell line derived from day 9 in vivo–produced sheep embryo. This significant study demonstrated that differentiated cells have the ability to originate a new individual and cells could be induced to enter a quiescent state. Finally, in 1997, the first mammal cloned from an adult, differentiated cell was named Dolly; it was produced by transferring a differentiated somatic cell into an enucleated mature oocyte [15]. Since Dolly, several other mammalian species have been cloned.

Different approaches have been conducted to modify donor cells and for nuclear transfer to improve the efficiency of the technique. These manipulations were focused on donor cells, including (a) the synchrony of the cell cycle stage among donor cells and the synchrony between donor cells and recipient oocytes; (b) the use of somatic cells from donors of various ages, tissue origins, passages, and culture conditions; (c) the transfer of stem cells with epigenetic marks; and (d) the drug-induced modification of the epigenetic marks in donor cells. The efficiency of nuclear transfer has been dramatically improved from the initial success rate, although none of these efforts eliminated the common problems associated with nuclear transfer, suggesting the need for further studies on nuclear reprogramming to better understand the underlying mechanisms of reprogramming and consequently to enhance the ability to reprogram differentiated somatic nuclei.

Metaphase II enucleated oocytes are most frequently used as recipient cytoplasm in SCNT studies. Two sources are available for these stage oocytes: (a) in vivo mature oocytes collected directly from animals and (b) aspirated oocytes collected from the ovaries of live or slaughtered animals and cultured in vitro for maturation (as described for microinjection). The selection of the source of recipient oocyte depends on the experimental species. In some species, the developmental competence is similar for both the oocyte sources. Conversely, in other species, the developmental competence is rather variable for in vivo or in vitro recipient oocytes.

The use of recombinant somatic cell lines for nuclear transfer shows several advantages: it allows the introduction of transgenes by traditional transfection methods, increases the efficiency of transgenic animal production to 100%, and overcomes the problem of founder mosaicism. Additionally, SCNT protocols may also be used to clone transgenic founders obtained by pronuclear microinjection. Several steps of SCNT technique to produce transgenic goats are described in Table 4.

Step	Description
1. Isolation of fibroblast cell line	The primary goat fibroblast cells used as karyoplast donors are obtained by biopsy. Tissues are minced, washed, and transferred into culture flasks. Cells are cultured in TCM 199 supplemented with 10% fetal bovine serum (FBS), nucleosides, 2 mM L-glutamine, 0.1 mM β-mercaptoethanol as an antioxidant, and 50 μg/mL gentamicin. After three subpassages, cells are frozen and stored in liquid nitrogen.
2. Oocyte recovery	Oocytes can be obtained from follicle-stimulating hormone (FSH)-stimulated animals (as described by laparoscopic oocyte collection) or slaughterhouse ovaries
3. Oocyte enucleation	Cumulus–oocyte complexes (COCs) are vortexed at 18 to 22 hours post in vitro maturation (IVM) in TL-HEPES containing 0.6 mg/mL hyaluronidase for removal of the cumulus cells. Oocytes are washed in modified TL-HEPES (TL-HEPES supplemented with 10% FBS), selected based on the presence of a polar body, and labeled with 3 μg/mL Hoechst 33342 for 1 min. Mature oocytes are enucleated in inverted microscope with micromanipulators and epifluorescent illumination.
4. Donor cell preparation	Fibroblast cells are trypsinized, washed, and held (<1.5 hours) in modified TL-HEPES immediately before transfer into the recently enucleated oocytes.
5. Reconstruction and fusion	A single fibroblast cell is injected into the perivitelline space of each enucleated oocyte. Karyoplast–cytoplast couplets are manually aligned between two stainless steel electrodes in a microslide fusion chamber filled with fusion buffer (0.3 M mannitol, 0.1mM $MgSO_4 7H_2O$, 0.05mM $CaCl_2$, 0.5mM HEPES and 4mg/mL BSA) and fused by a single DC pulse (1.30 kV/cm for 25 μsec) delivered by electrofusion equipment. Fused couplets are activated by a 5-min exposure to 5 μM ionomycin, washed extensively in modified TL-HEPES, and then incubated for 3 hours in 2 mM 6-dimethylaminopurine. Following activation, reconstructed embryos are cultured for 36 hours prior to transfer to recipient females.

Table 4. Brief methodology for the use of somatic cell nuclear transfer (SCNT) technique to produce transgenic goats.

There are no major changes in the technical characteristics of SCNT methods for farm animals since the pioneer description of the first successful embryonic cloning [31]. Researchers working in embryology consider the *zona pellucida* as important for the correct embryonic development. However, this technique requires the use of micromanipulators, highly skilled labor, and very expensive equipment. Some years ago, other forms of the technique (zona-free procedures) were described [36]. For clear distinction, the term hand-made cloning (HMC) was used for the micromanipulation-free SCNT method. Several publications with zona-free NT techniques agree that these methods are easier to learn and do not require skilled workforce, originating embryos with a final quality comparable with that of traditional cloning [37, 38]. Our group obtained a transgenic female goat founder by pronuclear microinjection, which expressed acceptable levels of the hG-CSF in the milk [14]. Currently, the objective of our group is to increase the number of transgenic animals by cloning techniques. Therefore, using skin fibroblast primary culture cells collected from the transgenic founder goat as nucleus donor cells, the HMC cloning procedure was used (Figure 6) to test the developmental potential of cloned transgenic goat embryos [39].

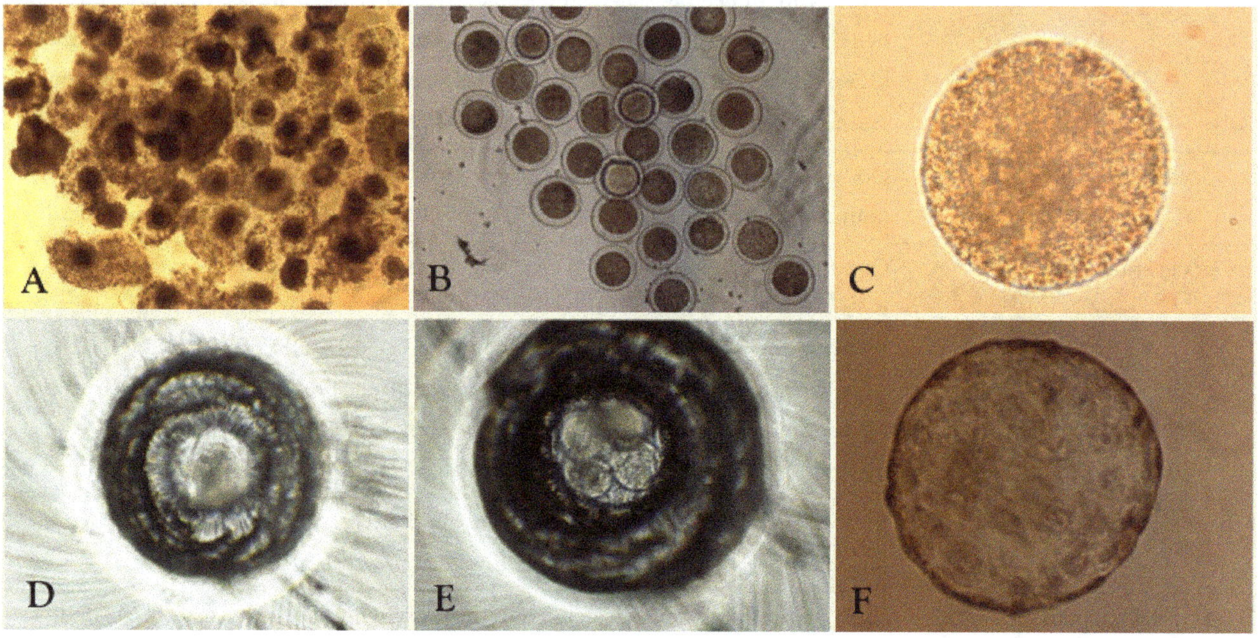

Figure 6. In vitro production steps of goat hand-made cloned embryos using donor cells from a transgenic founder goat. A: *Cumulus*–oocyte complexes after 20 hours of in vitro maturation (200×). B: In vitro–matured oocytes based on evaluation of polar body (200×). C: Zona-free oocyte after removal of the *zona pellucida* (600×). D: Transgenic one-cell stage cloned embryo on day 1 of culture into a microwell (400×). E: Transgenic four-cell stage cloned embryo on day 2 of culture into a microwell (400×). F: Transgenic cloned blastocyst on day 7 of development (600×).

4. Embryo transfer and pregnancy diagnosis

Crossbred recipient goats can be selected from those exhibiting estrus 48–72 hours prior to the scheduled embryo transfer. Nuclear transfer embryos are transferred into oviduct's recipients. For the embryo transfer procedure, we used the same general anesthesia protocol as for oocyte collection. Likewise, goats are placed in dorsal recumbence for midventral laparotomy procedure. The ovaries are then examined for evidence of ovulation; thereafter, the uterus is exteriorized, and embryos are transferred into the oviduct (2-day embryos) or the uterus (7-day embryos) ipsilateral to the ovary with the most ovulation points. Embryos (5–10/recipient) are transferred via a small plastic catheter through the tubal ostium, the catheter gently advanced into the oviduct.

The need for early diagnosis of pregnancy in goats, especially for valuable embryos, as in the case of transgenesis and cloning, lead to the common application of transrectal or abdominal real-time B-mode ultrasonography (US) for pregnancy examination. The distension of the uterus, the evidence of embryonic vesicles or the embryo, the fetal heartbeat, and placentomes are positive signals for pregnancy. Ideal time for abdominal scanning is between 40 and 60 days. Prior to 40–45 days, the transducer may have to be placed higher in the inguinal region, but if transrectal US is used the diagnosis is anticipated up to 25–30 days. In our studies, transrectal US examinations of recipients begin on day 25 of gestation, whereas additional US assessments are performed at intervals of 7 and 10 days until unequivocal pregnancy diagnosis. The fetal heartbeat is normally used to diagnose pregnancy and fetus viability. Goats with pregnancy confirmed are monitored at 2-week intervals until day 90, when they are separated from the others and placed in groups (two or three) per pen. Monitoring of pregnant recipients past day 45 is performed through abdominal US (Figure 7).

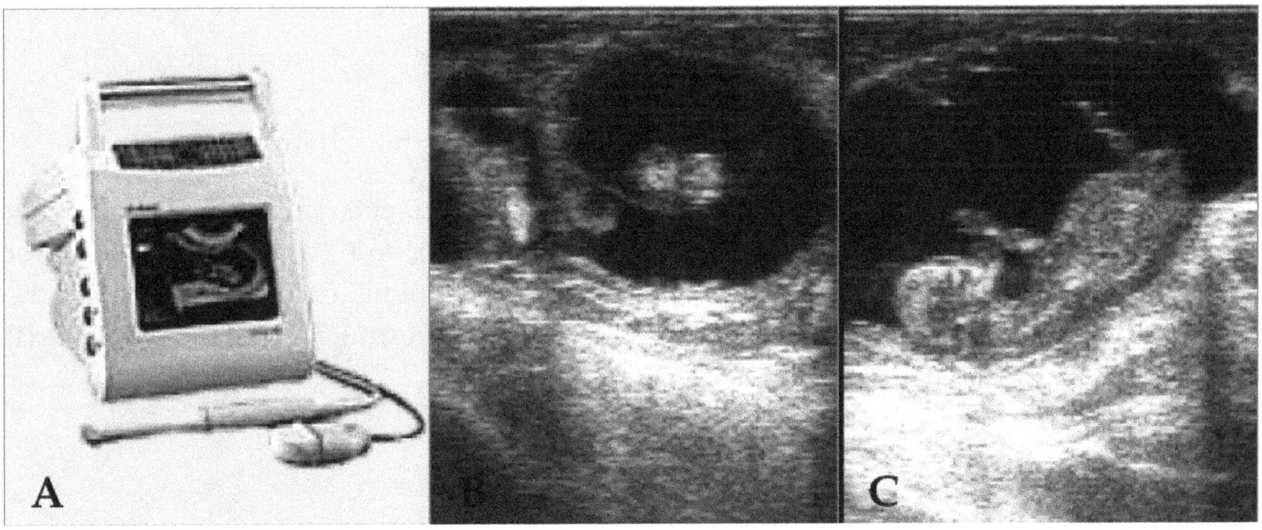

Figure 7. Pregnancy diagnosis in recipient goats. A: Real-time ultrasonography (US) equipment and its probes for transrectal and abdominal examination. Pregnant recipients at 25 (B) and 45 days (C) after embryo transfer.

Real-time US is also currently used for other kind of assessments, such as to verify and monitor the ovarian follicular dynamics, the time of ovulation, and to accurately estimate the number of corpora lutea at days 6–7 after induced superovulation in goats. Nevertheless, studies focusing the superovulation response in goats, particularly, to verify early ovulatory response (*corpora haemorraghica* (CH)), are very limited, and the accuracy of the method with this objective has not been assessed. Thus, our group proposed to assess the efficiency of real-time US as a noninvasive method to estimate the ovulation ratio after superovulation in goats by counting the number of CH's in the ovaries. The sensitivity, specificity, and total efficiency of the ultrasonographic estimation of ovulations in a goat were 100.0% [40]. Thus, transrectal real-time US is a valuable method for assessing the ovulatory response and can be used to select donor goats in a transgenesis program or to estimate embryo yield after superovulation.

5. Conclusions

Research in producing transgenic animals currently uses goats for its lower cost and precise genomic alterations. This statement is reinforced by the approval and the commercialization of the first human biological drug from the milk of transgenic goats.

Nuclear transfer protocols have evolved, which make it possible to obtain transgenic goats efficiently. Also, it seems that goats generated by SCNT do not suffer from the health problems reported in sheep and cattle. Moreover, the HMC technique does not require the use of micromanipulators and accessories. Also, the HMC procedure does not request micromanipulators or related equipment, making the method accessible to laboratories with limited funding.

Acknowledgements

Technical information in this chapter was derived from experiments funded by the following Brazilian funding agencies: CNPq (Brasília), CAPES (Brasília), and FUNCAP (Fortaleza) for providing the conditions for the experiments. The authors would like to thank the staff of the Laboratory of Physiology and Control of Reproduction for the technical assistance and animal care.

Author details

Vicente J. F. Freitas[1*], Luciana M. Melo[1], Dárcio I.A. Teixeira[1], Maajid H. Bhat[1], Irina A. Serova[2], Lyudmila E. Andreeva[3] and Oleg L. Serov[2]

*Address all correspondence to: vicente.freitas@uece.br

1 Laboratory of Physiology and Control of Reproduction, Faculty of Veterinary, State University of Ceará, Fortaleza, Brazil

2 Institute of Cytology and Genetics, Russian Academy of Sciences, Novosibirsk, Russia

3 Institute of Molecular Genetics, Russian Academy of Sciences, Moscow, Russia

References

[1] Gordon JW, Scangos GA, Plotkin DJ, Barbosa JA, Ruddle FH. Genetic transformation of mouse embryos by microinjection of purified DNA. Proc Natl Acad Sci USA. 1980;77,7380-7384.

[2] Palmiter RD, Brinster RL, Hammer RE, Trumbauer ME, Rosenfeld MG, Birnberg NC, Evans RM. Dramatic growth of mice that develop from eggs microinjected with metallothionein-growth hormone fusion genes. Nature. 1982;300,611-615.

[3] Hammer RE, Pursel VG, Rexroad CE Jr, Wall RJ, Bolt DJ, Ebert KM, Palmiter RD, Brinster RL. Production of transgenic rabbits, sheep and pigs by microinjection. Nature. 1985;315,680-683.

[4] Capecchi MR. Altering the genome by homologous recombination. Science. 1989;244,1288-1292.

[5] Houdebine LM. Production of pharmaceutical proteins by transgenic animals. Comp Immunol Microbiol Infect Dis. 2009;32,107-121.

[6] Bösze Z, Baranyi M, Whitelaw CB. Producing recombinant human milk proteins in the milk of livestock species. Adv Exp Med Biol. 2008;606,357-393.

[7] Schmidt C. Belated approval of first recombinant protein from animal. Nat Biotechnol. 2006;24,877. DOI: 10.1038/nbt0806-877.

[8] Lavine G. FDA approves first biological product derived from transgenic animal. Am J Health Syst Pharm. 2009;66,518.

[9] Ebert KM, Selgrath JP, DiTullio P, Denman J, Smith TE, Memon MA, Schindler JE, Monastersky GM, Vitale JA, Gordon K. Transgenic production of a variant of human tissue-type plasminogen activator in goat milk: generation of transgenic goats and analysis of expression. Nat Biotechnol. 1991;9,835-838.

[10] Krimpenfort P, Rademakers A, Eyestone W, van der Schans A, van den Broek S, Kooiman P, Kootwijk E, Platenburg G, Pieper F, Strijker R, Boer, H. Generation of transgenic dairy cattle using 'in vitro' embryo production. Nat Biotechnol. 1991;9,844-847.

[11] Zhang J, Lao W, Cheng Y, Cheng G, Xu S. Expression of HBsAg gene in transgenic goats under direction of bovine alpha-S1 casein control sequence. Chin J Biotechnol. 1997;13,99-104.

[12] Maga EA, Shoemaker CF, Rowe JD, Bondurant RH, Anderson GB, Murray JD. Production and processing of milk from transgenic goats expressing human lysozyme in the mammary gland. J Dairy Sci. 2006;89,518-524.

[13] Zhang J, Li L, Cai Y, Xu X, Chen J, Wu Y, Yu H, Yu G, Liu S, Zhang A, Cheng G.

Expression of active recombinant human lactoferrin in the milk of transgenic goats. Protein Expr Purif. 2007;57,127-135. DOI: 10.1016/j.pep.2007.10.015.

[14] Freitas VJF, Serova IA, Moura RR, Andreeva LE, Melo LM, Teixeira DIA, Pereira AF, Lopes-Jr ES, Dias LPB, Nunes-Pinheiro DCS, Sousa FC, Alcântara-Neto AS, Albuquerque ES, Melo CHS, Rodrigues VHV, Batista RITP, Dvoryanchikov GA, Serov OL. The establishment of two transgenic goat lines for mammary gland hG-CSF expression. Small Rum Res. 2012;105,105-113.

[15] Wilmut I, Schnieke AE, McWhir J, Kind AJ, Campbell KH. Viable offspring derived from fetal and adult mammalian cells. Nature. 1997;385,810-813.

[16] Baguisi A, Behboodi E, Melican DT, Pollock JS, Destrempes MM, Cammuso C, Williams JL, Nims SD, Porter CA, Midura P, Palacios MJ, Ayres SL, Denniston RS, Hayes ML, Ziomek CA, Meade HM, Godke RA, Gavin WG, Overstrom EW, Echelard Y. Production of goats by somatic cell nuclear transfer. Nat Biotechnol. 1999;17,456-461.

[17] Baldassarre H, Wang B, Keefer CL, Lazaris A, Karatzas CN. State of the art in the production of transgenic goats. Reprod Fertil Dev. 2004;16,465-470.

[18] Moura RR, Lopes-Junior ES, Teixeira DIA, Serova IA, Andreeva LE, Melo LM, Freitas VJF. Pronuclear embryo yield in Canindé and Saanen goats for DNA microinjection. Reprod Domest Anim. 2010;45,e101-e106.

[19] Freitas VJF, Serova IA, Andreeva LE, Melo LM, Teixeira DIA, Pereira AF, Moura RR, Lopes-Jr ES, Souza-Fabjan JMG, Batista RITP, Serov OL. The comparison of two embryo donor breeds for the generation of transgenic goats by DNA pronuclear microinjection. Anim Prod Sci. 2014;54,564-568.

[20] Freitas VJF, Melo LM, Batista RITP, Souza-Fabjan JMG, Teixeira DIA, Serova IA, Andreeva LE, Burkov IA, Serov OL. Goats (*Capra hircus*) as bioreactors for production of recombinant proteins interesting to pharmaceutical industry. Clon Transgen. 2014,3,3-9. DOI: 10.4172/2168-9849.1000130.

[21] Cognié Y, Poulin N, Locatelli Y, Mermillod P. State of-the-art production, conservation and transfer of in-vitro-produced embryos in small ruminants. Reprod Fertil Dev. 2004;16,437-445.

[22] van Wagtendonk-de Leeuw AM. Ovum pick up and in vitro production in the bovine after use in several generations: a 2005 status. Theriogenology. 2006;65,914-925.

[23] Gibbons A, Pereyra Bonnet F, Cueto MI, Salamone D, Catala M. Recovery of sheep and goat oocytes by laparoscopy. Acta Sci Vet. 2008;36,223-230.

[24] Avelar SRG, Moura RR, Sousa FC, Pereira AF, Almeida KC, Melo CHS, Teles-Filho ACA, Baril G, Melo LM, Teixeira DIA, Freitas VJF. Oocyte production and in vitro maturation in Canindé goats following hormonal ovarian stimulation. Anim Reprod. 2012;9,1-7.

[25] Sanchez DJD, Melo CHS, Souza-Fabjan JMG, Sousa FC, Rocha AA, Campelo IS, Teixeira DIA, Pereira AF, Melo LM, Freitas VJF. Repeated hormonal treatment and laparoscopic ovum pick-up followed by in vitro embryo production in goats raised in the tropics. Livest Sci. 2014;165,217-222.

[26] Souza JMG, Duffard N, Bertoldo MJ, Locatelli Y, Corbin E, Fatet A, Freitas VJF, Mermillod, P. Influence of heparin or the presence of cumulus cells during fertilization on the in vitro production of goat embryos. Anim Reprod Sci. 2013;138,82-89.

[27] Souza-Fabjan JMG, Panneau B, Duffard N, Locatelli Y, Figueiredo JR, Freitas VJF, Mermillod P. In vitro production of small ruminant embryos: late improvements and further research. Theriogenology. 2014;81,1149-1162.

[28] Souza-Fabjan JMG, Locatelli Y, Duffard N, Corbin E, Touzé JL, Perreau C, Beckers JF, Freitas VJF, Mermillod P. In vitro embryo production in goats: Slaughterhouse and laparoscopic ovum pick up-derived oocytes have different kinetics and requirements regarding maturation media. Theriogenology. 2014;81,1021-1031.

[29] Spermann H. Embryonic Development and Induction. New York: Hafner; 1938. p. 210-211.

[30] Hoppe PC, Illmensee K. Full-term development after transplantation of parthenogenetic embryonic nuclei into fertilized mouse eggs. Proc Natl Acad Sci USA. 1982;79,1912-1916.

[31] Willadsen SM. The development capacity of blastomeres from 4- and 8-cell sheep embryos. J Embryol Exp Morphol. 1981;65,165-172.

[32] Willadsen SM. Nuclear transplantation in sheep embryos. Nature. 1986;320,63-65.

[33] Campbell KH, McWhir J, Ritchie WA, Wilmut I. Sheep cloned by nuclear transfer from a cultured cell line. Nature. 1996;380,64-66. DOI: 10.1038/380064a0.

[34] Cibelli JB, Stice SL, Golueke PJ, Kane JJ, Jerry J, Blackwell C, Ponce de León FA, Robl JM. Cloned transgenic calves produced from nonquiescent fetal fibroblasts. Science. 1998;280,1256-1258.

[35] Murakami H, Nagashima H, Takahagi Y, Fujimura T, Miyagawa S, Okabe M, Seya T, Shigehisa T, Taniguchi N, Shirakura R, Kinoshita T. Production of transgenic pigs expressing human DAF (CD55) regulated by the porcine MCP gene promoter. Transplant Proc. 2000;32,2505-2506.

[36] Zhang P, Liu P, Dou H, Chen L, Chen L, Lin L, Tan P, Vajta G, Gao J, Du Y, Ma RZ. Handmade cloned transgenic sheep rich in omega-3 Fatty acids. PLoS One. 2013;8,e55941. DOI: 10.1371/journal.pone.0055941.

[37] Vajta G, Lewis IM, Trounson AO, Purup S, Maddox-Hyttel P, Schmidt M, Pedersen HG, Greve T, Callesen H. Handmade somatic cell cloning in cattle: analysis of factors contributing to high efficiency in vitro. Biol Reprod. 2003;68,571-578.

[38] Vajta G, Callesen H. Establishment of an efficient somatic cell nuclear transfer system for production of transgenic pigs. Theriogenology. 2012;77,1263-1274.

[39] Liu H, Li Y, Wei Q, Liu C, Bolund L, Vajta G, Dou H, Yang W, Xu Y, Luan J, Wang J, Yang H, Staunstrup NH, Du Y. Development of transgenic minipigs with expression of antimorphic human cryptochrome 1. PLoS One. 2013;8,e76098.

[40] Pereira AF, Feltrin C, Almeida KC, Carneiro IS, Avelar SRG, Alcântara Neto AS, Sousa FC, Melo CHS, Moura RR, Teixeira DIA, Bertolini LR, Freitas VJF, Bertolini M. Analysis of factors contributing to the efficiency of the in vitro production of transgenic goat embryos (*Capra hircus*) by handmade cloning (HMC). Small Rum Res. 2013;109,163-172.

[41] Teixeira DIA, Lopes-Júnior ES, Sousa FC, Pinheiro ESP, Serova IA, Andreeva LE, Freitas VJF. The use of real-time ultrasonography to select embryo donors participating in a transgenesis goat programme. Small Rum Res. 2008;76,215-219.

7

Major Components in Limiting Litter Size

María-José Argente

Abstract

The litter size is an important trait in prolific species such as rabbits and pigs. However, selection on litter size has had limited success in these species because of its low heritability and sex-limited expression. The litter size is a complex physiological trait in prolific species, affected by several components that are expressed sequentially, for example, ovulation, fertilization, embryo development, and fetal survival. The selection for ovulation rate or/and prenatal survival has been proposed to improve litter size indirectly. However, these alternative methods have not reached the expected response rate. Implantation is also a critical point in successful gestation, one-third to one-half of prenatal mortality occurring during peri-implantation. The uterus must provide an adequate microenviroment for the growth and development of embryo and for receptivity to implantation. There are multitudes of cellular events involved in crosstalk between embryo and maternal uterus during peri-implantation. A better understanding of molecular mechanisms affecting the implantation process could help to propose new strategies for litter size improvement in prolific species.

Keywords: Ovulation rate, embryonic survival, litter size, candidate gene, quantitative trait loci (QTL)

1. Introduction

Litter size is a complex physiological trait in prolific species, being affected by several component traits representing sequential events, e.g., ovulation, fertilization, embryo development, and fetal survival. Fertilization rate is usually high, exceeding 90 to 95% in rabbit [1], pigs [2], and mice [3]. Therefore, prenatal survival is considered a limiting factor of litter size. Uterine capacity is an important component in prenatal survival [4; for a review address to 5]. Whenever the ovulation is not a limiting factor, this trait depicts the ability of the uterus to support embryo development through gestation [6]. Approximately 30 to 40% of ova shed do not result in fetuses at term in prolific species such as rabbits, pigs, and mice (see review [7]) due to potentially viable embryos exceeds uterine capacity. One-third to one-half of these losses occurs during peri-implantation in rabbits [8], pigs [9], and mice [10]. Preimplantation embryo losses are mainly associated with embryonic viability [4], including chromosomal abnormalities [11] and oviductal or uterine environment, particularly in relation with the suitability of oviductal or uterine secretions [12]. The oviduct must provide an adequate milieu for sperm capacitation, gamete fertilization, and the first embryo cleavages until the embryo enters the uterus [13]. Many proteins that may contribute to these functions have been identified in the oviduct, including the insulin-like growth factor 1 (IGF1) [14], oviductin (OVGP1, known also as MUC9) [15], tissue inhibitor of metalloproteinase (TIMP1) [16], plasminogen activator inhibitor 1 (PAI1) [17], uteroglobin [18], and leptin [19]. A progesterone-primed uterus coordinates the embryo survival and receptivity. The asynchrony between embryo development and uterine environment increases the number of dead embryos in the peri-implantation gestational stage [20, 21].

In rabbits, the period between pregnancy days 8 to 17 is critical for fetal survival, corresponding to the stage when the hemochorial placenta of rabbit is established and the control of the fetal nutrition is transferred to the placenta [22]. A secondary critical period for fetal survival occurs between pregnancy days 17 and 24, accompanying the uterine enlargement, the increased tension exerted on the spherical conceptus, and the reduction of blood flowing through the maternal vessels of the uterus [23]. In this moment, the placenta should compensate these limiting factors and increase its surface area for fetomaternal exchanges [24] and stimulate the development of an adequate vascular network [25, 26]. In pigs, the distribution of mortality after implantation is slightly different throughout pregnancy [9], the first peak occurring between days 30 and 40 of gestation (10-15%) while the second peak is observed in the last two weeks of gestation (5-10%). Despite the existing species differences, it is now accepted that each embryo requires a minimum space in uterus to attach, survive, and develop. Thus, a decrease in the availability of uterine space increases prenatal mortality in pluriparous species, such as rabbits [4] or pigs [27], despite that the factors involved in the process may differ among species. In pigs, the area of attachment between the placenta and endometrium is a limiting factor of uterine capacity, due to its noninvasive placentation [9]. Thus, it was proposed that uterine capacity in pigs could be defined more correctly as the total amount of placental mass or surface area that a dam can support to term [28]. Earlier studies in pigs indicate that the limiting influence of the uterine capacity is generally exerted after the day 30 of gestation and that the effects of moderate crowding of embryos before day 30 could be compensated by increased placental efficiency later in gestation (see review [29]).

2. Selection experiments for components of litter size

Litter size is an important economic trait in prolific species such as rabbits or pigs. However, direct selection for litter size has not presented the success expected for these species (see review [7, 30]), which may be due to the fact that it is a female sex-limited trait with low heritability. The selection response for litter size has been established around 0.1 young per generation in rabbits [31, 32] and pigs [33-36]. This response has been much lower than that reported in mice: 0.15 to 0.20 young per generation [37-40]. Recently, in pigs, a selection experiment for litter size at day 5 after farrowing obtained a selection response around to 0.25 young per generation [41]. However, additional information on following generations will be required to confirm this trend.

The leading components of litter size are the ovulation rate and the prenatal survival; these parameters are also the limiting factors for the litter size improvement. For this reason, selection for ovulation rate and prenatal survival has been proposed as an indirect approach for increasing litter size. Selection for ovulation rate in prolific species, namely rabbits [42], pigs [43-45], and mice [40, 46], were in fact successful to improve the ovulation rate, but it was not convoyed by a corresponding increase in litter size in either pigs or rabbits, which was attributed to an increase in prenatal loss.

The selection for prenatal survival in pigs [45] and mice [40] allowed to increase litter size in both pigs and mice, but it was not more advantageous compared with the direct selection for litter size. Besides, prenatal survival might be limited by uterine capacity, defined as the maximum number of fetuses that a dam can support at birth when ovulation rate is not a limiting factor [6]. However, the ability to establish this trait is not easily performed across species, as it is dependent on the species physiology. In pigs, litter size in unilaterally ovario-hysterectomized females may be considered an indirect measure of the uterine capacity in pigs [6], since the remaining ovary nearly doubled its ovulation rate originating the fetal over-crowding in the ipsilateral uterine horn. Conversely, in unilaterally ovariectomized female rabbits and mice, which possess a duplex uterus that impedes intercornual transmigration, the number of total fetuses would represent their uterine capacity [47, 48].

Selection experiments for increased uterine capacity failed to obtain the expected success on litter size, in rabbits [8, 49], pigs [44], or mice [50]. Considering that uterine capacity in pigs would be more appropriately measured using the total amount of placental mass or the surface area that a dam can support to term as variables [28], a divergent selection experiment for placental efficiency developed in this species achieved success [51]. However, selection for increased placental efficiency will unlikely result in correlated increase in litter size [52].

A joint selection for ovulation rate and prenatal survival using an index would expectably show a greater response on litter size, since these parameters are optimally weighted [53, 54]. In pigs [55] and mice [56], the use of this joint selection successfully increased litter size, but the gain was lower than expected, probably due to a low precision of the estimated genetic correlations or the use of inappropriate economic weights [57]. Alternatively, a two-stage selection was proposed, which would be less affected by the precision of the genetic parameters; its application to rabbits [58] and pigs [59] obtained greater response on litter size than the observed in the other experiments of selection for litter size [31-36].

Selection for the environmental variability in litter size has been recently proposed as an alternative method for increased litter size. A reduction in litter size environmental variability would increase litter size heritability, and consequently its response to selection [60]. In rabbits, the environmental variance of a doe was estimated as within-doe variance of litter size.

Selection for environmental variance of litter size was successful [61], and as a consequence of reducing litter size variability the litter size was increased due to higher embryo survival [61].

3. Genetic control before implantation

The maternal genome controls virtually all aspects of early embryo development, through several maternal gene products such as mRNA and protein, which are loaded into the egg during oogenesis. As development proceeds, two processes subsequently lead to the maternal-to-zygotic transition (MZT) during which developmental control is transferred to the zygotic genome: first, a subset of the maternal mRNAs is degraded; second, the embryonic genome is transcriptionally activated. These maternal gene products play an important role in the regulation of the first cleavages until embryonic genome is activated [62]. Zygotic genome activation (ZGA) is a critical event determining the transition from maternal to embryonic control of development. Disruption of these critical events by specific chemicals or environmental factors results in irreversible arrest of embryo development [63]. ZGA has been shown to be a species-specific phenomenon, occurring at 2-cell stage in mice [64], 4-cell stage in pigs [65], and 8- to 16- cell stages in sheep, cows, or rabbits [66]. Many maternal-effect genes have been identified initially in mouse during the MZT, and several of whom have been detected posteriorly in rabbits and pigs.

Genome-wide gene activation in the zygote (ZGA) is regarded as crucial for preimplantation embryonic development. Multiple maternal factors were identified on the regulation of ZGA, which are listed in Table 1. Ablation of the gene encoding for these proteins results in embryonic arrest at cleavage-stage development.

These factors play critical roles in the regulation of embryo preimplantation development. For example, DICER1 enzyme is required for completion of oocyte mitotic maturation, and oocytes are arrested in metaphase of meiosis II when DICER1 gene is deleted [67, 68]. HSF1 protein is required for oocyte maturation. Embryos produced in knockout females for this gene are unable to proceed into the 2-cell stage after fertilization, possibly due to mitochondrial damage and altered redox homeostasis [69, 70]. UCHL1 is an import factor in blocking polyspermy [71, 72]. AGO2 is involved in the destruction of maternally inherited transcripts and activation of zygotic gene expression; knockout female for this factor is infertile because embryos fail to undergo the first cleavage [73, 74]. Moreover, embryos from MATER, ZAR1, PADI6, and SEBOX knockout females do not develop beyond the 2-cell stage embryo [75-82], and embryos lacking SMARCA4, DNMT1, DNMT3A, TET, and KLF4 are unable to reach the 8-cell stage [83-91]. BCLXL, HDAC1, and C-MYC exhibit maximum expression in 8-cell rabbit embryos coinciding with start of ZGA [90]. Hence the peak expression of transcripts at ZGA might be a requirement for embryo development. OCT4, NANOG, and SOX2 were co-expressed in epiblasts, and the combinatorial expression of these three genes is critical for the embryo development [92, 93].

Symbol gene	Name	Effect	References in	
			Mice*	Rabbits and pigs
DICER1	Endoribonuclease Dicer or helicase with RNase motif	Metaphase II	[67]	[68]
HSF1	Heat shock factor 1	Zygote	[69]	[70]
UCHL1	Ubiquitin carboxyl-terminal hydrolase L1	Zygote	[71]	[72]
AGO2	Argonaute 2	Zygote	[73]	[75]
MATER or *NLRP5*	Maternal antigen that embryos require	2-cell stage	[75]	[76]
ZAR1	Zygote arrest 1	2-cell stage	[77]	[78]
PADI6	Peptidylarginine deiminase type 6	2-cell stage	[79]	[80]
SEBOX	Skin-embryo-brain-oocyte homeobox	2-cell stage	[81]	[82]
BRG1 or *SMARCA4*	Brahma-related gene 1	4-cell stage	[83]	[84]
DNMT1	DNA cytosine methyltransferase 1	4-cell stage	[85]	[86]
DNMT3A	DNA cytosine methyltransferase 3 alpha	4-cell stage	[87]	[86]
TET1, 2 and *3*	Ten-eleven translocation (Tet) dioxygenases	4-cell stage		[88]
KLF4	Kruppel-like factor 4	4-cell stage		[89, 90]
BCLXL	B-cell lymphoma-extra large	8-cell stage		[89]
HDAC1	Histone deacetylase 1	8-cell stage		[89]
C-MYC	Avian myelocytomatosis viral oncogene homolog	8-cell stage		[90]
NANOG	The homeoprotein Nanog	16-cell stage	[91]	[84, 89, 93]
OCT4 or *POU5F1*	Octamer-binding protein 4	Morula	[92]	[89, 90, 93]
SOX2	SRY-box containing gene 2	Blastocyst		[84, 89, 93]

* Using gene-knockout mouse models.

Table 1. Maternal genes acting on early fetal embryogenesis.

4. Genetic control during implantation

Implantation requires a complex interaction among the developing embryo, decidualizing endometrium and developing maternal immune tolerance. For the successful implantation, it is of upmost importance the synchronization between the acquisition of implantation competency by the blastocyst and a receptive state in the uterine endometrium, for which the concurrence of the ovarian steroid hormones dynamics is crucial. The ovarian hormones determine a complex interplay of locally produced molecules in the endometrium, including cytokines, growth factors, homeobox transcription factors, lipid mediators, and morphogen genes that are involved in the complex process of implantation. The crosstalk between the blastocyst and the uterus is limited in most mammal species for a brief period, named as *window of implantation*. During this short period, in response to a viable embryo, the endometrium responds through species-specific transformation of the superficial tissue architecture, a process known as decidualization, allowing the organ to accommodate embryonic growth and placentation. The decidua will function like a barrier, protecting the embryo against the maternal immune system [see review 94].

Steroid hormones

The progesterone receptor (PR) was identified as one of the molecules that genetic polymorphisms were associated with the risk of implantation failure. The PR is encoded by *PGR* gene, and has two isoforms, PRA and PRB, both of them are expressed in the uterus [94]. Studies in mice showed that a deletion on PRA provokes severe abnormalities in ovarian and uterine function and impairs implantation [94]. A study in rabbits reported that favorable allele of PRG had an additive effect of 0.25 for implanted embryos and kits for litter size [95].

Estrogen receptor (ER) was also implied in uterine receptivity for embryo implantation. ER presents two isoforms, known as ERα (encoded by *ESR1* gene) and ERβ (encoded by *ESR2* gene). In mice, knocking out the ERα gene leads to unsuccessful implantation [94]. In pigs, favorable allele of ESR shows an additive effect between 0.45 and 0.75 piglets for litter size [96, 97]. In another hand, it has been shown that the ERBB receptor feedback inhibitor 1 (*ERRFI1*) gene is involved with successful implantation, which was associated with its suppression of *ESR1* activity in the uterine epithelium, a crucial event for embryo implantation. Despite that the ovaries of ERRFI1 knockout female mice show a normal morphology and steroidogenesis function, its uterine horns do not develop an implantation site [98]. The steroid receptor coactivators (SRC1 and SRC2) present distinct physiological functions in the female reproductive system. For example, female mice lacking either *SCR1* or *SCR2* show progesterone resistance and compromised decidualization, whereas deletion of both *SCR1* and *SCR2* genes provoke infertility in female due to a complete blockage of decidualization [99], suggesting that one of the steroid receptor coactivators may, at least in part, compensate the absence of the other. Also the blockage of the repressor of estrogen receptor activity (REA), a significant modulator of estrogen responsiveness, was reported to induce implantation failures [100].

Moreover, prostaglandins (PGs) play an important role in various reproductive processes, including ovulation and implantation [101]. Cyclo-oxygenases (COX-1 and COX-2) are crucial enzymes in the synthesis of various PGs. Females lacking COX-1 and COX-2 are infertile, due to abnormalities in ovulation, fertilization, implantation, and decidualization [102]. A study in pigs showed that favorable allele of COX-2 had an additive effect of 0.3 piglets for litter size [103].

Cytokines

During the embryo implantation, the endometrium undergoes a dramatic transformation into a specialized transitory tissue known as the decidua in species with invasive hemochorial placenta, such as rodents and lagomorphs. In other species with no invasive placenta as pigs, the changes in endometrium ought to allow the trophoblast and its supporting layer of extraembryonic mesoderm to contact successfully with the uterine epithelium. The placenta surrounding the developing embryo facilitates the nutrient transfer and limits trophoblast invasion. The endometrium is recognized as an important site of production of cytokines and their receptors, which are also potential regulators of the phenotype and activation status of the uterine-resident leukocytes. Leukocytes infiltrated in the endometrium are required for the immunotolerance pathways that allow the maternal organism to accommodate the conceptus during implantation and placental development. Several cytokines, such as the macrophage colony-stimulating factor 1 (CSF1, also known as M-CSF), granulocyte colony-stimulating factor (CSF3, also known as G-CSF), and granulocyte-macrophage colony-stimulating factor (CSF2, also known as GM-CSF), are implicated in the recruitment and phenotypic regulation of the abundant populations of endometrial macrophages, granulocytes, and dendritic cells. These cytokines appear to be related to the successful embryo implantation and placental growth [104]. Also leukemia inhibitory factor (LIF), interleukin-6 (IL-6) and interleukin-11 (IL-11), and its receptor (IL-11Ra) have unquestionable roles in the implantation process [105-107]. Studies in pig have reported an association between certain polymorphisms in LIF gene with the number of piglets [108-111].

Growth factors

The epidermal growth factor (EGF), heparin-binding epidermal growth factor-like growth factor (HB-EGF), the vascular endothelial growth factors (VEGFs), the IGF-I and IGF-II as well as the IGF-binding protein-1 (IGFBP-1) are important factors for implantation. It was shown that EGF plays a critical role in trophoblast invasion, differentiation, and proliferation. EGF deficiency during pregnancy causes intrauterine growth retardation or abortion [112], while the deletion of epidermal growth factor receptor (*EGFR* or *ERB1)* gene causes failure in the embryo development and the placenta formation [113]. Likewise, other factors involved in the implantation process, such as HB-EGF, are expressed in endometrial stromal and epithelial cells. It has been demonstrated that HB-EGF regulates endometrial cell proliferation, glandular epithelial secretion, and decidual transformation [113]. Gene knockout studies reveal that deletion of *HBEGF* reduces litter size [114].

VEGF is an endothelial-cell specific mitogen *in vitro*, and it is the main factor responsible for de novo blood vessel formation (vasculogenesis) and angiogenesis in vivo [115]. Proper level of VEGF expression is required for implantation [116]. Many critical cell responses, including mitogenesis, proliferation, growth, differentiation, and angiogenesis, are mediated by IGF-I and IGF-II [117]. Both IGF-I and IGF-II are necessary to maintain normal embryonic growth rates [117]. In addition, higher expression of *IGF1* mRNA has been observed during the peri-implantation period in mouse uterus [118]. Also in rabbit, it has been observed that IGF-II receptor plays an important role in embryo development and its implantation [119].

Transcription factors

Homeobox genes are transcriptional regulators evolutionarily conserved that control embryonic morphogenesis and differentiation [120]. *Homeobox A* (*HOXA10* and *HOXA11*) and *H6 homeobox 3* (*HMX3*) genes are expressed in uterine stromal cells during the receptivity period, and upregulated upon decidualization in response to steroid hormone stimulus. Ablation of *HOXA10, HOXA11,* or *HMX3* genes leads to implantation defects [121-123]. *MSX1* (also known as *HOX7.1*) and *MSX2* also belong to *homeobox* genes, and deletion of both genes results in female infertility due to altered uterine polarity and integrity of the surface epithelium [124]. The Kruppel-like transcription factors (*KLFs*) are implicated in diverse cellular processes, including proliferation, differentiation, and apoptosis. The Kruppel-like factor 5 (*KLF5*), one of these transcription factors, is essential for the establishment of uterine receptivity [125], and its depletion induces implantation failure [126]. *KLF9* is another KLF that plays an important role in blastocyst attachment; its loss reduces female fertility due to defective implantation [127]. The transcription factor named heart and neural crest derivatives-expressed transcript 2 (*HAND2*) and its protein are present in endometrial stromal cells adjacent to the surface epithelium in the uterus prior to the onset of implantation [128], suggesting that they may play a key role in uterine receptivity in mice [129]. Forkhead box protein A2 (FOXA2) is only expressed in the glandular epithelium of the uterus, and FOXA2 deficiency affect the endometrial gland formation and decidualization [130]. Deletion of chicken ovalbumin upstream promoter transcription factor II (*COUP-TFII*), which is mainly expressed in uterine stromal cells, originates implantation failure due to disrupted uterine receptivity associated with high estrogen activity [131].

Morphogen genes, lipid mediatiors, integrins, mucins, and others molecules

Proteins belonging to the transforming growth factor beta superfamily (TGF-β), Wingless (WNT), Hedgehog, and Notch have been identified as morphogens. Morphogens act directly on angiogenesis, cell growth, pattern formation, embryo development, metabolic regulation, cell migration, and tissue repair, while also presenting neurotropic effects. Five TGF-βs have been identified, of which TGF-β1, β2, β3 are abundant in mammals. However, only TFG-β1 appears to limit the number of implanted embryos [132]. Activins are also members of the TGF-β superfamily that participate in the regulation of several biological processes, including cell differentiation and proliferation, apoptosis, and the immune response [133]. Activin A plays an important role in the implantation of embryos in rabbits and mice, promoting decidualization and preventing the activation of T cells [134]. Among all bone morphogenetic proteins expressed in the uterus, only BMP2 shows intense expression in the stromal cells surrounding the implanted embryo under response to progesterone [135]. *In vitro* studies in undifferentiated stromal cells demonstrated that silencing the expression of BMP2 efficiently blocks the decidualization [136]. Deletion of *NODAL* gene was accompanied by severe malformation of the maternal decidua basalis during placentation and increasing fetal losses before birth [137]. Several components of WNTs signaling pathway are spatiotemporally regulated in the peri-implantation uterus and are crucial to implantation. The absence of WNT4, WNT5a, and WNT7a in the uterus induces defective embryo implantation and subsequent decidualization failure [138-140], while deletion of *WNT7b* gene provokes fetal losses during mid-gestation, due to failure of chorioallantoic fusion [141]. Indian hedgehog (IHH) is a member of the Hedgehog family. It has been reported that conditional deletion IHH protein in the uterus results in implantation failure [142]. Activation of smoothened (SOM), another member of the Hedgehog family, provokes hypertrophy in uterus along and consequently failure to decidual response [143]. The *NOTCH1* is responsible for cell survival, cell-to-cell communication, differentiation, and all fundamental processes for successful decidualization [144].

The blastocyst has a significant number of cannabinoid receptors (CB1) that are activated by the anandamide (AEA) produced in the uterus. It has been found that the levels of AEA are lower in the receptive uterus and at implantation sites than in the non-receptive uterus or at inter-implantation sites. These findings suggest the need for low AEA levels to activate uterine receptivity [145].

The integrin family of cell adhesion molecules is a major class of receptors for the extracellular matrix. They have many functions in cellular processes including differentiation, apoptosis, and attachment [146]. Previous studies have demonstrated that integrins exhibit distinctive expression patterns in different phases of uterine receptivity. Both α4β1 and αvβ3 integrins are present in uterus at the time of implantation, and intrauterine inhibition of these two molecules results in defective implantation [147, 148].

Symbol gene	**Name**	**Effects**	**References**
Morphogens			
TFG-β1	Transforming growth factor beta 1	Failures in immunotolerance during embryo implantation	[132, 133]
INHBA	Activin A	Limiting decidualization, and no preventing activation of T cells	[134]
BMP2	Bone morphogenetic protein 2	Block the decidual reaction	[135, 136]
NODAL	NODAL	Abnormal decidua basalis at midgestation and aberrant placental development	[137]
WNT4, WNT5A, WNT7A and *WNT7B*	Wingless-related MMTV integration site 4, 5a, 7a, 7b.	Implantation and decidualization failures	[138-141]
IHH	Indian hedgehog	Implantation failure	[142]
SMO	Smoothened	Uterine hypertrophy with the reduction in the number of uterine glands and impaired decidualization	[143]
NOTCH1	Notch1	Failure in decidualization	[144]
Lipid mediators			
AEA	Anandamide	Uterine receptivity	[145]
Integrins			
ITGA4/ITGB1	α4β1 integrin	Implantation and decidualization failures	[146, 147]
ITGAV/ITGB3	αvβ3 integrin	Implantation and decidualization failures	[146, 148]
Mucins			
MUC1	Mucin 1	Embryo attachment failure	[149, 151]
OVGP1 or *MUC9*	Oviductal glycoprotein 1	Fertilization and implantation failures	[152]
Other molecules			
CTNNB1	β-catenin	Defects in embryonic ectoderm cell layer	[153-154]
CX43 or *CJA1*	Connexin 43	Comprised decidualization; neovascularization defects	[155]

Table 2. Genes considered as critical to implantation.

Mucins also participate in the decidualization process. For example, mucin (MUC1) has been identified as an effective barrier that prevents embryo attachment to the uterine epithelium. Uterine MUC1 expression declines to undetectable levels prior to blastocyst attachment, reinforcing the impression that loss of MUC1 contributes to the establishment of a receptive uterus [149-151]. Other oviductal glycoprotein, as OVGP1 or MUC9, seems to affect the differentiation of endometrial and fetal cells by paracrine pathway, inhibiting the implantation and fetal development [152]. The active beta-catenin (CTNNB1) is only detected in morula and early blastocyst stages, its signal disappearing as soon as the blastocyst hatches from the zona pellucida [153]. Furthermore, deletion of *CTNNB1* gene provokes severe gastrulation defects that results in embryonic lethality [154]. Connexin 43 (CX43, also known as GJA1) is a major gap junction protein that is markedly expressed in the uterine stromal cells surrounding the implanted embryo during early pregnancy. Deletion of *CX43* gene leads to aberrant differentiation of uterine stromal cells, preventing the secretion of angiogenic factors, such as the VEGF. As consequence, the development of new blood vessels within the uterine stromal compartment suffers a striking impairment, resulting in the arrest of embryo growth and early pregnancy loss [155]. All these critical genes on successful implantation are listed on Table 2.

5. Approaches for identifying genes in pigs and rabbits

In pigs and rabbits, genetic markers associated with reproductive traits have been identified through two complementary approaches. The first approach has been performed through unbiased genome scans with anonymous DNA markers, such as microsatellites and more recently with thousands of single nucleotide polymorphisms (SNPs), which have been used to identify quantitative trait loci (QTL) with effects on reproductive traits. Genome-wide scanning usually proceeds without any presuppositions regarding the importance of specific functional features of the investigated traits. Until now, a total of 28 suggestive QTL have been reported on pig chromosomes (SSC) 2, 6, 7, 8, 11, 12, 14, 15, 16, 17, and 18 for litter size [156-165], on SSC 1, 2, 3, 4, 5, 6, 7, 8, 9, 10, 13, 14, 15, 16, 17, and 18 for ovulation rate [165-171], and on SSC8 for the uterine capacity and prenatal loss [159, 166, 168].

In the second approach, the physiological role of candidate gene is known, and the gene is scanned for polymorphisms and associations to variations within the trait. Numerous genes have been evaluated as candidate genes affecting litter size in pigs, such as the estrogen receptor (*ESR*) [95-97], retinol binding protein 4 (*RBP4*) [172], gonadotrophin-releasing hormone receptor (*GNRHR*) [173], osteopontin (*OPN*) [174], folate-binding protein (*FBP*) [175], mitogen-activated protein kinase 3 (*MAP3K3*) [176], vascular endothelial growth factor receptor (*KDR*) [176], ERBB2 interacting protein (*ERBB2IP*) [176], and peroxisome proliferator-activated receptor delta (*PPARD*) [176]. Another candidate genes have been found for progesterone receptor (*PRG*), *TIMP1*, oviductal glycoprotein 1 (*OVPG1*), hydroxysteroid-17-beta- dehydrogenase 4 (*HSD17B4*), endoplasmic oxidoreductin-1-like protein (*ERO1L*), and octamer-binding transcription factor 4 (*OCT4*) in rabbits [177-181]. However, associations of these genes with litter size are always population specific, and the causative mutations underlying litter size remain unexplored.

DNA microarray is a new powerful tool for studying the molecular basis of interactions on a scale that is impossible using conventional analysis, making possible to examine the expression of thousands of genes simultaneously. In order to expand the understanding of the biological processes involved in the success of female reproduction, several studies in gene expression were developed in pigs targeting to identify the changes in ovaries [182, 183] and the endometrium at implantation [184-191]. For example, after selecting for 11 generations using an index of ovulation rate and embryonic survival, followed by 7 generations of selection for litter size, a total of 71 differentially expressed genes were identified in ovarian tissues of the selected and control lines at days 2–6 of the follicular stage of the estrous cycle [182]. Many of these genes had not been previously associated with reproduction. From these genes, 59 were homologous to genes of known function, 5 had no known matches in GenBank, and 7 were homologous to sequences of unknown function. Among the differentially expressed genes identified were those associated with the transport of cholesterol in ovarian follicles and the synthesis of steroids, such as collagen type I receptor (*CD36L1*, also known as scavenger receptor class B type I). The experiment also showed the importance of studying the expression of all these genes at different times of estrous cycle. For instance, genes of steroidogenic acute regulatory protein (*STAR*), 3-β-hydroxysteroid dehydrogenase (*3βHSD*), were overexpressed in higher producing pig ovaries at day 2 of analysis, while were underexpressed at day 3. In contrast, plasminogen activator inhibitor 1 (*PAI1*) and cytochrome P450 17-α-hydroxylase (*CYP17*) were overexpressed at day 3.

In a different study, 189 genes were found to be differentially expressed in the ovaries of pregnant pigs with high and low prolificacy, of which 72 were overexpressed in the high prolificacy group, while 133 of them were overexpressed in the low prolificacy group [183]. These genes appear to cluster in three main biological processes: the first group would be related to the immune system response activation against external stimulus, the second group included integrated genes that regulate maternal homeostasis by complement and coagulation cascades, and the third was involved in lipid and fatty acid enzymes of metabolic processes of the steroidogenesis pathway. Among validated genes, 2-5-oligoadenylate synthetase 1 (*OAS1*) was found overexpressed in high prolificacy females, while a family with sequence similarity with 46 member C (*FAM46C*), secreted phosphoprotein 1 (*SPP1*), thiosulfate sulfurtransferase (*TST*), and vitronectin (*VTN*), were reported overexpressed in low prolificacy females.

Recent microarray analysis revealed more than 2000 differentially expressed genes in endometrium between pregnant and cyclic pigs at the time of implantation, i.e., on days 12 [186], 13 [191], 14 [187], 15 [188], 16 [192], 18 [191], or 24 [191] of gestation. Most genes were involved in cell motility as well as apoptosis, transporter activity, calcium ion binding, lipid metabolic processes, hormone activity, vascular development and proteolysis, immune response. The identified and validated genes that are upregulated included ADAM metallopeptidase with thrombospondin type 1 motif 20 (*ADAMTS20*), mucin 4 (*MUC4*), leukemia inhibitory factor receptor alpha (*LIFR*), interleukin 6 receptor (*IL6R*), interferon regulatory factor 1 (*IRF1*), immunoresponsive 1 homolog (IRG1), secreted phosphoprotein 1 (*SPP1*), osterocrin (*OSTN*), nuclear receptor interacting protein 1 (*NRIP1*), proteolipid protein 1 (*PLP1*), signal transducer and activator of transcription 1 (*STAT1*), serpin peptidase inhibitor, clade B (ovalbumin), member 7 (*SERPINB7*), s100 calcium binding protein A 9 (*S100A9*), Erb-B2 receptor tyrosine kinase 3 (*ERBB3*), and fibroblast growth factor 9 (*FGF9*). Contrasting, mucin 5AC, oligomeric mucus/gel-forming (*MUC5AC*), interleukin 11 receptor alpha (*IL11RA*), interleukin 24 (*IL24*), brain and acute leukemia cytoplasmic (*BAALC)*, defensin beta 1 (*PBD-2*), defensin beta 1 (*PBD-2*), cadherin 17 li (*CDH17*), FXYD domain containing ion transport regulator 4 (*FXYD4*), G protein-coupled receptor 83 (*CPR83*), and fibroblast growth factor receptor 3 (*FGFR3*) are downregulated [182-187, 191]. Litter size is controlled by a large number of genes.

Improvement in litter size has become one main objective of selection in pig and rabbit breeding programs. However, litter size is a complex trait, because it is controlled by numerous genes in complicated physiological networks such as those affecting ovulation rate, embryo survival, and uterine capacity. The genomic approaches, both QTL mapping and candidate gene analysis, have helped increase understanding in genetic control of litter size. Moreover newly developed tools based on DNA microarray techniques appear to be useful for in-depth understanding of the genetics of litter size in pigs and rabbits. A better understanding of genetic mechanisms controlling litter size could help to design more efficient selection strategies in improvement of this trait.

Author details

María-José Argente*

Address all correspondence to: mj.argente@umh.es

Departamento de Tecnología Agroalimentaria, Universidad Miguel Hernández de Elche, Orihuela, Alicante, Spain

References

[1] Peiró R, Gallego M, Blasco A, Santacreu MA. The effect of unilateral ovariectomy on early embryonic survival and embryo development in rabbits. World Rabbit Science. 2014; 22: 123–127.

[2] Geisert RD, Schmitt RAM. Early embryonic survival in the pig: Can it be improved? Journal of Animal Science. 2002; 80: E54–E65.

[3] Wilmut I, Sales DI, Ashworth CJ. Maternal and embryonic factors associated with prenatal loss in mammals. Journal of Reproduction and Fertility. 1986; 76: 851–864.

[4] Argente MJ, Santacreu MA, Climent A, Blasco A. Effects of intrauterine crowding on available uterine space per fetus in rabbits. Livestock Science. 2008; 114: 211–219.

[5] Ford SP. Embryonic and fetal development in different genotypes in pigs. Journal of Reproduction and Fertility. Supplement. 1997; 52: 165–176.

[6] Christenson RK, Leymaster KA, Young LD. Justification of unilateral hysterectomy–ovariectomy as a model to evaluate uterine capacity in swine. Journal of Animal Science. 1987; 67: 738–744.

[7] Blasco A, Bidanel JP, Bolet G, Haley CS, Santacreu MA. The genetics of prenatal survival of pigs and rabbits: a review. Livestock Production Science. 1993; 37: 1–21.

[8] Santacreu MA, Mocé ML, Climent A, Blasco A. Divergent selection for uterine capacity in rabbits. II. Correlated response in litter size and its components estimated with a cryopreserved control population. Journal of Animal Science. 2005; 83: 2303–2307.

[9] Ford SP, Vonnahme KA, Wilson ME. Uterine capacity in the pig reflects a combination of uterine environment and conceptus genotype effects. Journal of Animal Science. 2002; 80 (E. Suppl. 1); E66–E73.

[10] Holt M, Vangen O, Farstad W. Components of litter size in mice after 110 generations of selection. Reproduction. 2004; 127: 587–592.

[11] Pope WF, Xie S, Broerman DM, Nephew KP. Causes and consequences of early embryonic diversity in pigs. Journal of Reproduction and Fertility. Supplement. 1990; 40: 251–260.

[12] Bazer FW, Thatcher WW, Martinant-Botte F, Terqui M, Lacroix MC, Bernard S, Revault M, Dubois DH. Composition of uterine flushings from Large White and prolific Chinese Meishan gilts. Reproduction Fertility and Development. 1991; 3: 51–60.

[13] Bhatt P, Kadam K, Saxena A, Natraj U. Fertilization, embryonic development and oviductal environment: Role of estrogen induced oviductal glycoprotein. Indian Journal of Experimental Biology. 2004; 42: 1043–1055.

[14] Herrler A, Krusche AC, Beier HM. Insulin and insuline-like growth factor-I promote rabbit blastocyst development and prevent apoptosis. Biology of Reproduction. 1998; 59: 1302–1310.

[15] Buhi WC. Characterization and biological roles of oviduct-specific, oestrogen-dependent glycoprotein. Reproduction. 2002; 123: 355–362.

[16] Buhi WC, Alvarez IM, Pickard AR, McIntush EW, Kouba AJ, Ashworth CJ, Smith MF. Expression of tissue inhibitor of metalloproteinase-1 protein and messenger ribonucleic acid by the oviduct of cyclic, early-pregnant, and ovariectomized steroid-treated gilts. Biology of Reproduction. 1997; 57: 7–15.

[17] Kouba AJ, Alvarez IM, Buhi WC. Identification and localization of plasminogen activator inhibitor-1 within the porcine oviduct. Biology of Reproduction. 2000; 62: 501–510.

[18] Riffo M, Gonzalez KD, Nieto A. Uteroglobin induces the development and cellular proliferation of the mouse early embryo. Journal of Experimental Zoology Part A: Ecological Genetics and Physiology. 2007; 307: 28–34.

[19] Zerani M, Boiti C, Dall'Aglio C, Pascucci L, Maranesi M, Brecchia G. Leptin receptor expression and in vitro leptin actions on prostaglandin release and nitric oxide synthase activity in the rabbit oviduct. Journal of Endocrinology. 2005; 185: 319–325.

[20] Wilde MH, Xie S, Day ML, Pope WF. Survival of small and large littermate blastocysts in swine after synchronous and asynchronous transfer procedures. Theriogenology. 1988; 30: 1069–1074.

[21] Xie S, Broermann DM, Nephew KP, Bishop MD, Pope WF. Relationship between oocyte maturation and fertilization on zygotic diversity in swine. Journal of Animal Science. 1990; 68: 2027–2033.

[22] Adams CE. Studies on prenatal mortality in rabbit, Oryctolagus cuniculus: The amount and distribution of loss before and after implantation. Journal of Endocrinology. 1960; 19: 325–344.

[23] Hafez ESE, Tsutsumi Y. Changes in endometrial vascularity during implantation and pregnancy in the rabbit. American Journal Anatomy. 1966; 118: 249–282.

[24] Argente MJ, Santacreu MA, Climent A, Blasco A. Relationships between uterine and fetal traits in rabbits selected on uterine capacity. Journal of Animal Science. 2003; 81: 1265–1273.

[25] Mocé ML, Santacreu MA, Climent A, Blasco A. The effect of divergent selection for uterine capacity on prenatal survival in rabbits: maternal and embryonic genetic effects. Journal of Animal Science. 2004; 82: 68–73.

[26] Argente MJ, Santacreu MA, Climent A, Blasco A. Influence of available uterine space per fetus on fetal development and prenatal survival in rabbits selected for uterine capacity. Livestock Science. 2006; 102: 83–91.

[27] Chen ZY, Dziuk, PJ. Influence of initial length of uterus per embryo and gestation stage on prenatal survival, development, and sex ratio. Journal of Animal Science. 1993; 71: 1895–1901.

[28] Wilson ME, Biensen NJ, Ford SP. Novel insight into the control of litter size in the pig, using placental efficiency as a selection tool. Journal of Animal Science.1999; 77: 1654–1658.

[29] Foxcroft GR, Dixon WT, Dyck MK, Novak S, Harding JC, Almeida FC. Prenatal programming of postnatal development in the pig. Society for Reproduction and Fertility Supplement. 2009; 66: 213–231.

[30] Blasco A, Bidanel JP, Haley C. Genetics and neonatal survival. In: Varley MA, editors.

The Neonatal Pig. Development and Survival. CAB International: Wallingford, Oxon, UK; 1995. 17–38 pp.

[31] Rochambeau H De, Duzert R, Tudela F. Long term selection experiments in rabbit. Estimation of genetic progress on litter size at weaning. Proceedings of 6th World Congress on Genetics Applied to Livestock Production. 1998; 26: 112–115.

[32] García ML, Baselga M. Estimation of genetic response to selection in litter size of rabbits using a cryopreserved control population. Livestock Production Science. 2002; 74: 45–53.

[33] Haley CS, Avalos E, Smith C. Selection for litter size in the pig. Animal Breeding Abstract. 1988; 56: 319–332.

[34] Bolet B, Bidanel JP, Olliver L. Selection for litter size in pigs. II. Efficiency of closed and open selection lines. Genetic Selection and Evolution. 2001; 33: 515–528.

[35] Estany J, Villalba D, Tibau J, Soler J, Babot D, Noguera JL. Correlated response to selection for litter size in pigs: I. Growth, fat deposition, and feeding behavior traits. Journal of Animal Science. 2002; 80: 2556–2565.

[36] Holl JW, Robinson OW. Result from nine generations of selection for increased litter size in swine. Journal of Animal Sicience. 2003; 81: 624–629.

[37] Bradford GE. Genetic control of ovulation rate and embryo survival in mice. I. Response to selection. Genetics. 1969; 61: 905–921.

[38] Falconer DS. Improvement of litter size in a strain of mice at a selection limit. Genetics Research. 1971; 17: 215–235.

[39] Bakker H, Wallinga JH, Politiek RD. Reproduction and body weight of mice after long term selection for large litter size. Journal of Animal Science. 1978; 46: 1572–1580.

[40] Gion JM, Clutter AC, Nielsen MK. Alternative methods of selection for litter size in mice: II. Response to thirteen generations of selection. Journal of Animal Science. 1990; 68: 3543–3556.

[41] Nielsen B, Su G, Lund MS, Madsen P. Selection for increased number of piglets at d 5 after farrowing has increased litter size and reduced piglet mortality. Journal of Animal Science. 2014; 91(6): 2575–2582.

[42] Laborda P, Mocé ML, Santacreu MA, Blasco A. Selection for ovulation rate in rabbits: Genetic parameters, direct response, and correlated response on litter size. Journal of Animal Science. 2011; 89: 2981–2987.

[43] Cunningham PJ, England ME, Young LD, Zimmerman DR. Selection for ovulation rate in swine: Correlated response in litter size and weight. Journal of Animal Science. 1979; 48: 509–516.

[44] Leymaster KA, Christenson RK. Direct and correlated responses to selection for ovulation rate or uterine capacity in swine. Journal of Animal Science, 2000; 78(Suppl 1): 68.

[45] Rosendo AT, Druet T, Gogué J, Bidanel JP. Direct responses to six generations of selection for ovulation rate or prenatal survival in Large White pigs. Journal of Animal Science. 2007; 85: 356–364.

[46] Land RD, Falconer DS. Genetic studies of ovulation rate in the mouse. Genetics Research.1969; 13: 25–46.

[47] Blasco A, Argente MJ, Haley CS, Santacreu MA. Relationships between components of litter size in unilaterally ovariectomized and intact rabbit does. Journal of Animal Science. 1994; 72: 3066–3072.

[48] Clutter AC, Nielsen MK, Johnson RK. Alternative methods of selection for litter size in mice: I. Characterization of base population and development of methods. Journal of Animal Science. 1990; 68: 3536–3542.

[49] Bolet G, Santacreu MA, Argente MJ, Climent A, Blasco A. Divergent selection for uterine capacity in unilaterally ovariectomized rabbits. I. Phenotypic and genetic parameters. Proceedings of 5th World Congress Genetic Applied Livestock Production. 1994; 19: 261–264.

[50] Clutter AC, Kochera Kirby YL, Nielsen MK. Uterine capacity and ovulation rate in mice selected 21 generations on alternative criteria to increase litter size. Journal of Animal Science. 1994; 72: 577–583.

[51] Mesa H, Safranski TJ, Fischer KA, Cammack KM, Lamberson WR. Selection for placental efficiency in swine: Genetic parameters and trends. Journal of Animal Science. 2005; 83: 983–991.

[52] Mesa H, Cammack KM, Safranski TJ, Green JA,Lamberson WR. Selection for placental efficiency in swine: conceptus development. Journal of Animal Science. 2012; 90: 4217–4222.

[53] Johnson RK, Zimmerman DR; Kittok RK. Selection for components of reproduction in swine. Livestock Production Science. 1984; 11: 541–558.

[54] Bennett GL, Leymaster KA. Integration of ovulation rate, potential embryonic viability and uterine capacity into a model of litter size in swine. Journal of Animal Science. 1989; 67: 1230–1241.

[55] Johnson RK, Nielsen MK, Casey DS. Responses in ovulation rate, embryonal survival and litter traits in swine to 14 generations of selection to increase litter size. Journal of Animal Science. 1999; 77: 541–557.

[56] Kirby YK, Nielsen MK, Alternative methods of selection for litter size in mice: III. Response to 21 generations of selection. Journal of Animal Science. 1993; 71: 571–578.

[57] Falconer DS, Mackay TFC. Introduction to Quantitative Genetics, 4th edn Harlow, UK: Addison Wesley Longman. 1996. 317–339 pp.

[58] Ziadi C, Mocé ML, Laborda P, Blasco A, Santacreu MA. Genetic selection for ovulation rate and litter size in rabbits: Estimation of genetic parameters and direct and correlated responses. Journal of Animal Science. 2013; 91: 3113–3120.

[59] Ruiz-Flores A, Johnson RK. Direct and correlated responses to two-stage selection for ovulation rate and number of fully formed pigs at birth in swine. Journal of Animal Science. 2001; 79: 2286–2299.

[60] Hill WG, Mulder HA. Genetic analysis of environmental variation. Genetics Research. 2010; 92: 381–395.

[61] Argente MJ, García ML, Muelas R, Blasco A. Effect of selection for residual variance of litter size on components of litter size in rabbits. Proceedings of 10th World Congress on Genetics Applied to Livestock Production. 2014; 149: 244–250.

[62] Schultz RM. The molecular foundations of the maternal to zygotic transition in the preimplantation embryo. Human Reproduction Update. 2002; 8: 323–331.

[63] Latham KE, Kutyna K, Wang Q. Genetic variation in trophectoderm function in parthenogenetic mouse embryos. Developmental Genetics. 1999; 24: 329–335.

[64] Schultz RM. Regulation of zygotic gene activation in the mouse. BioEssays. 1993; 8: 531–538.

[65] Jarrell VL, Day BN, Praher RS. The transition from maternal to zygotic control of development occurs during the 4-cell stage in the domestic pig, Sus scrofa: Quantitative and qualitative aspects of protein synthesis. Biology of Reproduction. 1991; 44: 62–68.

[66] Telford NA, Hogan A, Franz CR, Schultz GA. Expression of genes for insulin and insulin-like growth factors and receptors in early postimplantation mouse embryos and embryonal carcinoma cells. Molecular Reproduction and Development. 1990; 27: 81–92.

[67] Murchison EP, Stein P, Xuan Z, Pan H, Zhang MQ, Schultz RM, Hannon GJ. Critical roles for Dicer in the female germline. Genes & Development. 2007; 21: 682–693.

[68] Stowe HM, Curry E, Calcatera SM, Krisher RL, Paczkowski M, Pratt SL. Cloning and expression of porcine Dicer and the impact of developmental stage and culture conditions on MicroRNA expression in porcine embryos. Gene. 2012; 501(2): 198–205.

[69] Le Masson F, Razak Z, Kaigo M, Audouard C, Charry C, Cooke H, Westwood JT, Christians ES. Identification of heat shock factor 1 molecular and cellular targets during embryonic and adult female meiosis. Molecular and Cellular Biology. 2011; 31(16): 3410–3423.

[70] Kohata C, Izquierdo-Rico MJ, Romar R, Funahashi H. Development competence and relative transcript abundance of oocytes derived from small and medium follicles of prepubertal gilts. Theriogenology. 2013; 80(9): 970–978.

[71] Mtango NR, Sutovsky M, Susor A, Zhong Z, Latham KE, Sutovsky P. Essential role of maternal UCHL1 and UCHL3 in fertilization and preimplantation embryo development. Journal of Cellular Physiology. 2012; 227(4): 1592–1603.

[72] Yi YJ, Manandhar G, Sutovsky M, Li R, Jonáková V, Oko R, Park CS, Prather RS, Sutovsky P. Ubiquitin C-terminal hydrolase-activity is involved in sperm acrosomal function and anti-polyspermy defense during porcine fertilization. Biology of Reproduction. 2007; 77(5): 780–793.

[73] Lykke-Andersen K, Gilchrist MJ, Grabarek JB, Das P, Miska E, Zernicka-Goetz M. Maternal Argonaute 2 is essential for early mouse development at the maternal-zygotic transition. Molecular Biology of the Cell. 2008; 19(10): 4383–4392.

[74] Krawczynski K, Najmula J, Bauersachs S, Kaczmarek MM. MicroRNAome of porcine conceptuses and trophoblasts: expression profile of micromas and their potential to regulate genes crucial for establishment of pregnancy. Biology of Reproduction. 2015; 92(1): 21.

[75] Tong ZB, Gold L, Pfeifer KE, Dorward H, Lee E, Bondy CA, Dean J, Nelson LM. Mater, a maternal effect gene required for early embryonic development in mice. Nature Genetics. 2000; 26(3): 267–268.

[76] Pisani LF, Ramelli P, Lazzari B, Braglia S, Ceciliani F, Mariani P. Characterization of maternal antigen that embryos require (MATER/NLRP5) gene and protein in pig somatic tissues and germ cells. Journal of Reproduction and Development. 2010; 56(1): 41–48.

[77] Wu X, Viveiros MM, Eppig JJ, Bai Y, Fitzpatrick SL, Matzuk MM. Zygote arrest 1 (Zar1) is a novel maternal-effect gene critical for the oocyte-to-embryo transition. Nature Genetics. 2003; 33: 187–191.

[78] Uzbekova S, Roy-Sabau M, Dalbiès-Tran R, Perreau C, Papillier P, Mompart F, Thelie A, Pennetier S, Cognie J, Cadoret V, Royere D, Monget P, Mermillod P. Zygote arrest 1 gene in pig, cattle and human: Evidence of different transcript variants in male and female germ cells. Reproductive Biology and Endocrinology. 2006; 4: 12.

[79] Esposito G, Vitale AM, Leijten FP, Strik AM, Koonen-Reemst AM, Yurttas P, Robben TJ, Coonrod S, Gossen JA. Peptidylarginine deiminase (PAD) 6 is essential for oocyte cytoskeletal sheet formation and female fertility. Molecular and Cellular Endocrinology. 2007; 273: 25–31.

[80] Xia X, Yan C, Wu W, Zhou Y, Hou L, Zuo B, Xu D, Ren Z, Xiong Y. Characterization of the porcine peptidylarginine deiminase type VI gene (PADI6) promoter: Sp1 regulates basal transcription of the porcine PADI6. Gene. 2015; 1119(15): 1151.

[81] Cinquanta M, Rovescalli AC, Kozak CA, Nirenberg M. Mouse Sebox homeobox gene expression in skin, brain, oocytes, and two-cell embryos. Proceedings of the National Academy of Sciences of the United States of America. 2000; 97: 8904–8909.

[82] Zheng Z, Zhao MH, Jia JL, Heo YT, Cui XS, Oh JS, Kim NH. Knockdown of maternal homeobox transcription factor SEBOX gene impaired early embryonic development in porcine parthenotes. Journal of Reproduction and Development. 2013; 59(6): 557–562.

[83] Bultman SJ, Gebuhr TC, Pan H, Svoboda P, Schultz RM, Magnuson T. Maternal BRG1 regulates zygotic genome activation in the mouse. Genes and Development. 2006; 20: 1744–1754.

[84] Magnani L, Cabot RA. Manipulation of SMARCA2 and SMARCA4 transcript levels in porcine embryos differentially alters development and expression of SMARCA1, SOX2, NANOG, and EIF1. Reproduction. 2009; 137(1): 23–33.

[85] Howell CY, Bestor TH, Ding F, Latham KE, Mertineit C, Trasler JM, Chaillet JR. Genomic imprinting disrupted by a maternal effect mutation in the Dnmt1 gene. Cell. 2001; 104(6): 829–838.

[86] Huan Y, Wang H, Wu Z, Zhang J, Liu Z, He H. The expression patterns of DNA methylation reprogramming related genes are associated with the developmental competence of cloned embryos after zygotic genome activation in pigs. Gene Expression Patterns. 2015; 18(1–2): 1–7.

[87] Okano M, Bell DW, Haber DA, Li E. DNA methyltransferases Dnmt3a and Dnmt3b are essential for de novo methylation and mammalian development. Cell. 1999; 99(3): 247–257.

[88] Lee K, Hamm J, Whitworth K, Spate L, Park KW, Murphy CN, Prather RS. Dynamics of TET family expression in porcine preimplantation embryos is related to zygotic genome activation and required for the maintenance of NANOG. Development Biology. 2014: 386(1): 86–95.

[89] Henderson GRW, Brahmasani SR, Yelisetti UM, Konijeti S, Katari VC, Sisinthy S. Candidate gene expression patterns in rabbit preimplantation embryos developed in vivo and in vitro. Journal of Assisted Reproduction and Genetics. 2014; 31(7): 899–911.

[90] Táncos Z, Bock I, Nemes C, Kobolák J, Dinnyés A. Cloning and characterization of rabbit POU5F1, SOX2, KLF4, C-MYC and NANOG pluripotency-associated genes. Gene. 2015; 566(2): 148–157.

[91] Yamaguchi S, Kurimoto K, Yabuta Y, Sasaki H, Nakatsuji N, Saitou M, Tada T. Conditional knockdown of Nanog induces apoptotic cell death in mouse migrating primordial germ cells. Development. 2009; 136: 4011–4020.

[92] Kehler J, Tolkunova E, Koschorz B, Pesce M, Gentile L, Boiani M, Lomeli H, Nagy A, McLaughlin KJ, Sholer HR, Tomilin A. Oct4 is required for primordial germ cell survival. EMBO Report. 2004; 5: 1078–1083.

[93] du Puy L, Lopes SM, Haagsman HP, Roelen BA. Analysis of co-expression of OCT4, NANOG and SOX2 in pluripotent cells of the porcine embryo, in vivo and in vitro. Theriogenology. 2011; 75(3): 513–526.

[94] Zhang S, Lin H, Kong S, Wang S, Wang H, Wang H, Armant DR. Physiological and molecular determinants of embryo implantation. Molecular Aspects of Medicine. 2013; 34(5): 939–980.

[95] Peiró R, Merchán M, Santacreu MA, Argente MJ, García ML, Folch JM, Blasco A. Identification of single-nucleotide polymorphism in the progesterone receptor gene and its association with reproductive traits in rabbits. Genetics. 2008; 180(3): 1699–1705.

[96] Short TH, Rothschild MF, Southwood OI, McLaren DG, de Vries A, van der Steen H, Eckardt GR, Tuggle CK, Helm J, Vaske DA, Mileham AJ, Plastow GS. Effect of the estrogen receptor locus on reproduction and production traits in four commercial pig lines. Journal of Animal Science. 1997; 75(12): 3138–3142.

[97] Rothschild MF, Jacobson C, Vaske DA, Tuggle C, Wang L, Short T, Eckardt G, Sasaki S, Vicent, A, McLaren D, Southwood O, van der Steen, H, Mileham A, Plastow G. The estrogen receptor locus is associated with a major gene influencing litter size in pigs. Proceedings of the National Academy of Science of the United States of America. 1996; 93(1): 201–205.

[98] Kim TH, Lee D-K, Franco HL, Lydon JP, Jeong JW. ERBB receptor feedback inhibitor 1 regulation of estrogen receptor activity is critical for uterine implantation in mice. Biology of Reproduction. 2010; 82(4): 706–713.

[99] Xu J, Li Q. Review of the in vivo functions of the p160 steroid receptor coactivator family. Molecular Endocrinology. 2003; 17(9): 1681–1692.

[100] [100] Park S, Yoon S, Zhao Y, Park SE, Liao L, Xu J, Lydon JP, DeMayo FJ, O'Malley BW, Bagchi MK, Katzenellenbogen BS. Uterine development and fertility are dependent on gene dosage of the nuclear receptor coregulator REA. Endocrinology. 2012; 153(8): 3982–3994.

[101] Salleh N. Diverse roles of prostaglandins in blastocyst implantation. Scientific World Journal. 2014; 968141.

[102] Lim H, Paria BC, Das SK, Dinchuk JE, Langenbach R, Trzaskos JM, Dey SK. Multiple female reproductive failures in cyclooxygenase 2-deficient mice. Cell. 1997; 91(2): 197–208.

[103] Sironen AI, Uimari P, Serenius T, Mote B, Rothschild M, Vilkki J. Effect of polymorphisms in candidate genes on reproduction traits in Finnish pig populations. Journal of Animal Science. 2010; 88(3): 821-827.

[104] Robertson AS, Allanson M, Mau VJ. Molecular regulation of uterine leukocyte recruitment during early pregnancy in the mouse. Placenta. 1998; 19(1): 101–119.

[105] Kamei Y, Tsutsumi O, Kuwabara Y, Taketani Y. Intra-uterine growth retardation and fetal losses are caused by epidermal growth factor deficiency in mice. American Journal of Physiology. 1993; 264(3): R597–R600.

[106] Robertson SA, O'Connell A, Ramsey A. The effect of interleukin-6 deficiency on implantation, fetal development and parturition in mice. Proceedings of the Australian Society for Reproductive biology. 2000; 31: 97.

[107] Robb L, Li R, Hartley L, Nandurkar H, Koentgen F, Begley CG. Infertility in female mice lacking the receptor for interleukin 11 is due to a defective uterine response to implantation. Nature Medicine. 1998; 4(3): 303–308.

[108] Stewart CL, Kaspar P, Brunet LJ, Bhatt H, Gadi I, Köntgen F, Abbondanzo SJ. Blastocyst implantation depends on maternal expression of leukaemia inhibitory factor. Nature. 1992; 359(6390): 76-79.

[109] Mucha A, Ropka-Molik K, Piórkowska K, Tyra M, Oczkowicz M. Effect of EGF, AREG and LIF genes polymorphisms on reproductive traits in pigs. Animal Reproduction Science. 2013; 137(1-2): 88-92.

[110] Lin HC, Liu GF, Wang AG, Kong LJ, Wang XF, Fu JL. Effect of polymorphism in the leukemia inhibitory factor gene on litter size in large white pigs. Molecular Biology Reports. 2009; 36(7): 1833-1838.

[111] Spötter A, Müller S, Hamann H, Distl O. Effect of polymorphisms in the genes for LIF and RBP4 on litter size in two German pig lines. Reproduction in Domestic Animal. 2009; 44(1): 100-105.

[112] Threadgill DW, Dlugosz AA, Hansen LA, Tennenbaum T, Lichti U, Yee D, LaMantia C, Mourton T, Herrup K, Harris RC, Barnard JA, Yuspa SH, Coffey RJ, Magnusom T. Targeted disruption of mouse EGF receptor: Effect of genetic background on mutant phenotype. Science. 1995; 269(5221): 230–234.

[113] Xie H, Wang H, Tranguch S, Iwamoto R, Mekada E, Demayo FJ, Lydon JP, Das SK, Dey SK. Maternal heparin-binding-EGF deficiency limits pregnancy success in mice. Proceedings of the National Academy of Sciences of the United States of America. 2007; 104(46): 18315-18320.

[114] Ferrara N. Vascular endothelial growth factor: Basic science and clinical progress. Endocrine Reviews. 2004; 25(4): 581–611.

[115] Halder JB, Zhao X, Soker S, Paria BC, Klagsbrun M, Das SK, Dey SK. Differential expression of VEGF isoforms and $VEGF_{164}$-specific receptor neuropilin-1 in the mouse

uterus suggests a role for $VEGF_{164}$ in vascular permeability and angiogenesis during implantation. Genesis. 2000; 26: 213–224.

[116] Lammers R, Gray A, Schlessinger J, Ullrich A. Differential signalling potential of insulin- and IGF-1-receptor cytoplasmic domains. EMBO Journal. 1989; 8(5): 1369–1375.

[117] Baker J, Liu JP, Robertson EJ, Efstratiadis A. Role of insulin-like growth factors in embryonic and postnatal growth. Cell. 1993; 75(1): 73–82.

[118] Kapur S, Tamada H, Dey SK, Andrews GK. Expression of insulin-like growth factor-I (IGF-I) and its receptor in the peri-implantation mouse uterus, and cell-specific regulation of IGF-I gene expression by estradiol and progesterone. Biology of Reproduction. 1992; 46(2): 208–219.

[119] Llobat L, Marcos-Jiménez F, Peñaranda DS, Thieme R, Navarrete A, Vicente JS. mRNA expression in rabbit blastocyst and endometrial tissue of candidate gene involved in gestational losses. Reproduction in Domestic Animals. 2012.; 47: 281-287.

[120] Krumlauf R. Hox genes in vertebrate development. Cell. 1994; 78(2): 191–201.

[121] Gendron RL, Paradis H, Hsieh-Li HM, Lee DW, Potter SS, Markoff E. Abnormal uterine stromal and glandular function associated with maternal reproductive defects in Hoxa-11 null mice. Biology of Reproduction. 1997; 56(5): 1097–1105.

[122] Bagot CN, Kliman HJ, Taylor HS. Maternal Hoxa10 is required for pinopod formation in the development of mouse uterine receptivity to embryo implantation. Developmental Dynamics. 2001; 222(3): 538–544.

[123] Wang W, Van De Water T, Lufkin T. Inner ear and maternal reproductive defects in mice lacking the Hmx3 homeobox gene. Development. 1998; 125(4): 621–634.

[124] Daikoku T, Cha J, Sun X, Tranguch S, Xie H, Fujita T, Hirota Y, Lydon J, DeMayo F, MaxsonR, Dey SK. Conditional deletion of MSX homeobox genes in the uterus inhibits blastocyst implantation by altering uterine receptivity. Developmental Cell. 2011; 21(6):1014-1025.

[125] Cha J, Sun X, Dey SK. Mechanisms of implantation: Strategies for successful pregnancy. Nature Medicine. 2012; 18: 1754-1767.

[126] Sun X, Zhang L, Xie H, Wan H, Magella B, Whitsett JA, Dey SK. Kruppel-like factor 5 (KLF5) is critical for conferring uterine receptivity to implantation. Proceedings of the National Academy of Sciences of the United States of America. 2012; 109(4): 1145–1150.

[127] Simmen RCM, Eason RR, McQuown JR, Linz AL, Kang TJ, Chatman L, Jr., Till SR, Fujii-Kuriyama Y, Simmen FA, Oh SP. Subfertility, uterine hypoplasia, and partial progesterone resistance in mice lacking the Kruppel-like factor 9/basic transcription element-binding protein-1 (Bteb1) gene. Journal of Biological Chemistry. 2004; 279(28): 29286–29294.

[128] Huyen D, Bany B. Evidence for a conserved function of heart- and neural crest derivatives-expressed transcript 2 (Hand2) in mouse and human decidualization. Reproduction (Cambridge, England). 2011; 142(2): 353-368.

[129] Li Q, Kannan A, DeMayo FJ, Lydon JP, Cooke PS, Yamagishi H, Srivastava D, Bagchi MK, Bagchi IC. The antiproliferative action of progesterone in uterine epithelium is mediated by Hand2. Science. 2011; 331(6019): 912–916.

[130] Jeong JW, Kwak I, Lee KY, Kim TH, Large MJ, Stewart CL, Kaestner KH, Lydon JP, DeMayo FJ. Foxa2 is essential for mouse endometrial gland development and fertility. Biology of Reproduction. 2010; 83(3): 396-403.

[131] Kurihara I, Lee D-K, Petit FG, Jeong J, Lee K, Lydon JP, Demayo FJ, Tsai MJ, Tsai SY. COUP-TFII mediates progesterone regulation of uterine implantation by controlling ER activity. Barsh G, ed. PLoS Genetics. 2007; 3(6): e102.

[132] Roelen BAJ, Lin HY, Knežević V, Freund E, Mummery C. Expression of TGF-βs and their receptors during implantation and organogenesis of the mouse embryo. Developmental Biology. 2002; 166(2): 716-728.

[133] Jones RL, Salamonsen LA, Findlay JK. Potential roles for endometrial inhibins, activins and follistatin during human embryo implantation and early pregnancy. Trends in Endocrinology and Metabolism. 2002; 13(4): 144–150.

[134] Jones RL, Findlay JK, Farnworth PG, Robertson DM, Wallace E, Salamonsen LA. Activin A and inhibin A differentially regulate human uterine matrix metalloproteinases: Potential interactionsduring decidualization and trophoblast invasion. Endocrinology. 2006; 147(2): 724–732.

[135] Paria BC, Ma W, Tan J, Raja S, Das SK, Dey SK, Hogan BL. Cellular and molecular responses of the uterus to embryo implantation can be elicited by locally applied growth factors. Proceedings of the National Academy of Sciences of the United States of America. 2001; 98(3): 1047–1052.

[136] Li Q, Kannan A, Wang W, Demayo FJ, Taylor RN, Bagchi MK, Bagchi IC. Bone morphogenetic protein 2 functions via a conserved signaling pathway involving Wnt4 to regulate uterine decidualization in the mouse and the human. Journal of Biological Chemistry. 2007; 282(43): 31725–31732.

[137] Park CB, DeMayo FJ, Lydon JP, Dufort D. NODAL in the uterus is necessary for proper placental development and maintenance of pregnancy. Biology of Reproduction. 2012; 86(6):194.

[138] Franco HL, Dai D, Lee KY, Rubel CA, Roop D, Boerboom D, Jeong JW, Lydon JP, Bagchi IC, Bagchi MK, DeMayo FJ. WNT4 is a key regulator of normal postnatal uterine development and progesterone signaling during embryo implantation and decidualization in the mouse. FASEB Journal. 2011; 25(4): 1176-1187.

[139] Cha J, Bartos A, Park C, Sun X, Li Y, Cha SW, Ajima R, Ho HY, Yamaguchi TP, Dey SK. Appropriate crypt formation in the uterus for embryo homing and implantation requires Wnt5a-ROR signaling. Cell Reports. 2014; 8(2): 382-392.

[140] Dunlap KA, Filant J, Hayashi K, Rucker EB 3rd, Song G, Deng JM, Behringer RR, DeMayo FJ, Lydon J, Jeong JW, Spencer TE. Postnatal deletion of Wnt7a inhibits uterine gland morphogenesis and compromises adult fertility in mice. Biology of Reproduction. 2011; 85(2): 386-396.

[141] Parr BA, Cornish VA, Cybulsky MI, McMahon AP. Wnt7b regulates placental development in mice. Developmental Biology. 2001; 237(2): 324–332.

[142] Lee K, Jeong J, Kwak I, Yu CT, Lanske B, Soegiarto DW, Toftgard R, Tsai MJ, Tsai S, Lydon JP, DeMayo FJ. Indian hedgehog is a major mediator of progesterone signaling in the mouse uterus. Nature Genetics. 2006; 38(10):1204–1209.

[143] Franco HL, Lee KY, Rubel CA, Creighton CJ, White LD, Broaddus RR, Lewis MT, Lydon JP, Jeong JW, DeMayo FJ. Constitutive activation of smoothened leads to female infertility and altered uterine differentiation in the mouse. Biology of Reproduction. 2010; 82(5): 991-999.

[144] Afshar Y, Jeong JW, Roqueiro D, DeMayo F, Lydon J, Radtke F, Radnor R, Miele L, Fazleabas A. Notch1 mediates uterine stromal differentiation and is critical for complete decidualization in the mouse. FASEB Journal. 2012; 26(1): 282-294.

[145] Schmid PC, Paria BC, Krebsbach RJ, Schmid HHO, Dey SK. Changes in anandamide levels in mouse uterus are associated with uterine receptivity for embryo implantation. Proceedings of the National Academy of Sciences of the United States of America. 1997; 94(8): 4188-4192.

[146] Desgrosellier JS, Cheresh DA. Integrins in cancer: biological implications and therapeutic opportunities. Nature Reviews Cancer. 2010; 10(1): 9-22.

[147] Illera MJ, Lorenzo PL, Gui YT, Beyler SA, Apparao KB, Lessey BA. A role alphavbeta3 integrin during implantation in rabbit model. Biology of Reproduction. 2003; 68: 766-771.

[148] Johnson GA, Robert C, Burghardt RC, Bazer FW. Osteopontin: A leading candidate adhesion molecule for implantation in pigs and sheep. Journal of Animal Science and Biotechnology. 2014; 5: 56-70.

[149] Hoffman LH, Olson Ge, Carson DD, Chilton BS. Progesterone and implanting blastocysts regulate Muc1 expression in rabbit uterine epithelium. Endocrinology. 1998; 139: 266-271.

[150] Ren Q, Guan S, Fu J, Wang A. Temporal and spatial expression of Muc1 during implantation in sows. International Journal of Molecular Sciences. 2010; 11: 2322-2335.

[151] Surveyor GA, Gendler SJ, Pemberton L, Das SK, Chakraborty I, Julian J, Pimental RA, Wegner CC, Dey SK, Carson DD. Expression and steroid hormonal control of Muc-1 in the mouse uterus. Endocrinology. 1995; 136(8): 3639-3647.

[152] Wang J, Huang ZP, Nie GY, Salamonsen LA, Shen QX. Immunoneutralization of endometrial monoclonal nonspecific suppressor factor beta (MNSFβ) inhibits mouse embryo implantation in vivo. Molecular Reproduction Development. 2007; 74(11): 1419–1427.

[153] Jha RK, Titus S, Saxena D, Kumar PG, Laloraya M. Profiling of E-cadherin, beta-catenin and Ca(2+) in embryo-uterine interactions at implantation. FEBS Letters. 2006; 580: 5653-5660.

[154] Haegel H, Larue L, Ohsugi M, Fedorov L, Herrenknecht K, Kemler R. Lack of beta-catenin affects mouse development at gastrulation. Development. 1995 121: 3529-3537.

[155] Laws MJ, Taylor RN, Sidell N, DeMayo FJ, Lydon JP, Gutstein DE, Bagchi MK, Bagchi IC. Gap junction communication between uterine stromal cells plays a critical role in pregnancy-associated neovascularization and embryo survival. Development. 2008; 135(15): 2659-2668.

[156] Yasue H, Mikawa S, Uenishi H, Wada Y. Analysis of allele segregation distortion in a swine resource family. Animal Biotechnology. 1999; 10(3): 147–152.

[157] Cassady JP, Johnson RK, Pomp D, Rohrer GA, Van Vleck LD, Spiegel EK, Gilson KM. Identification of quantitative trait loci affecting reproduction in pigs. Journal of Animal Science. 2001; 79(3): 623–633.

[158] de Koning DJ, Rattink AP, Harlizius B, Groenen MAM, Brascamp EW, van Arendonk JAM. Detection and characterization of quantitative trait loci for growth and reproduction traits in pigs. Livestock Production Science. 2001; 72(3): 185–198.

[159] King AH, Jiang Z, Gibson JP, Haley CS, Archibald AL. Mapping quantitative trait loci affecting female reproductive traitson porcine chromosome 8. Biology of Reproduction. 2003; 68(6): 2172–2199.

[160] Holl JW, Cassady JP, Pomp D, Johnson RK. A genome scan for quantitative trait loci and imprinted regions affecting reproduction in pigs. Journal of Animal Science. 2004; 82(12): 3421–3429.

[161] Tribout T, Iannuccelli N, Druet T, Gilbert H, Riquet J, Gueblez R, Mercat MJ, Bidanel JP, Milan D, Le RP. Detection of quantitative trait loci for reproduction and production traits in large white and French Landrace pig populations. Genetics Selection Evolution. 2008; 40(1): 61–78.

[162] Li K, Ren J, Xing Y, Zhang Z, Ma J, Guo Y, Huang L. Quantitative trait loci for litter size and prenatal loss in a White DurocxChinese Erhualian resource population. Animal Genetics. 2009; 40(6): 963–966.

[163] Onteru SK, Fan B, Nikkilä MT, Garrick DJ, Stalder KJ, Rothschild MF. Whole-genome association analyses for lifetime reproductive traits in the pig. Journal of Animal Science. 2011; 89(4): 988-995.

[164] Onteru SK, Fan B, Du ZQ, Garrick DJ, Stalder KJ, Rothschild MF. A whole-genome association study for pig reproductive traits. Animal Genetics. 2012; 43(1): 18–26.

[165] Hernandez SC, Finlayson HA, Ashworth CJ, Haley CS, Archibald AL. A genome-wide linkage analysis for reproductive traits in F_2 large white × Meishan cross gilts. Animal Genetics. 2014; 45(2): 191-197.

[166] Wilkie PJ, Paszek AA, Beattie CW, Alexander LJ, Wheeler MB, Schook LB. A genomic scan of porcine reproductive traits reveals possible quantitative trait loci (QTLs) for number of corpora lutea. Mammalian Genome. 1990; 10(6): 573–578.

[167] Rathje TA, Rohrer GA, Johnson RK. Evidence for quantitative trait loci affecting ovulation rate in pigs. Journal of Animal Science. 1997; 75(6): 1486–1494.

[168] Rohrer GA, Ford JJ, Wise TH, Vallet JL, Christenson RK. Identification of quantitative trait loci affecting female reproductive traits in a multigeneration Meishan-White composite swine population. Journal of Animal Science. 1999; 77(6): 1385–1391.

[169] Campbell EM, Nonneman D, Rohrer GA. Fine mapping a quantitative trait locus affecting ovulation rate in swine on chromosome 8. Journal of Animal Science. 2003; 81(7): 1706–1714.

[170] Bidanel JP, Rosendo A, Iannuccelli N, Riquet J, Gilbert H, Caritez JC, Billon Y, Amigues Y, Prunier A, Milan D. Detection of quantitative trait loci for teat number and female reproductive traits in Meishan × Large White F2 pigs. Animal. 2008; 2(6): 813-820.

[171] Schneider JF, Nonneman DJ, Wiedmann RT, Vallet JL, Rohrer GA. Genomewide association and identification of candidate genes for ovulation rate in swine. Journal of Animal Science. 2014; 92(9): 3792-803.

[172] Rothschild MF, Messer L, Day A, Wales R, Short T, Southwood O, Plastow G. Investigation of the retinol-binding protein 4 (RBP4) gene as a candidate gene for increased litter size in pigs. Mammalian Genome. 2000; 11(1): 75–77.

[173] Jiang Z, Gibson JP, Archibald AL, Haley CS. The porcine gonadotropin-releasing hormone receptor gene (GNRHR): Genomic organization, polymorphisms, and association with the number of corpora lutea. Genome. 2001; 44(1): 7–12.

[174] Korwin-Kossakowska A, Kamyczek M, Cieslak D, Pierzchala M, Kuryl J. The effect of the polymorphism of leptin (LEP), leptin receptor (LEPR) and osteopontin (OPN) genes on selected reproduction traits of synthetic Line 990 sows. Animal Science Papers and Reports. 2002; 20(3): 159–168.

[175] Vallet JL, Freking BA, Leymaster KA, Christenson RK. Allelic variation in the secreted folate binding protein gene is associated with uterine capacity in swine. Journal of Animal Science. 2005; 83(8): 1860–1867.

[176] Spötter A, Hamann H, Müller S, Distl O. Effect of polymorphisms in four candidate genes for fertility on litter size in a german pig line. Reproduction in Domestic Animals. 2010; 45(4): 579–584.

[177] Peiró R, Herrler A, Santacreu MA, Merchán M, Argente MJ, García ML, Folch JM, Blasco A. Expression of progesterone receptor related to the polymorphism in the PGR gene in the rabbit reproductive tract. Journal of Animal Science. 2010; 88: 421-427.

[178] Argente MJ, Merchán M, Peiró R, García ML, Santacreu MA, Folch JM, Blasco A. Candidate gene analysis for reproductive traits in two lines of rabbits divergently selected for uterine capacity. Journal of Animal Science. 2010; 88(3): 828-836.

[179] García ML, Peiró R, Argente MJ, Merchán M, Folch JM, Blasco A, Santacreu MA. Investigation of the oviductal glycoprotein 1 (OVGP1) gene associated with embryo survival and development in the rabbit. Journal of Animal Science. 2010; 88(5): 1597-1602.

[180] Estellé J, Sastre Y, Merchán M, Peiró R, Santacreu MA, Folch JM. TIMP-1 as candidate gene for embryo survival in two divergent lines selected for uterine capacity in rabbits. Molecular Reproduction and Development. 2006; 73: 678-684.

[181] Ballester M, Castelló A, Peiró R, Argente MJ, Santacreu MA, Folch JM. Identification of differentially expressed genes in the oviduct of two rabbit lines divergently selected for uterine capacity using suppression subtractive hybridization. Animal Genetic. 2013; 44: 296-304.

[182] Caetano AR, Johnson RK, Ford JJ, Pomp D. Microarray profiling for differential gene expression in ovaries and ovarian follicles of pigs selected for increased ovulation rate. Genetics. 2004; 168: 1529-1537.

[183] Fernandez-Rodriguez A, Muñoz M, Fernandez A, Pena RN, Tomas A, Noguera JL, Ovilo C, Fernandez AI. Differential gene expression in ovaries of pregnant pigs with high and low prolificacy levels and identification of candidate genes for litter size. Biology of Reproduction. 2011; 84(2): 299–307.

[184] Østrup E, Bauersachs S, Blum H, Wolf E, Hyttel P. Differential endometrial gene expression in pregnant and nonpregnant sows. Biology of Reproduction. 2010; 83(2): 277-285.

[185] Kim M, Seo H, Choi Y, Shim J, Kim H, Lee CK, Ka H. Microarray analysis of gene expression in the uterine endometrium during the implantation period in pigs. Asian-Australasian Journal of Animal Sciences. 2012; 25(8): 1102-1116.

[186] Samborski A, Graf A, Krebs S, Kessler B, Bauersachs S. Deep sequencing of the porcine endometrial transcriptome on day 14 of pregnancy. Biology of Reproduction. 2013; 88(4): 84.

[187] Samborski A, Graf A, Krebs S, Kessler B, Reichenbach M, Reichenbach HD, Ulbrich SE, Bauersachs S. Transcriptome changes in the porcine endometrium during the preattachment phase. Biology of Reproduction. 2013; 89(6): 134.

[188] Franczak A, Wojciechowicz B, Kotwica G. Transcriptomic analysis of the porcine endometrium during early pregnancy and the estrous cycle. Reproduction of Biology. 2013; 13(3): 229-237.

[189] Gu T, Zhu MJ, Schroyen M, Qu L, Nettleton D, Kuhar D, Lunney JK, Ross JW, Zhao SH, Tuggle CK. Endometrial gene expression profiling in pregnant Meishan and Yorkshire pigs on day 12 of gestation. BMC Genomics. 2014; 15: 156.

[190] Su L, Liu R, Cheng W, Zhu M, Li X, Zhao S, Yu M. Expression patterns of MicroRNAs in Porcine endometrium and their potential roles in embryo implantation and placentation. PLoS ONE. 2014; 9(2): e87867.

[191] Chen X, Li A, Chen W, Wei J, Fu J, Wang A. Differential gene expression in uterine endometrium during implantation in pigs. Biology of Reproduction. 2015; 92(2): 52.

[192] Franczak A, Wojciechowicz B, Kolakowska J, Zglejc K, Kotwica G. Transcriptomic analysis of the myometrium during peri-implantation period and luteolysis–The study on the pig model. Functional & Integrative Genomics. 2014; 14(4): 673-682.

8

Proliferative Endometrial Lesions Hidden Behind the Feline Pyometra

Maria dos Anjos Pires, Hugo Vilhena, Sónia Miranda, Miguel Tavares Pereira, Fernanda Seixas and Ana Laura Saraiva

Abstract

The literature refers to pyometra as the most important pathology in the feline uterus, which is often associated with cystic endometrial disease (cystic endometrial hyperplasia/ pyometra complex or CEH-Pyo). The etiology of pyometra is complex and probably multifactorial, but hormonal influences are suggested to play an important role in the pathogenesis. Progestagen-based contraceptives may be risk factors for the CEH-Pyo syndrome, for endometrial adenocarcinoma and also to mammary tumors in this species.

The histopathological descriptions of pyometra include an enlarged uterus containing purulent fluid, variable endometrial infiltration of neutrophils and bacterial colonization. The degree of hyperplasia of endometrial glands is variable, and frequently the endometrium becomes atrophic. The severity of endometritis is variable. Thereby, the type of inflammatory cells infiltrating the uterine wall or lumen varies accordingly and may include neutrophils, macrophages, plasma cells and lymphocytes.

The clinical diagnosis of pyometra is often based on the clinical signs and the physical examination, supported by ultrasound findings. The surgical excision of the uterus is the recommended treatment when the animal is not intent for breeding, as most pyometra clinical signs resolve after ovariohysterectomy.

Nevertheless, our clinical practice demonstrated that, in cats, pyometra often masks other uterine conditions that may present a worst prognosis and may interfere with the expected outcome. Thus, although seldom requested, the pathological analysis of the uterus with pyometra should be performed following surgery, even if significant macroscopic alterations are not visible, as one frequent finding in pyometra specimens is the co-existence of feline endometrial adenocarcinoma (FEA).

FEA is usually described as a rare pathology in cats, but recent descriptions suggest that it may be more frequent than thought. Some morphological and clinical features of FEA, as well as molecular markers, have been recently described. Moreover, age is not an adequate factor for triage, since some FEA cases were described in young animals, prompting pathologists, clinicians and researchers into this new reality.

Therefore, this chapter proposes to highlight aspects that reinforce the need for careful observation of all the cases of CEH-Pyo, to exclude the co-existence of FEA that can worsen the prognosis.

Keywords: Cat, cystic hyperplasia, endometrial adenocarcinoma, pyometra, queen

1. Introduction

The cystic endometrial hyperplasia-pyometra (CEH-Pyo) complex is the most frequent and important uterine disorder in queens [1–5]. Ovarian hormones are considered the main factors in CEH-Pyo complex development, and progesterone is considered the principal component in its pathogenesis. Nevertheless, estrogen's effects in uterus have also been implicated as causing hyperplasia of the endometrium and cystic dilation of endometrial glands, with concomitant increased secretion of fluid that favors the progression of CEH to pyometra [6, 7]. Pyometra is characterized by uterine inflammation and infiltration of the endometrium and uterine lumen by neutrophils and bacteria, leading to the development of the clinical signs [8]. Ultimately, the condition may originate sepsis [2, 9, 10].

In queens, pyometra has been described co-existing with other uterine conditions, such as a disorder of sexual development [11], uterine torsion [12], in ovariectomized (but not hysterectomized) cats [13] and in the uterine stump of neutered queens with ovarian remnant syndrome [14, 15]. Moreover, several cases of uterine neoplasia associated with pyometra have been reported in queens [16–18]. The factors involved in the development of feline endometrial adenocarcinomas (FEA) are still unclear. However, it seems probable that the endogenous or exogenous ovarian steroid hormones that are associated with CEH and pyometra development might also influence endometrial carcinogenesis and tumor progression [1, 2, 12, 17, 18]. Moreover, one could also hypothesize that the feline uterus reacts to any irritative stimulus by enhancing the inflammatory response, like it happens in dogs, which would lead to pyometra [8].

2. Pathophysiology of the cystic endometrial hyperplasia-pyometra complex

Female cats are a polyestrous seasonal species (Figure 1). The photoperiod is a major factor influencing the onset and duration of seasonal ovarian activity [19]. During the breeding season, queens may show ovulatory or anovulatory estrous cycles. In cats, spontaneous ovulation seldom occurs. Cats are considered a species with induced ovulation in which a physical repetitive stimulus, as the one associated with coitus or mechanical stimulation of the vagina, is required to trigger a consistent pulse of pre-ovulatory luteinizing hormone (LH). This will stimulate the ovulation of large ovarian follicles, which occurs in approximately 30 to 50 hours after the LH surge [6]. However, less frequently, ovulation can also occur without mating and is designated as spontaneous ovulation. Frequent grooming, self-grooming and presence of a tomcat are putative pheromonal, tactile and visual stimuli that may trigger LH secretion and induce ovulation [20–22].

In anovulatory cycles, recurrent follicular (estrogen-dominated) phases develop and in absence of a luteal phase, they are separated only by a short period corresponding to follicular atresia and emergence of a new follicular wave. Consequently, in non-ovulatory cycles, the uterus is not exposed to progesterone [6, 19, 22].

In ovulatory cycles, follicular and luteal (progesterone-dominated) phases alternate. The follicular stage is similar to that of non-ovulatory cycles. After ovulation, progesterone secretion by corpora lutea initiates within 24 to 48 hours. The luteal phase lasts approximately 30 days in non-pregnant cycles (diestrous or luteal phase), but it may also last up to 50 days if a pseudopregnancy develops, and in pregnant cycles the progesterone dominancy lasts about 60 days [6, 19, 22].

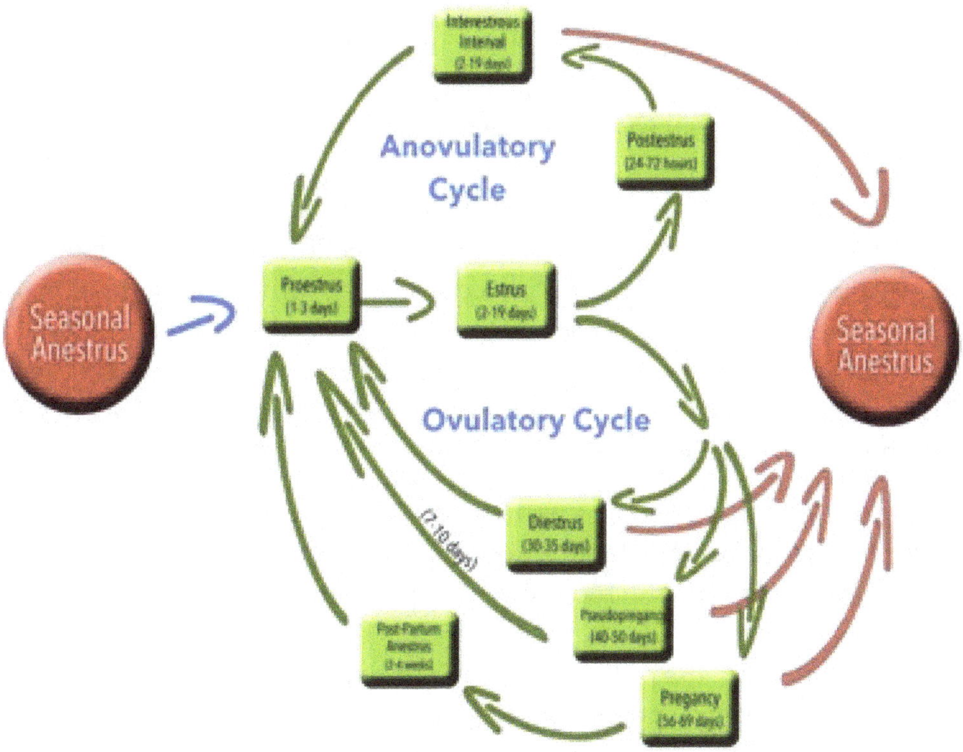

Figure 1. Drawing of the queen's reproductive cycle—the estrous cycles begin at puberty or after a seasonal anestrous. The figure compares the alternacy of the stages of the cycle in the anovulatory (top) or ovulatory (bottom) estrous cycles. The season onset always originates a new cycle whereas, at the end of season, the female can enter anestrous issued from an anovulatory cycle or at the end of a luteal phase (diestrous, pseudopregnancy or pregnancy) if the cycle was ovulatory.

The effects of progesterone in the uterus are major players in the pathogenesis of the CEH-Pyo complex. Progesterone plays diverse physiological roles designed to facilitate embryo survival and the success of a term pregnancy. These include the following: (1) the increase in coiling and in the secretory activity of endometrial glands (which expand the surface and the amount of fluid produced), (2) the decreased myometrial contractility, (3) the closure of the uterine cervix and (4) a depression in the uterine immune response [6, 21]. Our studies showed that T lymphocytes are the most common immune cells found in the cyclic feline endometrium, and we noted a variation of these cells and macrophage localization in the uterus between follicular and luteal stages, with an apparent migration of these cell populations from the surface layers to deeper layers of the endometrium during the luteal stages, which favor the embryo implantation [23].

Repeated exposure to consecutive progesterone cycles or the exogenous administration of progestagens for contraception may aggravate these effects, and stimulate the proliferation and cystic dilation of the endometrial glands leading to CEH development, luminal fluid accumulation, and uterine distention [24]. These changes in the uterine microenvironment, in particular the accumulation of a mucinous fluid in the uterus and the immune suppression, predispose to colonization of the uterine content by ascending bacteria from vaginal microbiota, originating pyometra [2, 6, 8, 21].

Pyometra is often diagnosed with functional ovarian corpora lutea. The progesterone is considered the main hormone implicated, although estrogens are also considered important in the CEH-Pyo complex pathogenesis. The disease has also been reported in cats with ovaries containing only follicles and basal serum progesterone concentrations or in cats with inactive ovaries [5, 25]. Estrogens cause dilation of the cervix during estrous, predisposing the uterus to ascending bacterial colonization by the normal vaginal biota [2, 19, 26]. In fact, the existence of a subclinical chronic low-grade bacterial uterine infection that may develop during the proestrous or estrous has also been implicated as a causative factor responsible for the endometrial proliferation occurring earlier in the pathogenesis of the CEH-Pyo complex [8]. Moreover, estradiol increases the estrogen and progesterone receptors in the uterus and enhances uterine response to continued stimulation with estradiol and concurrent or subsequent stimulation by progesterone [21].

It is questionable who comes first in the pyometra pathology: the colonization by bacteria that promotes inflammation or the instability or anomalous morphology/physiology of the endometrium (as CEH or FEA), which endorse the proliferation of opportunist bacteria that arrive from the cervix and vagina [6]. The most common bacteria isolated from feline pyometra is *Escherichia coli*, but other agents of the normal vaginal flora and of suspected fecal contamination have been also detected, including *Streptococcus* spp., *Staphylococcus* spp., *Pasteurella* spp., *Klebsiella* spp., *Proteus* spp., *Pseudomonas* spp., *Moraxella* spp., and *Tritrichomonas foetus* [6, 19, 25, 27, 28]. Sterile pyometra may also be observed [2].

3. Epidemiology

Cystic endometrial hyperplasia/pyometra complex is considered to be frequent in female cats. However, information on prevalence of feline pyometra is scarce. The few published studies suggest that it is less frequent than in dogs [3], which is in accordance with our experience (unpublished data). This difference has been attributed to two main factors. The first is related to the fact that queens are induced ovulators, and therefore, the feline uterus is less frequently exposed to progesterone influence [20, 24]. The second is related with the habit to neuter queens at younger ages to prevent unwanted pregnancies and heat behavior [15].

The prevalence of feline CEH-Pyo complex increases with age and is considered a common condition in cats older than 5 years of age and in cats receiving exogenous progestogens [1, 3, 5, 29–32]. Age effects in CEH-Pyo complex are associated to a cumulative effect of repeated hormonal stimulation in subsequent estrous cycles. Spontaneous ovulations and progestogen treatments predispose the uterus to abnormal endometrial proliferation, which leads to CEH development [25]. However, pyometra can also develop in younger animals, presumably due to other etiological process not yet described in the literature.

Feline breeds like the Sphynx, Siberian cat, Ocicat, Korat, Siamese, Ragdoll, Maine Coon, and Bengal seem to present a higher rate of pyometra, according to a study performed in Sweden [3]. Age at first mating, age at first parturition, and the number of parturitions do not appear to influence pyometra development [6].

Although feline pyometra is associated with a high morbidity, the mortality rate is relatively low, ranging from 5.7 to 8.0% of cases [3, 4].

4. Diagnosis

CEH is usually a silent pathology, with infertility being the main symptom in the early stages of the disease [6, 21, 33]. At this stage, queens are usually asymptomatic, and the physical examination, results from hematology, serum biochemistry, and urinalysis are unremarkable in most cases [5]. In cases of marked hyperplasia, or when mucometra develops, abdominal distention can be observed and an enlarged uterus can be detected on abdominal palpation [6]. In some cases, the uterus can be visualized on abdominal radiographs, but this diagnostic procedure presents low sensitivity in CEH. Abdominal ultrasound and histopathology are the most reliable methods for CEH diagnosis [19].

Pyometra is commonly associated with overt clinical symptoms and the abnormalities detected on complementary diagnostic exams vary with the severity of the disease, which typically develop one week to two months after estrous [2, 34]. The most common clinical symptoms include a mucous-purulent to hemorrhagic vulvar discharge (Figure 2), anorexia, lethargy, abdominal distention, dehydration, a palpable uterus, fever, and leukocytosis [4, 19]. The vulvar discharge is present in cases of open-cervix pyometra, but it can be unapparent due to the fastidious grooming habits of queens [6]. Queens with closed-cervix pyometra often have a significant abdominal distention and signs of severe illness [19].

Diagnosis is based on the history and clinical signs, and it is usually confirmed by abdominal radiography and/or ultrasound [13]. On abdominal radiology, an enlarged uterus with dorsal and cranial displacement of the small intestine can be detected, but these are also found in

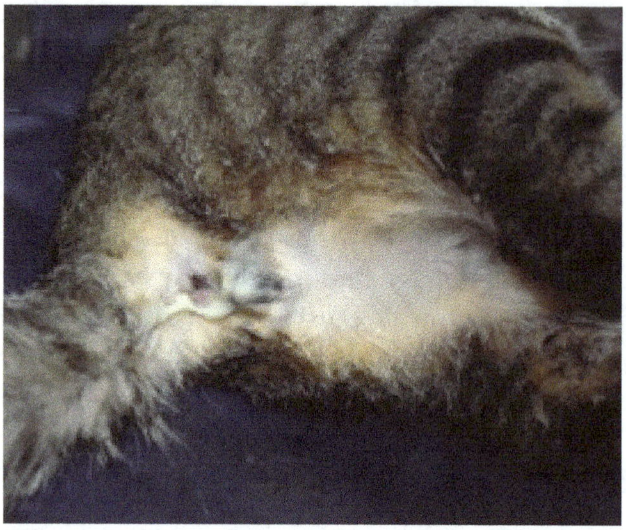

Figure 2. Purulent vulvar discharge in a queen with open-cervix pyometra.

early pregnancy [6]. In the abdominal ultrasound, it is possible to observe an enlarged uterus containing hypoechoic to anechoic intraluminal fluid (Figures 3A and B). This is the diagnostic procedure of choice in most cases [25].

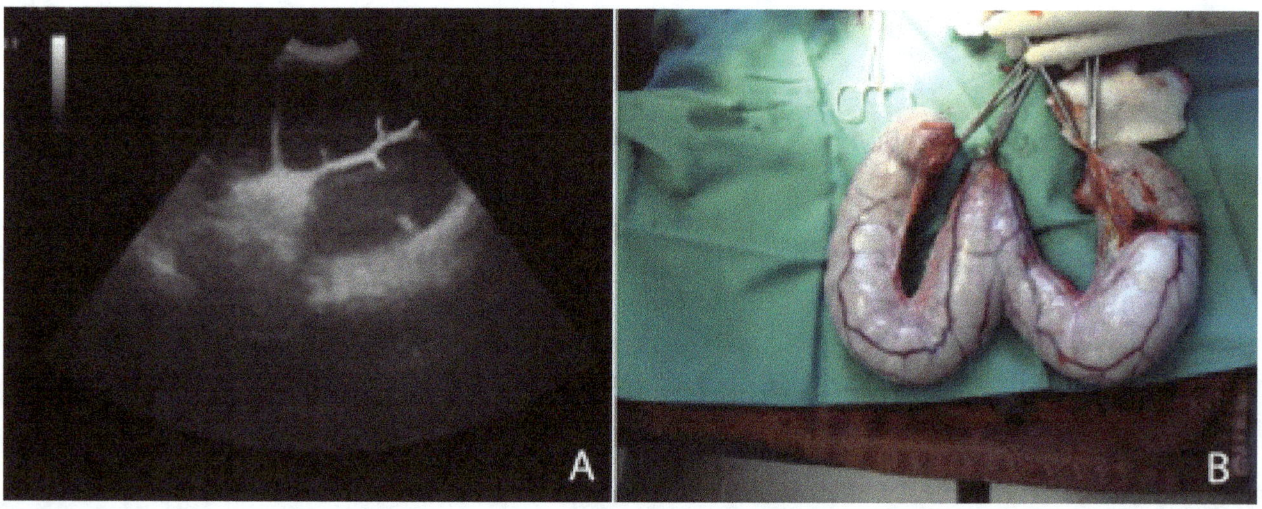

Figure 3. Pyometra in a female cat. (A) Abdominal ultrasound of a queen with pyometra showing a marked uterine distention with a narrow uterine wall, and the presence of hypoechoic intraluminal fluid; the loss of definition of the uterine contour and the unfolding of uterine walls are also observed. (B) Macroscopic aspect of the uterus after surgical excision.

5. Treatment

Pyometra is considered an emergency; thus, treatment should be rapid and aggressive due to the risk of septicemia, endotoxemia, azotemia, uterine rupture, peritonitis and shock [9, 10, 24,

34]. Surgical treatment (ovariohysterectomy—Figures 3B and 4) associated with supportive therapy is considered the treatment of choice [25, 26]. Support treatment includes intravenous fluid therapy and antibiotics administration [19]. Fluid therapy is directed to correct fluid deficits and electrolytic and acid-base imbalances, to correct azotemia and to maintain adequate tissue perfusion [6]. Whenever possible, antibiotics should be chosen based on bacterial culture and sensitivity. However, these procedures may take longer than the time available to recover the female from a life-threatening situation. Therefore, if not possible or during pending culture results, wide broad bactericide antibiotics effective against the most common bacteria are recommended, including ampicillin, amoxicillin + clavunalate, trimethoprim sulfonamide, cephalosporins and enrofloxacin [6]. Ovariohysterectomy rapidly and permanently eliminates the site of infection, and most cats recover successfully [19].

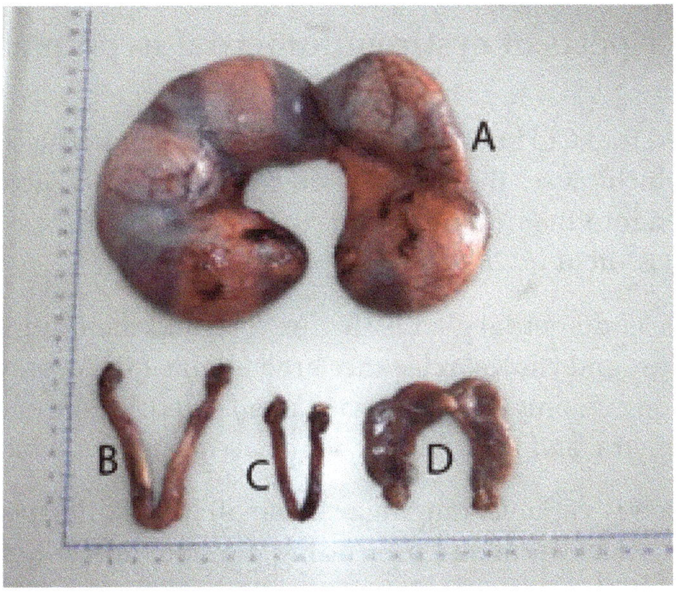

Figure 4. Different morphological aspects and sizes of queen's uterus collected at ovariohysterectomy. Each specimen represents a different clinical condition: (A) late-pregnant uterus; (B) uterus in a follicular stage; (C) uterus in anestrous; (D) uterus with pyometra.

Medical treatment could be considered in queens with reproductive added value, which are clinically stable and present an open-cervix pyometra [24, 35]. Medical treatment is directed to avoid synthesis or effects of progesterone, and to evacuate the uterine content.

Treatment with prostaglandins (natural or synthetic prostaglandin F2α, such as cloprostenol) or with aglepristone (an anti-progestagen) has proved to be effective in feline pyometra [36–39]. However, care should be taken before starting a treatment with prostaglandins as they should not be used in closed-cervix pyometra due to the risk of uterine rupture [24, 35, 38]. Prostaglandins promote luteolysis, consequently decreasing plasma concentrations of progesterone, and stimulate myometrial contraction leading to uterine evacuation [24, 40]. Mild and transient side effects, such as vocalization, panting, restlessness, grooming, tenesmus, salivation, diarrhea, mydriasis, emesis, urination and lordosis, can develop after prostaglandin administration [38, 39].

Aglepristone binds to the progesterone receptors and consequently inhibits progesterone effects. Once progesterone influences are withdrawn, the cervix will open and allow the elimination of the uterine content [37]. No side effects were observed in cats treated with aglepristone [37]. This compound may be used to treat a closed pyometra if the animal is stabilized and the risks of septicemia are residual. Prostaglandins and aglepristone can be used in association [25]. One good approach is to start the treatment with aglepristone and to associate the prostaglandins 24 to 48 h after the first aglepristone administration [37].

Queens submitted to medical treatment will be at risk for recurrence of pyometra at every ovulation after treatment if they do not become pregnant. To prevent it, they should be bred on the next cycle and then spayed when no longer needed for breeding [34].

6. Morphological aspects of cystic endometrial hyperplasia

It is well acknowledged that CEH pathogenesis involves the hyperplasia of the endometrium with cyst formation, which causes the accumulation of endometrial secretions [8], first into the glands lumen and then into the uterine lumen, with variable amount of serous, mucous, or purulent (neutrophils) content [6, 26].

Cysts develop from the endometrial glands (Figure 5A), and their number, size, distribution, histological morphology, and clinical relevance are variable [8]. As previously stated, CEH is usually a silent pathology, mainly if the cysts are small-sized. Cystic endometrial hyperplasia can progress into pyometra, and may or may not be associated with bacterial infection [41].

The morphology of CEH is variable, ranging from small gland dilatations to cystic structures occupying the entire endometrium and protruding into the uterine lumen. It is possible that the increased pressure of the cystic fluid modifies the morphology of glandular epithelium from cubic to squamous. In our experience (unpublished data), the size and number of cysts are not related with the grade of endometritis (mainly composed of a plasma cell and macrophages infiltrate) or with pyometra.

Dow [26] classified the severity of the CEH in four grades, being grade 1 the less severe, characterized by an endometrium with approximately normal dimensions and the presence of small cysts; and grade 4, the most severe, characterized by a significantly thickened endometrium or an atrophic endometrium with large ulcers associated to pyometra [26]. Different degrees of inflammation exist, ranging from a discreet mononuclear endometrium infiltration around the cysts in grade 2 to a severe infiltration of neutrophils, macrophages, and plasma cells in grade 4 [8].

The stimulus for connective tissue deposition around small aggregates of endometrial cystic glands can cause expansion and protrusion of a small portion of the endometrium, originating endometrial polyps, which are considered as morphologic variations of CEH. Queens can present single or multiple endometrial polyps, with variable dimensions [8], which can be spread along the uterine horns due to estrogen or progesterone influence [7].

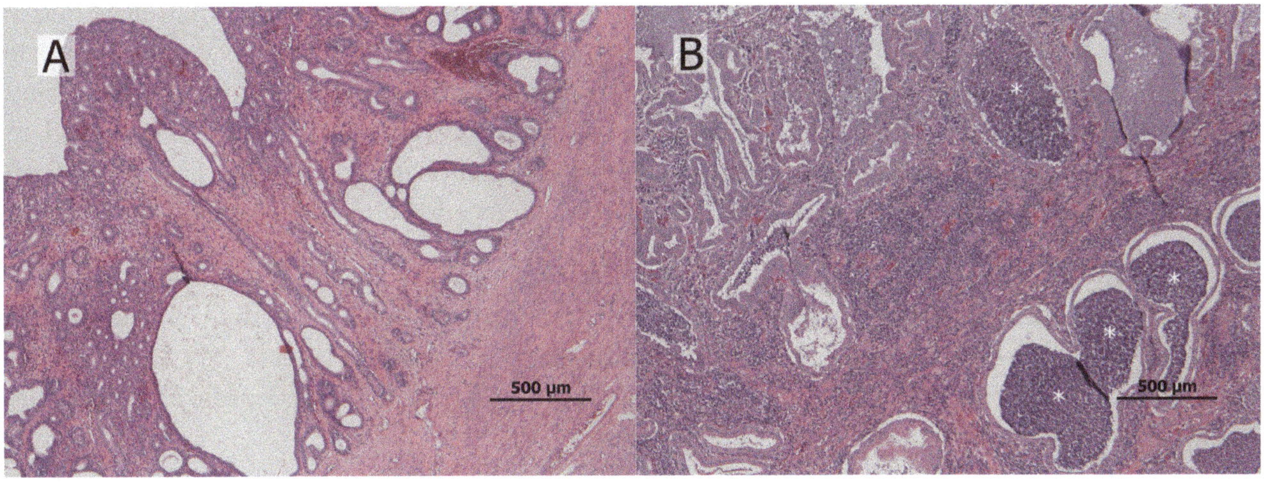

Figure 5. (A) Histological aspect of grade 2 CEH showing a hyperplastic endometrium with differently sized cysts. (B) Histological aspect of pyometra in a queen. Note the endometrium with cystic structures filled with neutrophils (*) opening into the lumen (arrow). An intense mononuclear infiltrate in endometrium characterizes endometritis (arrowhead) in a hyperplastic endometrium when concomitant CEH exists. Hematoxylin and Eosin. Bar = 500 μm.

7. Morphological aspects of pyometra

Morphological aspects of pyometra in queens are similar to those observed in the bitch and other species. The enlargement of the uterus is variable, ranging from a subtle increase to a severe enlargement in the endometrium dimension when associated to harsh glands and epithelium hyperplasia. The major feature of pyometra is the presence of neutrophils in the lumen of the uterus [7]. The glands are filled with mucous and neutrophils (Figure 5B) in a hyperplastic endometrium, evolving to the rupture of cysts into the lumen. Contrasting to that observed in the healthy endometrium, a raise of macrophages and B lymphocytes in the lumen and in the endometrium is observed following an initial rise in neutrophils, but only minor changes on the values for T lymphocytes are reported [23]. An asymptomatic pyometra often evolves to an atrophic and ulcerated endometrium, the final phase of this disease. The myometrium becomes thinner and uterine rupture can be eminent in more severe cases.

Endometritis is a concomitant feature with pyometra (Figure 5B). In the beginning of the process, neutrophils are the predominant cell type in the endometrium stroma, but macrophages and mainly plasma cells become more intense in chronic cases, presenting ulcerated and atrophic endometrium [7].

8. Feline endometrial adenocarcinoma: frequency, morphological and clinical aspects

Adenocarcinoma of the uterus is a malignant neoplasm, which arises from the endometrial epithelia [42]. Although reported in the rabbit, cow, rat, mouse, guinea pig, horse, dog, and cat [43], it is considered to be rare in domestic animals [44, 45]. In the literature, feline endometrial adenocarcinoma (FEA) is referred as an infrequent tumor [44]. However, recent reports of FEA suggest that the disease might be more common than assumed, and its incidence may be underestimated [18, 47, 46]. Feline endometrial adenocarcinoma is usually reported in geriatric queens, older than nine years of age [42, 48–50], but recently Cho et al. [51] and Payan at al. [18] reported FEA in young queens, less than two and one years old, respectively [18, 51].

The common practice of elective ovariohysterectomy in cats is being referred as protective from uterine neoplasia [11, 52]. Notwithstanding, the rarity of these tumors might be related to inadequate post-surgical or post-mortem evaluation of the genital tract [7, 53]. In fact, most clinicians do not require a histopathological evaluation of genital apparatus based on the assumption that "it is normal" or "only a pyometra".

Data related to the factors involved in the development of feline endometrial adenocarcinomas (FEA) are still insufficient. However, it seems probable that the endogenous or exogenous ovarian steroid hormones, associated to pyometra development, might also influence endometrial carcinogenesis and tumor progression [1, 18]. As recently reported, according to the database of our laboratory in a 13-year period, feline endometrial adenocarcinoma was diagnosed in 20.30% of cases ($n = 37/197$) while 33.5% (66/197) presented uterine inflammatory disease [12]. However, in 41.16% (81/197), the queen failed to manifest signs of uterine dysfunction or history of infertility. FEA either evolves as a silent disease [3, 47] or is associated to nonspecific and vague clinical signs that can be associated to other uterine disorders, including CEH, pyometra, or abortion. Therefore, there is a strong possibility that these tumors may be underdiagnosed [16, 18, 47].

Uterine tumors are more commonly found in sexually intact queens [54], but there are three reported cases of FEA in the uterine stump of ovariohysterectomized cats [44, 43, 54]. As such, these cases represent unusual complications following an incomplete ovariohysterectomy of the uterine body in a previously undiagnosed diseased uterus [44, 54].

Saraiva et al. [47] described three different histotypes of FEA based on cell features, growth pattern, and invasiveness, which included papillary serous carcinoma, clear cell carcinoma and "in situ" (non-invasive) carcinoma [16]. Microscopically, this classification resembles the human endometrial carcinoma classification [55]. Macroscopically, these patterns are indistinguishable, presenting as a multiple whitish papillary masses or a diffuse thickening of the endometrium along both uterine horns and corpus, which are better appreciable in a longitudinal section of this organ. Occasionally, the uterine wall may be thinner (atrophic) and when a concomitant pyometra is present, a purulent exudate may be observed in the lumen. Tumor invasion of the myometrium may be detected, and serosa rupture can occur promoting peritonitis and peritoneal carcinomatosis. Distant metastases in lungs or liver were seldom reported [54].

Regarding the cyclic uterus, in the presence of tumors we found higher numbers of macrophages and T lymphocytes. An increased number of B cells was only observed in pyometra cases and in FEA cases associated with pyometra. These findings suggest that B lymphocytes may be more relevant in uterine inflammation and that T cells may be more important in endometrial tumor oncobiology. These findings deserve more investigation to ascertain the real function of the various cell types on the etiopathogeny of uterine inflammation and neoplasia [23].

9. Hidden behind feline pyometra

In our experience, pyometra may not be the only uterine pathology observed in a queen that presents a mucous, hemorrhagic, or purulent vulvar discharge accompanied with other clinical signs compatible with pyometra.

Chronic inflammation associated with pyometra has been suggested to have a tumorigenic effect, and the inflammatory process may itself mask an underlying neoplasm in an under-evaluated uterus [56]. Association of pyometra with FEA was also suggested by some authors reporting cases of endometrial carcinoma in queens with pyometra [1, 16–18, 44]. Data related to the etiological factors involved in the development of feline endometrial adenocarcinomas are still scarce; however, it seems possible that the endogenous or exogenous ovarian steroids hormones associated with pyometra development might also influence endometrial carcinogenesis and tumor progression [1, 18].

Our laboratory records show that approximately 44% (29/66) of diagnosed pyometra co-existed with FEA, and that 24.2%(16/66) had concurrent CEH.

The clinical anamnesis is an important tool that must be explored to exclude the possibility of gestation or abortion. Whenever ovariohysterectomy is the adopted solution, the uterus must be totally excised, including the body and the cervix in the excised specimen. The histopathology is the unique complementary analysis that can distinguish and discard pyometra from other uterine lesions. Thus, it is of utmost importance to submit these organs to histopathological examination. At surgery, if any part of the uterine horns or the uterus is left in place (partial hysterectomy), it can bring problems in a near future, as some description of FEA [25] and pyometra were reported in uterine stumps of spayed queens [5, 14, 47, 54]. So, the complete and careful surgery associated with histopathological observation, alongside to a correct clinical monitoring, are the most important points to discard possible lesions that could be hidden behind a pyometra.

Acknowledgements

This work was sponsored/financed/founded by the Portuguese Science and Technology Foundation (FCT) under the Project PEst-OE/AGR/UI0772/2011, PEst-OE/AGR/UI0772/2014 and UID/CVT/00772/2013.

The authors would like to thank Mrs. Lígia Lourenço (UTAD) for her excellent contribution processing the histology samples. We also thank to all colleagues who sent feline OVH specimens for diagnosis to the Laboratory of Histology and Anatomical Pathology of the University of Trás-os-Montes and Alto Douro (UTAD).

Author details

Maria dos Anjos Pires[1,2*], Hugo Vilhena[1,3,4], Sónia Miranda[1,3,4], Miguel Tavares Pereira[5], Fernanda Seixas[1,2] and Ana Laura Saraiva[2,3]

*Address all correspondence to: apires@utad.pt

1 CECAV, Animal and Veterinary Research Centre, Universidade de Trás-os-Montes e Alto Douro, Vila Real, Portugal

2 Veterinary Sciences Department, Universidade de Trás-os-Montes e Alto Douro, Portugal

3 Department of Veterinary Medicine, Escola Universitária Vasco da Gama, Coimbra, Portugal

4 Baixo Vouga Veterinary Hospital, Águeda, Portugal

5 Vet4, Estarreja, Portugal

References

[1] Keskin A, Yilmazbas G, Yilmaz R, Ozyigit MO, Gumen A. Pathological abnormalities after long-term administration of medroxyprogesterone acetate in a queen. J Feline Med Surg. 2009; 11:518-21.

[2] Johnson CA. Medical management of feline pyometra. In: Kirk´s Current Veterinary Therapy XI, ed. Kirk RW and Bonagura JD, 1992, pp. 969-71. Elsevier Saunders, Pennsylvania.

[3] Hagman R, Ström Holst B, Möller L, Egenvall A. Incidence of pyometra in Swedish insured cats. Theriogenology. 2014; 82:114-20.

[4] Kenney KJ, Matthiesen DT, Brown NO, Bradley RL. Pyometra in cats: 183 cases (1979-1984). J Am Vet Med Assoc. 1987; 191:1130-2.

[5] Potter K, Hancock DH, Gallina AM. Clinical and pathologic features of endometrial hyperplasia, pyometra, and endometritis in cats: 79 cases (1980-1985). J Am Vet Med Assoc. 1991; 198:1427-31.

[6] Agudelo CF. Cystic endometrial hyperplasia-pyometra complex in cats. A review. Vet Quart. 2005; 27(4):173-82.

[7] Schlafer DH, Miller RB. Female genital system. In: Jubb, Kennedy & Palmer's Pathology of Domestic Animals. Edited by M. Grant Maxie. Vol 3. Chap 4. 2007; pp. 460-473.

[8] Schlafer DH, Gifford AT. Cystic endometrial hyperplasia, pseudo-placentational endometrial hyperplasia, and other cystic conditions of the canine and feline uterus. Theriogenology. 2008; 70:349-58.

[9] Costello MF, Drobatz KJ, Aronson LR, King LG. Underlying cause, pathophysiologic abnormalities, and response to treatment in cats with septic peritonitis: 51 cases (1990-2001). J Am Vet Med Assoc. 2004; 225(6):897-902.

[10] Brady CA, Otto CM, Van Winkle TJ, King LG. Severe sepsis in cats: 29 cases (1986-1998). J Am Vet Med Assoc. 2000;217:531-5.

[11] Schulman J, Levine SH. Pyometra involving uterus masculinus in a cat. J Am Vet Med Assoc. 1989; 194:690-1

[12] Stanley SW, Pacchiana PD. Uterine torsion and metabolic abnormalities in a cat with a pyometra. Can Vet J. 2008; 49:398-400.

[13] de Faria VP, Norsworthy GD. Pyometra in a 13-year-old neutered queen. J Feline Med Surg. 2008; 10:185-7.

[14] Rota A, Pregel P, Cannizzo FT, Sereno A, Appino S. Unusual case of uterine stump pyometra in a cat. J Feline Med Surg. 2011; 13:448-50.

[15] Demirel MA, Acar DB. Ovarian remnant syndrome and uterine stump pyometra in three queens. J Feline Med Surg. 2012; 14:913-8.

[16] Saraiva AL, Payan-Carreira R, Gärtner F, Pires MA. Feline Endometrial Adenocarcinomas. In: MA Longoria and JI Alcalá (Eds.) Adenocarcinoma: Pathogenesis, Treatment and Prognosis. Series Cancer Etiology, Diagnosis and Treatments. Nova Science Publishers, Hauppauge, NY, 2012; pp. 175–189.

[17] Sontas BH, Erdogan Ö, Apaydin Enginler SÖ, Turna Yilmaz Ö, Şennazli G, Ekici H (2013) Endometrial adenocarcinoma in two young queens. J Small Anim Pract 2013; 54:156-9.

[18] Payan-Carreira R, Saraiva A, Santos T, Vilhena H, Sousa A, Santos C, Pires MA. Feline endometrial adenocarcinoma in females < 1 year old: a description of four cases. Reprod Domest Anim. 2013; 48:70-7.

[19] Feldman EC, Nelson RW. Feline reproduction. In: Canine and Feline Endocrinology and Reproduction, 3rd Ed, ed Feldman EC and Nelson RW, 2004, pp. 1016-1045. Elsevier Saunders, Missouri.

[20] Lawler DF, Johnston SD, Hegstad RL, Keltner DG, Owens SF. Ovulation without cervical stimulation in domestic cats. J Reprod Fertil Suppl. 1993; 47:57-61.

[21] von Reitzenstein M, Archbald LF, Newell SM. Theriogenology question of the month. Pyometra, hydrometra, or mucometra. J Am Vet Med Assoc. 2000; 216:1221-3.

[22] England GC. Physiology and endocrinology of the female. In: BSAVA Manual of Canine and Feline Reproduction and Neonatology, 2nd Ed, ed. England G and von Heimendahl A, 2010; pp. 1-12. British Small Animal Veterinary Association, Gloucester.

[23] Tavares Pereira M. Comparison of Macrophages and Lymphocytes in Non-diseased Endometrium and Feline Endometrial Adenorcarcinomas, 2013; Master Thesis, UTAD, Vila Real, Portugal.

[24] Wiebe VJ, Howard JP. Pharmacologic advances in canine and feline reproduction. Top Companion Anim Med. 2009; 24:71-99

[25] Axnér E. Clinical approach to conditions of the non-pregnant and neutered queen. In: BSAVA Manual of Canine and Feline Reproduction and Neonatology, 2nd Ed, ed. England G and von Heimendahl A, 2010; pp. 185-190. British Small Animal Veterinary Association, Gloucester.

[26] Dow C. The cystic hyperplasia-pyometra complex in the cat. Veterinary Record, 1962; 74:141.

[27] Dahlgren SS, Gjerde B, Pettersen HY. First record of natural Tritrichomonas foetus infection of the feline uterus. J Small Anim Pract. 2007; 48:654-7.

[28] Majoy SB, Sharp CR, Dickinson AE, Cunningham SM. Septic pericarditis in a cat with pyometra. J Vet Emerg Crit Care (San Antonio). 2013; 23:68-76.

[29] Lawler DF, Evans RH, Reimers TJ, Colby ED, Monti KL. Histopathologic features, environmental factors, and serum estrogen, progesterone, and prolactin values associated with ovarian phase and inflammatory uterine disease in cats. Am J Vet Res. 1991; 52:1747-53.

[30] Chatdarong K, Rungsipipat A, Axnér E, Linde Forsberg C. Hysterographic appearance and uterine histology at different stages of the reproductive cycle and after progestagen treatment in the domestic cat. Theriogenology. 2005; 64:12-29.

[31] Bellenger CR, Chen JC. Effect of megesterol acetate on the endometrium of the prepubertally ovariectomised kitten. Res Vet Sci. 1990; 48:112-8.

[32] Romagnoli S. Progestins to control feline reproduction: historical abuse of high doses and potentially safe use of low doses. J Feline Med Surg. 2015; 17:743-52.

[33] Baldwin, CJ. Pregnancy Loss in the Queen. In: Kirk´s Current Veterinary Therapy XIV, ed. Bonagura JD and Twedt DC, 2009; pp. 1041-1045. Elsevier Saunders, Missouri.

[34] Traas AM. Feline reproduction. In: Textbook of Veterinary Internal Medicine, 7th Ed, ed. Ettinger SJ and Feldman EC, 2010, pp. 1729-1746. WB Saunders, Philadelphia.

[35] Biddle D, Macintire DK. Obstetrical emergencies. Clin Tech Small Anim Pract. 2000; 15:88-93.

[36] Arnbjerg J, Flagstad A. Prostaglandin F2 alpha treatment of feline open pyometra. Nord Vet Med. 1985; 37:286-90.

[37] Nak D, Nak Y, Tuna B. Follow-up examinations after medical treatment of pyometra in cats with the progesterone-antagonist aglepristone. J Feline Med Surg. 2009; 11:499-502.

[38] García Mitacek MC, Stornelli MC, Tittarelli CM, Nuñez Favre R, de la Sota RL, Stornelli MA. Cloprostenol treatment of feline open-cervix pyometra. J Feline Med Surg. 2014; 16:177-9.

[39] Davidson AP, Feldman EC, Nelson RW. Treatment of pyometra in cats, using prostaglandin F2 alpha: 21 cases (1982-1990). J Am Vet Med Assoc. 1992; 200:825-8.

[40] Verstegen JP, Onclin K, Silva LD, Donnay I. Abortion induction in the cat using prostaglandin F2 alpha and a new anti-prolactinic agent, cabergoline. J Reprod Fertil Suppl. 1993; 47:411-7.

[41] De Bosschere H, Ducatelle R, Vermeirsch R, Van Den BroeckW, Coryn M. Cystic endometrial hyperplasia-pyometra complex in the bitch: should the two entities be disconnected? Theriogenology 2001; 55:1509-19.

[42] Preiser H, 1964. Endometrial Adenocarcinoma in a Cat Pathol Vet, 1964; 1: 485

[43] Meier H (1956). Carcinoma of the uterus in the cat: two cases. Cornell vet 46: 188-200. as quoted by McEntee (1990).

[44] Miller MA, Ramos-Vara JA, Dickerson MF, Johnson GC, Pace LW, Kreeger JM, Turnquist SE, Turk JR. Uterine neoplasia in 13 cats. J Vet Diagn Invest. 2003; 15:515-22.

[45] Cotchin E, Spontaneous uterine cancer in animals. Br J Cancer, 1964; 18:209-27.

[46] Saraiva AL, Payan-Carreira R, Gartner F, Santana I, Rema A, Lourenço LM, Pires MA. Immunohistochemical expression of cyclooxygenase-2 (COX-2) in feline endometrial adenocarcinoma and in normal and hyperplastic endometria. Reprod Domest Anim. 2015; 50(2):333-40.

[47] Saraiva AL, Payan-Carreira R, Gärtner F, Fortuna da Cunha M, Rêma A, Faria F, Lourenço LM, Pires MA. An immunohistochemical study on the expression of sex steroid receptors, Ki-67 and cytokeratins 7 and 20 in feline endometrial adenocarcinomas. BMC Vet Res, 2015; 11:204.

[48] Belter LF, Crawford EM, Bates HR. Endometrial adenocarcinoma in a cat. Pathol Vet. 1968; 5:429-31.

[49] McEntee K. The uterus: atrophic, metaplastic and proliferative lesions – neoplasia. In: Reproductive Pathology of Domestic Animals, Academic Press, San Diego, CA, 1990; pp. 179–190.

[50] Klein MK, Tumours of the female reproductive system: uterine tumours. In: Withrow S, Vail D (eds), Withrow and MacEwen's Small Animal Clinical Oncology, 4th Ed, Saunders/Elsevier Health Sciences, St Louis, Missouri, USA, 2007; pp. 613-614.

[51] Cho S-J, Lee H-A, Hong S, Kim O. Uterine adenocarcinoma with feline leukaemia virus infection. Lab Anim Res. 2011; 27:347–51.

[52] Taylor KH. Female reproductive tumours. In: Henry CJ, Higginbotham ML (Eds.), Cancer Management in Small Animal Practice [Part V: Specific Tumours. Saunders, Elsevier, St. Louis, Missouri, 2010; pp. 268–274.

[53] Sontas BH, Erdogan Ö, Apaydin Enginler SÖ, Turna Yilmaz Ö, Şennazli G, Ekici H (2013) Endometrial adenocarcinoma in two young queens. J Small Anim Pract. 2013; 54:156-9.

[54] Anderson C, Pratschke K. Uterine adenocarcinoma with abdominal metastases in an ovariohysterectomised female cat. J Fel Med Surg. 2011; 13(1): 4-47.

[55] Horn L-C, Meinel A, Handzel R, Einenkel J. Histopathology of endometrial hyperplasia and endometrial carcinoma – an update. Annal Diag Pathol. 2007; 11(4): 297-311.

[56] McEntee K. The uterus: atrophic, metaplastic and proliferative lesions – neoplasia. In: Reproductive Pathology of Domestic Animals, Academic Press, San Diego, CA, 1990; pp. 179-190.

9

The Primordial to Primary Follicle Transition — A Reliable Marker of Ovarian Function

Duda Malgorzata, Grzesiak Malgorzata, Knet-Seweryn Malgorzata and Zbigniew Tabarowski

Abstract

In many mammalian species, including humans, folliculogenesis begins in fetal life and progresses throughout adulthood. The growing follicles progress from a reserve of primordial follicles that constitute the pool of female gametes for the entire life. Primordial follicles may begin to grow either immediately after forming or at clearly defined species-specific gap. Alternatively, some follicles may become quiescent before they either degenerate or resume growth several months or years afterwards. The rate of follicular assembly and the primordial to primary follicle transition is a critical step in female fertility. Therefore, disturbed coordination of the formation of primordial follicles and activation of their growth may entail some reproductive disorders. A poor initial reserve or the precocious primordial follicle depletion will result in infertility that, in women, is escorted by a shortened reproductive lifespan and early menopause. Therefore, it seems necessary to reach a profounder understanding of the molecular and cellular mechanisms controlling follicular development during preantral transition. *In vitro* growth of isolated immature ovarian follicles (IVGF) appears as an emerging technology, allowing to expand the fertility options in particular ovarian disorders or after cancer treatment

Keywords: Preantral follicles, folliculogenesis, organ cultures, *in vitro*, growth of isolated immature ovarian follicles (IVGF)

1. Introduction

In the ovary of a mammalian female, the process of folliculogenesis begins during fetal life and proceeds until the end of reproductive capacity, which is manifested in cell proliferation and differentiation [1,2]. Folliculogenesis, involving growth and development of ovarian follicles from primordial to preovulatory stages, is a complex phenomenon requiring multidirectional regulation. The ovarian follicle plays an essential role in securing optimal conditions for oocyte maturation and its release during ovulation, for which it will provide an appropriate microenvironment based on locally produced molecules, such as sex steroids and peptide hormones, growth factors, and cytokines, while also providing the appropriate communication among particular compartments of an ovarian follicle [3–5]. Sex steroids produced by follicular cells are known to play one of the main roles in the regulation of ovarian function [6]. These steroids present in the systemic circulation actively participate in the regulation of pituitary gonadotropin secretion. On the other side, sex steroids present in the ovarian microenvironment act as paracrine factors important for the maintenance of follicular development [3]. The majority of information about the role of sex steroids in ovarian functioning has been obtained in studies directed at the action of estrogens [7] and progestagens [8]. Nowadays, increasing attention is being devoted to the action of androgens because the activation of androgen receptors (ARs) located in follicular cells [9,10] modulates the expression and activity of many genes vital for the maintenance of follicular development [11–13].

From the initial pool of ovarian follicles recruited to grow, only a few reach a preovulatory stage. Less than 1% of follicles elude the process of atresia at various stages of development, and the preantral to early antral transition is the most susceptible to this process. The pool of primordial follicles established in fetal life constitutes a reserve that will not increase during the postnatal period. The initial stages of folliculogenesis, including the accumulation of primordial follicles, the recruitment of primordial follicles from the resting pool, and their transition into primary follicles, are crucial for the female reproduction regardless of the species [14]. Improper coordination of the formation of primordial follicles and activation of their growth may entail disturbed folliculogenesis in mature individuals manifested by a reduction of fertility. Recent research has revealed that primordial and primary follicles might not die by classical apoptosis. It is therefore possible that, in the immature ovary, other mechanisms are involved in follicular atresia [15].

The main factor determining the selection of follicles into the antral stage is their ability to respond to gonadotropins, especially follicle-stimulating hormone (FSH). Preantral follicles display an increase in the number of FSH receptors (FSHR) that, when activated, stimulate granulosa cell proliferation, antrum formation, and biosynthesis of estradiol after the activation of aromatase enzyme. There is quite ample evidence that follicle development is dependent on their granulosa layer, the functioning of which is influenced by endocrine, paracrine, and autocrine mechanisms. Granulosa cells are involved in the control of oocyte maturation and proper execution of ovulation and participate in early embryogenesis, maintenance of corpus luteum function, and production of chemotactic factors and those involved in angiogenesis [16]. Sustained oocyte growth depends on the effective communication and crosstalk between granulosa cells and the oocyte, because granulosa cells remain the major source of nutrients for the gamete through homologous and heterologous gap junctional contacts [17].

The tool that allows studying the function of ovarian follicles irrespective of its complicated structure is the model of whole organ culture, which reflects the conditions and complicated interactions occurring *in vivo*. These kinds of cultures constitute very sensitive objects to test the biological activity of various factors; they allow to observe the responses to increased or decreased steroid hormone secretion, the induction or inhibition of cell proliferation, or the induction or inhibition of apoptosis. The technological revolution in reproductive biology that started with artificial insemination and embryo transfer technologies during the last 30 years has continued with oocyte *in vitro* maturation (IVM), *in vitro* fertilization (IVF), or *in vitro* embryo culture (IVC), to name only a few. IVM has particular significance, providing the technology platform for the abundant supply of mature, good quality oocytes for diverse applications, such as reducing the generation interval in important species or studying *in vitro* human reproduction. Despite the convenience of IVM, we still do not understand the precise factors and conditions occurring *in vivo*, which yield the highest-quality mature oocytes for successful fertilization and embryo development outcomes; hence, we cannot completely imitate these conditions. Thus, *in vitro* growth of isolated immature ovarian follicles (IVGF) appears as an emerging technology allowing to expand the fertility options, particularly in young cancer patients [18–20], and may serve as a potential source of fertilizable gametes. Thus, assisted reproductive technologies allied to a profound understanding of granulosa/oocyte interactions can benefit from the capability to sustain primordial and primary follicle growth *in vitro* while supporting the acquisition of oocyte competence.

On this basis, the objective of this chapter is to review relevant data concerning the molecular factors crucial to the regulation of early stages of folliculogenesis and to provide basic information to the design of future culture strategies promoting the *in vitro* development of ovarian follicles.

2. Development of the primordial follicle

In the mammalian embryo, ovarian development begins between 3 and 6 weeks after conception. During this period, ovarian rudiment is massively colonized by mesonephric cells, which are regarded as the follicular cell precursors, and the primordial germ cells (PGC) migrate into the genital ridge; hence, other events take place, such as the differentiation of the gonads according to gender, proliferation, and apoptosis [14,21,22]. Oocyte development begins in the mammalian female fetus together with the differentiation of PGC. Proliferating PGC migrate towards the nascent genital ridges, where they differentiate into oogonia, before entering the first meiotic division to become primary oocytes [26].

Mammalian oocytes develop and reach ovulatory maturity inside the follicles where they are covered at first by pre-granulosa and then by granulosa cells [23] (Figure 1). Over the lengthy process of follicle development, granulosa cells proliferate and the theca layer is formed [24], allowing the follicle to take advantage of blood supply. Then, follicles pass through the succeeding stages of development before reaching full maturation and the ability to ovulate [25]. Primary oocytes, which are arrested at diplotene of the first meiotic prophase since late prenatal life in most mammal species, are the organizing centers of primordial follicles. The oocyte is considered to play the most important role in follicular organization during folliculogenesis. It is assumed that the oocyte controls both the proliferation and the differentiation of granulosa cells into cells capable of secreting steroids and various proteins. On the contrary, several oocyte features, such as growth, differentiation, meiosis, cytoplasmic maturation, or control of transcriptional activity, are dependent on the presence and contact with granulosa cells [27]. Interestingly, when the oocyte reaches a certain size threshold, it secretes factors that inhibit the ability of granulosa cells to promote its own growth [4], which suggests that the oocyte may determine not only its own growth but also the growth of the whole follicle.

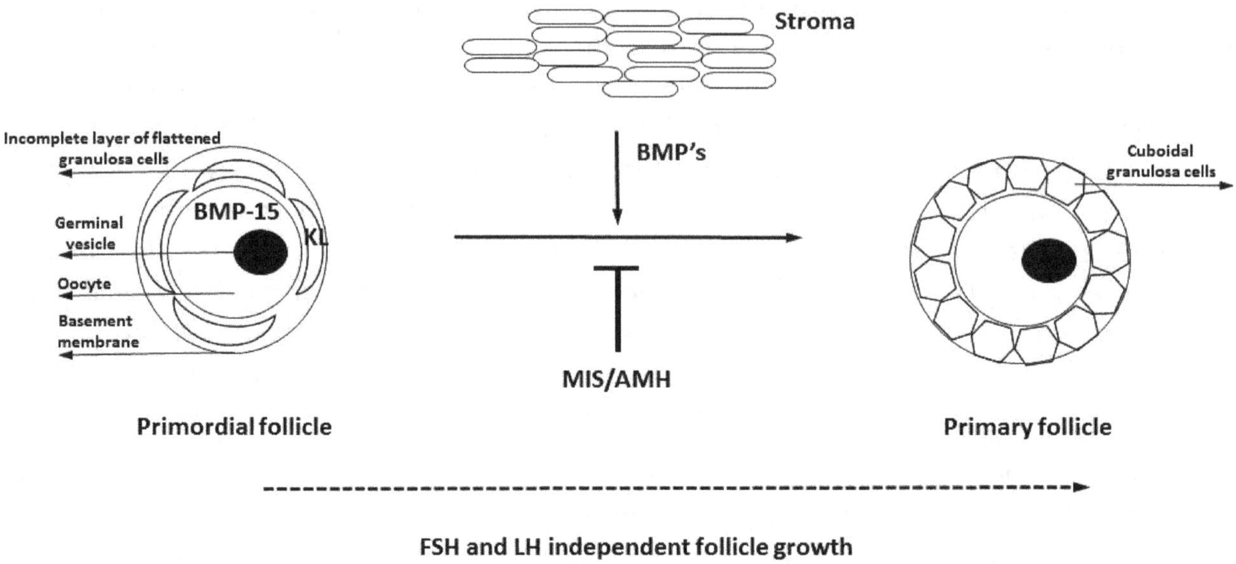

Figure 1. Simplified representation of early stages of mammalian folliculogenesis. KL, kit ligand; BMPs, bone morphogenic proteins; MIS/AMH, Müllerian inhibitory substance.

The assembly of primordial follicles, also described as the primordial follicle formation, demands for individual oocytes to associate with developing pre-granulosa cells, in a complex process that involves the breakdown of oocyte nests, oocyte separation, and subsequent recruitment of somatic (i.e. pre-granulosa) cells, which are regulated by circulating hormones and factors produced by the oocyte and somatic cells [28]. Newly created primordial follicles give rise to primary follicles that, in a series of transitions coordinated by gonadotropins, steroids, and other intraovarian factors, transform into preantral and then antral follicles and finally preovulatory follicles [6]. Among the signaling pathways that are important for primordial follicle assembly, apoptosis and autophagy are crucial in determining cellular fate. After nest separation, a large number of germ cells are lost by apoptosis; the mechanisms regulating cyst breakdown and germ cell death are still unclear. Indeed, much attention has been focused on germ cell elimination by apoptosis and the role of Bcl-2 gene family in regulating the balance between survival and death of oocytes before the formation of primordial follicles [29]. To date, autophagy has also been proposed to contribute in the mechanisms of prenatal and neonatal oocyte demise [30]. Increasing evidences showed that, in the immediate hours of the postnatal life, many tissues and organs evidence up-regulation of autophagy pathways, possibly acting as an adaptive response of the newborn organism to nutritional stress associated with the deprivation of placental nutrients [31]. The balance between quiescence and activation of the primordial follicle reserve seems to depend on a number of molecules. The phosphatidylinositol 3-kinase (PI3K) pathway was proposed as a key pathway playing a crucial integrative role by bridging the action of multiple factors in the balance between follicle growth suppression, activation, and the maintenance of healthy quiescence [32]. In mammalian ovaries, postnatal depletion of oocytes occurs also by atresia of follicles. Follicular atresia is directed by granulosa cell apoptosis and affects all stages of follicular development. Interestingly, recent evidence from studies on rats shows that autophagy of germinal cells is an alternative route to induce follicular atresia in the ovary [33]. This implies the importance of autophagy in cellular elimination within the ovary.

3. Primordial to primary follicle transition

The concept of primordial follicle activation refers to the process by which primordial follicles gradually exit the nongrowing follicle pool and enter the intermediate or primary follicle stage [23]. The clarification of the mechanisms that regulate primordial follicle activation is an important issue for the success of assisted reproduction [4]. Whereas the primordial follicle activation seems to depend mainly on signals originating in the ovary, pituitary and metabolism-related hormones are required for folliculogenesis to proceed past the primary or secondary stage [34]. In the ovary, the crosstalk between oocytes and somatic cells (i.e. granulosa or theca cells) occurs at an early stage of follicular development [4]. The activation of primordial follicles is associated with oocyte growth, and simultaneous differentiation of the adjacent pre-granulosa cells occurs. During the transition into primary follicles (showing a complete layer of cuboidal granulosa cells), pre-granulosa cells change into a cuboidal shape [23], and in the process, they form an intermediate form of follicles presenting both cuboidal and flattened pre-granulosa cells [35]. The proliferation of granulosa cells allows to originate multiple layers of cells, and follicle develops to secondary, antral, and further advanced follicle stages [23,36,37].

Recent research revealed that factors secreted by the oocytes regulate the initiation of primordial follicle growth [38] (Figure 2). The tyrosine kinase receptor Kit (c-Kit) and two different isoforms of its ligand (kit ligand, KL), localized in oocytes and granulosa cells, stimulate oocyte growth and maintain it in meiotic arrest depending on FSHR levels. The up-regulated expression of KL, triggered by low concentrations of FSH, promotes a reduction in the ratio of KL/c-Kit and stimulates oocyte growth, whereas high concentrations of FSH enhance follicle development but impair oocyte growth [5]. Other important regulators of follicle growth are activin [39] and oocyte-derived growth differentiation factor-9 (GDF-9) [14,40]. GDF-9 promotes follicular survival and growth during transition due to suppression of granulosa cell apoptosis and follicular atresia, whereas activin promotes FSH release, antral cavity formation, and granulosa cell proliferation. Bone morphogenetic protein-15 (BMP15) has been shown to promote granulosa cell growth by stimulation of the proliferation of undifferentiated granulosa cells in an FSH-independent manner. It was shown that two markers of proliferation, Ki-67 and proliferating cell nuclear antigen (PCNA), are regulated by these oocyte-derived factors (for review, see Ref. [41]). It was also suggested that PCNA could act as a key regulator of the development of ovarian follicles. The time expression of PCNA in oocytes is coincident with the initiation of primordial follicle formation. By promoting the apoptosis of oocytes, PCNA can also regulate primordial follicle assembly in neonatal mouse ovaries [42]. Moreover, proliferation of granulosa cells is increased by insulin-like growth factor-I (IGF-I), which was also associated with the regulation of follicular growth from the primordial stage [43]. The anti-Müllerian hormone (AMH) is synthesized early in the follicle formation, by the cuboidal granulosa cells of primordial follicles. This factor, subsequently produced by preantral and antral follicles, inhibits the initial recruitment of primordial ones as well as their further FSH-dependent growth [44].

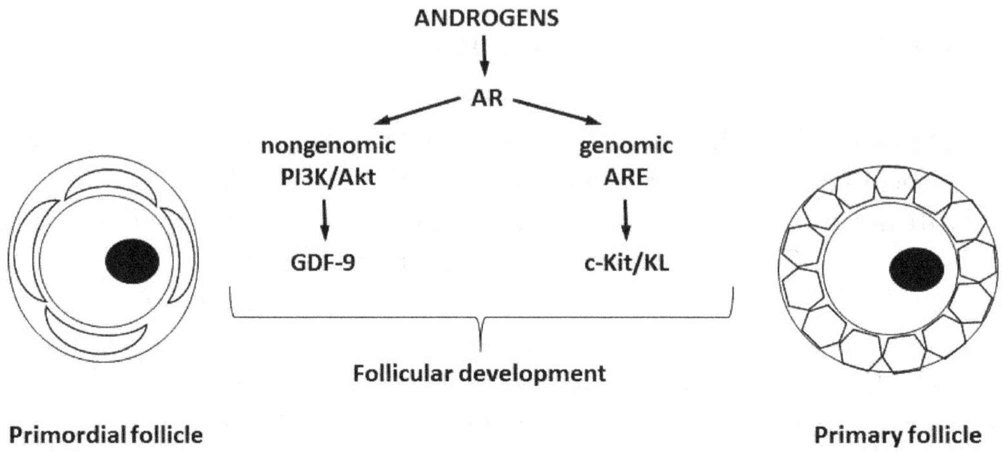

Figure 2. Mechanisms of androgen actions in follicular development. Physiological functions of androgens during primordial follicle recruitment are mediated through androgen response element (ARE)-dependent genomic actions and/or via PI3K/Akt nongenomic signaling pathway.GDF-9, growth differentiation factor-9; AR, androgen receptor.

The idea that androgens might regulate follicular development initially started with studies indicating AR expression in the different compartments of follicles throughout most stages of folliculogenesis [45–48]. However, AR expression pattern may differ between cell types, and in most species, AR is abundant in the preantral/antral stages of follicular development but declines as a follicle matures to the preovulatory stage [49–51]. Based on these observations, it was suggested that androgens might differentially regulate various stages of follicular development through an autocrine and/or paracrine way. It is generally accepted that androgens primarily affect preantral follicles and that their activities are important for preantral follicle growth and prevention of follicular atresia. Moreover, it seems possible that androgens are involved in the activation of primordial follicles [52–54] (Figure 2). How androgens influence primordial follicle recruitment and whether this is a primary or secondary response to androgens are still open-ended questions needing further investigation.

The mechanisms of primordial follicle activation can be studied using *in vitro* culture methods. However, until now, the success of primordial follicles culture as a method of oocyte growth has been limited to mice. Eppig and O'Brien [55] were the first to obtain mouse offspring derived from oocytes acquired from cultured primordial follicles. As to other species, several studies carried out in farm animals and primates showed that the transition of primordial into primary follicles in culture of cortical strips from caprine [56], bovine [57], baboon [58], and human [59] ovaries is possible. A confirmation of the normality of follicle development *in vitro* was obtained through the changes in follicle morphology and cell number as well as from the stage-specific follicular responsiveness to above-mentioned factors or the development of steroidogenic capacity.

4. Primordial and primary follicle isolation

In vitro follicle growth is a promising fertility preservation strategy [19,60] despite that, in some mammalian species, including humans and pigs, the success has been limited when the process started with primordial follicles. This could be explained by the fact that adequate isolation methods and culture strategies have not yet been fully established, thereby impairing the ability to obtain mature gametes from the culture of isolated primordial follicles in those species. The manipulation of primordial follicles is a challenge due to their small size and the existing physical connections between the oocyte and the surrounding squamous granulosa cells, which are also poorly studied. Conversely, the conditions that support their activation and growth are not well defined. Several studies have indicated that primordial and primary follicles rapidly degenerate in cultures carried under multiple conditions [61–63]. For example, primordial follicles isolated from human ovarian tissue using collagenase digestion and subsequently cultured in collagen gels resulted in the degeneration of the follicles within 24 hours [64].

The species and the reproductive age of the ovarian tissue affect preantral follicle yield in the ovary because of the existence of a larger number of follicles and the easiness of the isolation method in neonatal and prepubertal ovaries compared with mature ovaries [65,66]. The success of either culture or transplantation of isolated follicles depends on the high quality of retrieved follicles. That is why an effective method for retrieving viable, preantral follicles is an essential condition. Different methods are currently available to isolate follicles for preantral follicle culture. The mechanical isolation methods include the use of fine-gauge needles or forceps to isolate follicles from mice [67], rats [68], pigs [69], cattle [70], and humans [64]; the combination of ovarian dissociation methods, such as grating or mincing, with sieving [71]; and the follicular dissection from the ovarian cortex using a skin-grafting knife and/or small scalpel blades [34,70]. The mechanical isolation methods have a main advantage, as they allow retrieving intact follicles, surrounded by the basement membrane and theca layers, although they are slow and laborious techniques that typically yield only a small number of follicles [72]. These technical problems can be avoided by the use of enzymes to aid follicle recovery. The incubation of ovarian tissue in collagenase and/or DNAse (e.g. Refs. [55,73]) softens and disaggregates the tissue matrix and allow detaching follicles from the surrounding stroma with the aid of needles. However, the degradation of the basement membrane and the absence of theca cell layers are the most common undesirable consequences of the use of enzymes in follicle isolation [74], as they foster the spontaneous loss of granulosa cells from the follicles in culture. Nevertheless, the time of enzyme exposure can be controlled to minimize the damage [75].

The ovarian stroma is dense and fibrous; thereby, it is more efficiently isolated using a combination of mechanical and enzymatic procedures that have been shown to preserve follicle viability [64,76–78]. Dolmans et al. [77] developed a new isolation protocol using Liberase Blendzyme 3. This blend of purified enzymes allowed the isolation of a high number of preantral follicles, which were viable as well as morphologically and ultrastructurally normal. However, this type of Liberase is no longer produced. Therefore, the second generation of Liberase DH (Dispase High) Research Grade has been successfully tested for the preparation of human ovarian follicles [79]. However, the efficiency of the mixture may vary with the species. Our group recommends the use of Liberase TH (Thermolysin High) Research Grade to obtain a high number of fully isolated primordial follicles from porcine ovarian cortex [80], as it presents a really fibrous tissue. Using prepubertal gilt ovaries, we applied different types of Liberase (DH, TM, and TH) Research Grade and treatment protocols to isolate primordial follicles (Figure 3). The quality of the isolated follicles was evaluated by their general morphology and viability upon routine hematoxylin and eosin (H&E) and fluorescent staining, whereas their ultrastructure was assessed by electron microscopy. Additionally, to determine the purity of isolated follicles, a germ cell-specific protein, MSY2, was used to recognize oocytes. Liberase TH Research Grade was the mixture presenting a very high proportion of retrieved viable follicles whose majority exhibited good morphology with a complete granulosa cell layer. In addition, primordial follicles stained with either Hoechst 33342 or H&E indicated that Liberase TH Research Grade only occasionally induced atresia. This was supported by ultrastructural studies revealing that the oolemma-follicular cell interface was well preserved, which would allow the complex to express the correct metabolic profile (Figure 4). The results obtained in those experiments also showed that almost all of the Liberase TH Research Grade-isolated primordial follicles were MSY2 positive. As shown in the literature [81,82], primordial and primary follicles may rapidly degenerate after isolation because of the loss of critical connections between the oocyte and the granulosa cells. It seems that Liberase is a promising alternative to collagenase treatment, allowing the use of isolated primordial follicles for further reproductive studies.

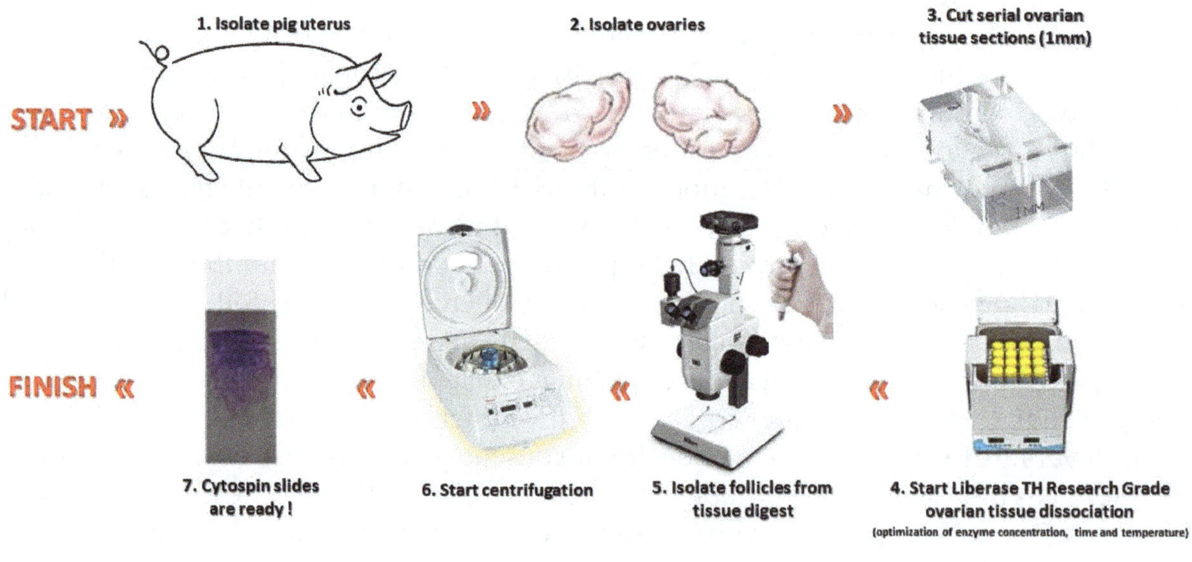

Figure 3. Pig ovarian medulla collection and preantral follicle isolation protocol. 1–3: Ovarian medulla collection (4- to 5-month-old prepubertal gilts); 4–6: isolation of primordial and primary follicles using different types of collagenase (types I, II, and IV) and Liberase (DH, TM, and TH); 7: evaluation of preantral follicles morphology (H&E staining). For details, see Ref. [80].

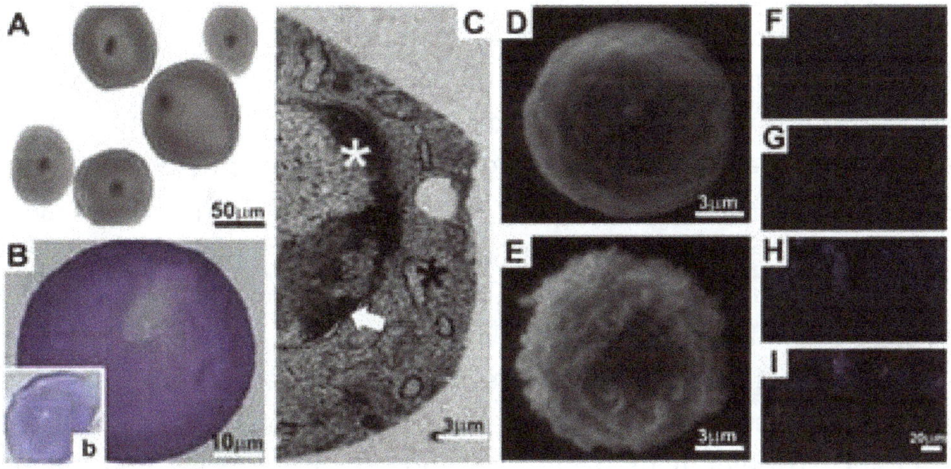

Figure 4. Morphology and ultrastructure of primordial and primary pig follicles isolated from ovarian medulla using Liberase TH-Research Grade. (A) Morphology of Liberase TH-treated pre-antral follicles (light microscopy), (B) morphology of Liberase TH-isolated pre-antral follicles stained with hematoxylin and eosin; (b) interrupted granulosa cells layer in pre-antral pig follicles isolated with collagenase (type II); (C) Transmission electron microscopy (TEM) showed a single uninterrupted layer of cuboid follicular cells (black asterisk) surrounding the oocyte (white asterisk), which was bordered by a continuous basal lamina (arrow). Scanning electron microscopy (SEM) of primordial (D) and primary (E) follicles isolated using Liberase TH. A continuous layer of cuboid follicular cells surrounds the oocyte in the primary follicle (E) while in the primordial follicle a flattened layer of cells covers the oocyte; immunoconfocal images recorded from three selected areas of centrifuged ovarian digest: follicles were stained for actin (F), MS2Y (G) and with DAPI (H), merged images (I).

5. Primordial and primary follicle culture

The clinical application of IVFG is still at the investigational stage, in a laboratory setting, although it stands a robust approach to study the basic biology of the ovary or the follicle under a controlled yet adjustable environment. Multiple culture systems have been developed to support the development of isolated preantral follicles [69,72,83], each one with its own advantages and providing useful insights into the follicle physiology. By this time, hydrogel-based follicle culture systems have been well characterized. The oocyte and the surrounding granulosa cells interact with each other and the environment, maintaining the same spatial location, connections, and dimensionality as in the intact ovary. The *in vitro* growth and development of mouse preantral follicles was successfully supported by alginate-based hydrogels, a substrate that was also applied to several large mammalian species, including dogs [84], rhesus monkeys [19], and humans [85], resulting in stage IV oocytes (human) [86], meiosis II (MII)-arrested eggs, and fertilized two-cell embryos (rhesus macaque) [87]. This developmental stage has not been reached in other systems.

It is commonly agreed that early follicular growth is largely independent of a gonadotropin stimulus; instead, it seems that it is controlled by paracrine and autocrine signals originating from several sources in the ovary, including stromal cells, macrophages, and other follicles [38]. Recent studies showed that these local factors may also play an important role in *in vitro* culture, supporting the growth of isolated preantral follicles: isolated primary ovarian follicles survived and grew when cocultured with purified ovarian stroma including theca-interstitial cells and macrophages [88] or with mouse embryonic fibroblasts (MEFs) as a feeder cell layer [89]. Coculture with MEFs resulted in an increased follicle survival, growth, and differentiation until antral follicles contained meiotically competent oocytes capable of reaching metaphase II in response to adequate hormone stimulation [89]. As suggested in those studies, individual primary follicles require factors beyond the standard culture media additives, including insulin, transferrin, selenium, fetuin, bovine serum albumin, and FSH [90], which can be supplied *in vitro* by coculture with stromal cells or MEFs; nevertheless, *in vivo* observations also suggest that follicles themselves may have a stimulatory effect on IVFG, because, in the mammalian ovarian cortex, the distinctive architecture and follicle distribution may influence follicle development: primordial and primary follicles are located close to the rigid, collagen-dense cortical stroma, whereas larger, growing follicles are typically closer to the interior medulla, which presents a less rigid stroma [91]. It has been shown, in a study examining the spatial relationship of follicles within ovaries, that follicles surrounded by growing follicles are more likely to be growing, suggesting the existence of an *in vivo* stimulatory effect of other follicles [92] that could be exerted by signals originating from both the oocyte and the growing follicles, which enhance the differentiation of preantral follicles.

6. Summary

In summary, the ability to sustain preantral follicle growth *in vitro* while supporting the acquisition of oocyte competence is of great scientific interest. This relies on supplying oocytes for assisted reproductive technologies and broadening our understanding of somatic cell/ oocyte interactions in species characterizing by prolonged follicular growth, such as humans and pigs. IVGF is becoming a useful tool to assess follicular development, offering also the potential to preserve reproductive options in cases of polycystic ovarian syndrome (PCOS), premature ovarian failure, or definitive sterility (post-oncotherapy). In addition, it is known that certain ovarian dysfunctions, such as PCOS and gonadotropin poor responsiveness, are consequences of deregulated follicle growth at this transitional stage. Therefore, the elucidation of molecular and cellular mechanisms involved in the control of follicular development during transition from preantral to early antral stage may provide an important insight into the pathophysiology and rational treatment of these disorders.

Acknowledgements

This work was supported by DEC-2013/09/B/NZ9/00226 from the National Polish Science Centre.

Author details

Duda Malgorzata[1*], Grzesiak Malgorzata[2], Knet-Seweryn Malgorzata[1] and Zbigniew Tabarowski[3]

*Address all correspondence to: duda.m@op.pl

1 Department of Endocrinology, Institute of Zoology, Jagiellonian University, Kraków, Poland

2 Department of Animal Physiology and Endocrinology, University of Agriculture, Kraków, Poland

3 Department of Experimental Hematology, Institute of Zoology, Jagiellonian University, Kraków, Poland

Conflict of interest

The authors declare that there is no conflict of interest that would prejudice the impartiality of this scientific work.

Author's contributions

Malgorzata Duda and Malgorzata Grzesiak prepared the paper, Malgorzata Knet-Seweryn collected articles for this chapter, and Tabarowski Zbigniew reviewed the paper. All authors read and approved the final manuscript version.

References

[1] McGee EA, Hsueh AJW. Initial and cyclic recruitment of ovarian follicles. Endocrine Reviews. 2000;21:200–214.

[2] Monniaux D, Clement F, Dalbies-Tran R, Estienne A, Fabre S, Mansanet C, Monget P. The ovarian reserve of primordial follicles and the dynamic reserve of antral growing follicles: what is the link? Biology of Reproduction. 2014;90:1–11.

[3] Gougeon A. Regulation of ovarian follicular development in primates: facts and hypotheses. Endocrine Reviews. 1996;17:121–155.

[4] Matzuk MM, Burns KH, Viveiros MM, Eppig JJ. Intercellular communication in the mammalian ovary: oocytes carry the conversation. Science. 2002;296:2178–2180. DOI: 10.1126/science.1071965

[5] Fortune JE, Ribera GM, Yang MY. Follicular development: the role of follicular microenvironment in selection of dominant follicle. Animal Reproduction Science. 2004;82:109–126.

[6] Drummond AE. The role of steroids in follicular growth. Reproductive Biology and Endocrinology. 2006;4:16.

[7] Huynh K, Jones G, Thouas G, Britt KL, Simpson ER, Jones MEE. Estrogen is not directly required for oocyte developmental competence. Biology of Reproduction. 2004;70:1263–1269.

[8] Stouffer RL. Progesterone as a mediator of gonadotropin action in the corpus luteum: beyond steroidogenesis. Human Reproduction Update. 2003;9:99–117.

[9] Hild-Petito S, West NB, Brenner RM, Stouffer RL. Localization of androgen receptor in the follicle and corpus luteum of the primate ovary during the menstrual cycle. Biology of Reproduction. 1991;44:561–568.

[10] Weil SJ, Vendola K, Zhou J, Adesanya OO, Wang J, Okafor J, Bondy CA. Androgen receptor gene expression in the primate ovary: cellular localization, regulation, and functional correlations. Journal of Clinical Endocrinology and Metabolism. 1998;83:2479–2485.

[11] Vendola K, Zhou J, Wang J, Bondy CA. Androgens promote insulin-like growth factor-I and insulin-like growth factor-I receptor gene expression in the primate ovary. Human Reproduction. 1999;14:2328–2332.

[12] Vendola K, Zhou J, Wang J, Famuyiwa OA, Bievre M, Bondy CA. Androgens promote oocyte insulin-like growth factor I expression and initiation of follicle development in the primate ovary. Biology of Reproduction. 1999;61:353–357.

[13] Lenie S, Smitz J. Functional AR signaling is evident in an *in vitro* mouse follicle culture bioassay that encompasses most stages of folliculogenesis. Biology of Reproduction. 2009;80:685–695.

[14] Skinner MK. Regulation of primordial follicle assembly and development. Human Reproduction Update. 2005;11:461–471.

[15] Tingen CM, Bristol-Gould SK, Kiesewetter SE, Wellington JT, Shea L, Woodruff TK. Prepubertal primordial follicle loss in mice is not due to classical apoptotic pathways. Biology of Reproduction. 2009;81:16–25.

[16] Vanderstichele H, Delaey B, de Winter J, de Jong F, Rombauts L, Verhoeven G, Dello C, van de Voorde A, Briers T. Secretion of steroids, growth factors and cytokines by immortalized mouse granulosa cells lines. Biology of Reproduction. 1994;50:1190–1202.

[17] Picton HM, Muruvi W, Jin P. Interaction of oocyte and somatic cells. In: Tan SL, Chian RC, Buckett WM, editors. In-Vitro Maturation of Human Oocytes—Basic Science to Clinical Application. Oxon, UK: Informa Health; 2007. p. 37–48.

[18] Jeruss JS, Woodruff TK. Preservation of fertility in patients with cancer. New England Journal of Medicine. 2009;360:902–911.

[19] Xu M, West-Farrell ER, Stouffer RL, Shea LD, Woodruff TK, Zelinski MB. Encapsulated three-dimensional culture supports development of nonhuman primate secondary follicles. Biology of Reproduction. 2009;81:587–594.

[20] Schmidt KT, Larsen EC, Andersen CY, Andersen AN. Risk of ovarian failure and fertility preserving methods in girls and adolescents with a malignant disease. BJOG: An International Journal of Obstetrics and Gynaecology. 2010;117:163–174.

[21] McNatty KP, Fidler AE, Juengel JL, Quirke LD, Smith PR, Heath DA, Lundy T, O'Connell A, Tisdall DJ. Growth and paracrine factors regulating follicular formation and cellular function. Molecular and Cellular Endocrinology. 2000;163:11–20.

[22] Hyttel P, Sinowatz F, Vejlsted M, Betteridge K. Essentials of Domestic Animal Embryology. 1st ed. Edinburgh: Elsevier; 2010.

[23] Picton HM. Activation of follicle development: the primordial follicle. Theriogenology. 2001;55:1193–1210.

[24] Young JM, McNeilly AS. Theca: the forgotten cell of the ovarian follicle. Reproduction. 2010;140:489–504.

[25] Binelli M, Murphy BD. Coordinated regulation of follicle development by germ and somatic cells. Reproduction, Fertility and Development. 2010;22:1–12.

[26] McLaughlin EA, McIver SC. Awakening the oocyte: controlling primordial follicle development. Reproduction. 2009;137:1–11.

[27] van den Hurk R, Zhao J. Formation of mammalian oocytes and their growth, differentiation and maturation within ovarian follicles. Theriogenology. 2005;63:1717–1751.

[28] Pepling ME. Follicular assembly: mechanisms of action. Reproduction. 2012;143:139–149.

[29] Fulton N, Martins da Silva SJ, Bayne RA, Anderson RA. Germ cell proliferation and apoptosis in the developing human ovary. The Journal of Clinical Endocrinology & Metabolism. 2005;90:4664–4670.

[30] Rodrigues P, Limback D, McGinnis LK, Plancha CE, Albertini DF. Multiple mechanisms of germ cell loss in the perinatal mouse ovary. Reproduction. 2009;137:709–720.

[31] Kuma A, Hatano M, Matsui M, Yamamoto A, Nakaya H, Yoshimori T, Ohsumi Y, Tokuhisa T, Mizushima M. The role of autophagy during the early neonatal starvation period. Nature. 2004;432:1032–1036.

[32] Kerr JB, Myers M, Anderson RA. The dynamics of the primordial follicle reserve. Reproduction. 2013;146:205–215.

[33] Escobar ML, Echeverría OM, Ortíz R, Vázquez-Nin GH. Combined apoptosis and

autophagy, the process that eliminates the oocytes of atretic follicles in immature rats. Apoptosis. 2008;13:1253–1266.

[34] Picton HM, Harris SE, Muruvi W, Chambers EL. The *in vitro* growth and maturation of follicles. Reproduction. 2008;136:703–15.

[35] Da Silva-Buttkus P, Jayasooriya GS, Mora JM, Mobberley M, Ryder TA, Baithun M, Stark J, Frank S, Hardy K. Effect of cell shape and packing density on granulosa cell proliferation and formation of multiple layers during early follicle development in the ovary. Journal of Cell Science. 2008;121:3890–3900.

[36] Lintern-Moore S, Moore GP. The initiation of follicle and oocyte growth in the mouse ovary. Biology of Reproduction. 1979;20:773–778.

[37] Hirshfield AN. Theca cells may be present at the outset of follicular growth. Biology of Reproduction. 1991;44:1157–1162.

[38] Peters H, Byskov AG, Himelstein-Braw R, Faber M. Follicular growth: the basic event in the mouse and human ovary. Journal of Reproduction and Fertility. 1975;45:559–566.

[39] Knox RV. Recruitment and selection of ovarian follicle for determination of ovulation rate in pig. Domestic Animal Endocrinology. 2005;29:385–397.

[40] McGrath SA, Esquela AF, Lee SJ. Oocyte-specific expression of growth differentiation factor-9. Molecular Endocrinology. 1995;9:131–136.

[41] Hsueh AJ, Kawamura K, Cheng Y, Fauser BC. Intraovarian control of early folliculogenesis. Endocrine Reviews. 2015;36:1–24.

[42] Xu B, Hua J, Zhang Y, Jiang X, Zhang H, Ma T, Zheng W, Sun R, Shen W, Sha J, Cooke HJ, Shin Q. Proliferating cell nuclear antigen (PCNA) regulates primordial follicle assembly by promoting apoptosis of oocytes in fetal and neonatal mouse ovaries. PLoS One. 2011;6:e16046.

[43] Qu J, Godin PA, Nisolle M, Donnez J. Expression of receptors for insulin-like growth factor-I and transforming growth factor-beta in human follicles. Molecular Human Reproduction. 2000;6:137–145.

[44] Visser JA, de Jong FH, Laven JSE, Themmen APN. Anti-Müllerian hormone: a new marker for ovarian function. Reproduction. 2006;131:1–9.

[45] Hillier SG, Tetsuka M, Fraser HM. Location and developmental regulation of androgen receptor in primate ovary. Human Reproduction. 1997;12:107–111.

[46] Szoltys M, Slomczynska M. Changes in distribution of androgen receptor during maturation of rat ovarian follicles. Experimental and Clinical Endocrinology & Diabetes. 2000;108:228–234.

[47] Slomczynska M, Tabarowski Z. Localization of androgen receptor and cytochrome P450 aromatase in the follicle and corpus luteum of the porcine ovary. Animal Reproduction Science. 2001;65:127–134.

[48] Juengel JL, Heath DA, Quirke LD, McNatty KP. Oestrogen receptor alpha and beta, androgen receptor and progesterone receptor mRNA and protein localization within the developing ovary and in small growing follicles of sheep. Reproduction. 2006;131:81–92.

[49] Cardenas H, Herrick JR, Pope WF. Increased ovulation rate in gilts treated with dihydrotestosterone. Reproduction. 2002;123:527–533.

[50] Hampton JH, Manikkam M, Lubahn DB, Smith MF, Garveric HA. Androgen receptor mRNA expression in the bovine ovary. Domestic Animal Endocrinology. 2004;27:81–88.

[51] Rice S, Ojha K, Whitehead S, Mason H. Stage-specific expression of androgen receptor, follicle stimulating hormone receptor, and anti-Mullerian hormone type II receptor in single, isolated human preantral follicles: relevance to polycystic ovaries. The Journal of Clinical Endocrinology and Metabolism. 2007;92:1034–1040.

[52] Abbott DH, Dumesic DA, Levine JE, Dunaif A, Padmanabhan V. Animal models and fetal programming of PCOS. In: Azziz R, Nestler JE, Dewailly D, editors. Contemporary Endocrinology: Androgen Excess Disorders in Women: Polycystic Ovary Syndrome and Other Disorders. 2nd ed. Humana Press Incorporation, Totowa, NY; 2006. p. 259–272.

[53] Smith P, Stickler TL, Veiga-Lopez A, Padmandabhan V. Developmental programming: differential effects of prenatal testosterone and dihydrotestosterone on follicular recruitment, depletion of follicular reserve and ovarian morphology in sheep. Biology of Reproduction. 2009;80:726–736.

[54] Magamage MPS, Zengyo M, Moniruzzaman M, Miyano T. Testosterone induces activation of porcine primordial follicles *in vitro*. Reproductive Medicine and Biology. 2011;10:21–30.

[55] Eppig JJ, O'Brien MJ. Development *in vitro* of mouse oocytes from primordial follicles. Biology of Reproduction. 1996;54:197–207.

[56] Silva JR, van den Hurk R, de Matos MH, dos Santos RR, Pessoa C, de Moraes MO, de Figueiredo. Influences of FSH and EGF on primordial follicles during *in vitro* culture of caprine ovarian cortical tissue. Theriogenology. 2004;61:1691–1704.

[57] Wandji SA, Srsen V, Voss AK, Eppig JJ, Fortune JE. Initiation *in vitro* of growth of bovine primordial follicles. Biology of Reproduction. 1996;55:942–948.

[58] Fortune JE, Kito S, Wandji SA, Srsen V. Activation of bovine and baboon primordial follicles *in vitro*. Theriogenology. 1998;49:441–449.

[59] Hovatta O, Silye R, Abir R, Krausz T, Winston RM. Extracellular matrix improves survival of both stored and fresh human primordial and primary ovarian follicles in long-term culture. Human Reproduction. 1997;12:1032–1036.

[60] Xu M, Barrett SL, West-Farrell ER, Kondapalli LA, Kiesewetter SE, Shea LD, Woodruff TK. *In vitro* grown human ovarian follicles from cancer patients support oocyte growth. Human Reproduction. 2009;24:2531–2540.

[61] Abir R, Fisch B, Nitke S, Okon E, Raz A, Ben Rafael Z. Morphological study of fully and partially isolated early human follicles. Fertility and Sterility. 2001;75:141–146.

[62] Hovatta O, Wright C, Krausz T, Hardy K, Winston RM. Human primordial, primary and secondary ovarian follicles in long-term culture: effect of partial isolation. Human Reproduction. 1999;14:2519–2524.

[63] O'Brien MJ, Pendola JK, Eppig JJ. A revised protocol for *in vitro* development of mouse oocytes from primordial follicles dramatically improves their developmental competence. Biology of Reproduction. 2003;68:1682–1686.

[64] Abir R, Roizman P, Fisch B, Nitke S, Okon E, Orvieto R, Ben Rafael Z. Pilot study of isolated early human follicles cultured in collagen gels for 24 hours. Human Reproduction. 1999;14:1299–1301.

[65] Figueiredo JR, Hulshof SC, Van den Hurk R, Ectors FJ, Fontes RS, Nusgens B, Bevers MM, Beckers JF. Development of a new mechanical method for the isolation of intact preantral follicles from fetal, calf and adult bovine ovaries. Theriogenology. 1993;40:789–799.

[66] Carambula SF, Goncalves PB, Costa LF, Figueiredo JR, Wheeler MB, Neves JP, Mondadori RG. Effect of fetal age and method of recovery on isolation of preantral follicles from bovine ovaries. Theriogenology.1999;52:563–571.

[67] Cortvrindt R, Smitz J, Van Steirteghem AC. *In vitro* maturation, fertilization and embryo development of immature oocytes from early preantral follicles from prepubertal mice in a simplified culture system. Human Reproduction. 1996;11:2656–2666.

[68] Zhao J, Dorland M, Taverne MA, Van Der Weijden GC, Bevers MM, Van Den Hurk R. *In vitro* culture of rat pre-antral follicles with emphasis on follicular interactions. Molecular Reproduction and Development. 2000;55:65–74.

[69] Wu J, Emery BR, Carrell DT. *In vitro* growth, maturation, fertilization, and embryonic development of oocytes from porcine preantral follicles. Biology of Reproduction. 2001;64:375–381.

[70] Gutierrez CG, Ralph JH, Telfer EE, Wilmut I, Webb R. Growth and antrum formation of bovine preantral follicles in long-term culture *in vitro*. Biology of Reproduction. 2000;62:1322–1328.

[71] Jewgenow K. Role of media, protein and energy supplements on maintenance of

morphology and DNA-synthesis of small preantral domestic cat follicles during short-term culture. Theriogenology. 1998;49:1567–1577.

[72] Telfer EE, Binnie JP, McCaffery FH, Campbell BK. *In vitro* development of oocytes from porcine and bovine primary follicles. Molecular and Cellular Endocrinology. 2000;163:117–123.

[73] Newton H, Picton HM, Gosden RG. *In vitro* growth of oocyte-granulosa cell complexes isolated from cryopreserved ovine tissue. Journal of Reproduction and Fertility. 1999;115:141–150.

[74] Nayudu PL, Fehrenbach A, Kiesel P, Vitt UA, Pancharatna K, Osborn SM. Progress towards understanding follicle development *in vitro*: appearances are not deceiving. Archives of Medical Research. 2001;32:587–594.

[75] Shuttleworth G, Broughton Pipkin F, Hunter MG. *In vitro* development of pig preantral follicles cultured in a serum-free medium and the effect of angiotensin II. Reproduction. 2002;123:807–818.

[76] Oktay K, Nugent D, Newton H, Salha O, Chatterjee P, Gosden RG. Isolation and characterization of primordial follicles from fresh and cryopreserved human ovarian tissue. Fertility and Sterility. 1997;67:481–486.

[77] Dolmans MM, Michaux N, Camboni A, Martinez-Madrid B, Van Langendonckt A, Nottola SA, Donnez J. Evaluation of Liberase, a purified enzyme blend, for the isolation of human primordial and primary ovarian follicles. Human Reproduction. 2006;21:413–420.

[78] Rice S, Ojha K, Mason H. Human ovarian biopsies as a source of pre-antral follicles. Human Reproduction. 2007;23:600–605.

[79] Kristensen SG, Rasmussen A, Byskov AG, Andersen CY. Isolation of pre-antral follicles from human ovarian medulla tissue. Human Reproduction. 2011;26:157–166.

[80] Duda M, Grzesiak M, Tabarowski Z, Tomanek M. Isolation of primordial and primary follicles from porcine ovarian medulla tissue. In: Reproduction Abstracts Vol. 1, World Congress of Reproductive Biology 2014; 2–4 September 2014, Edinburgh, UK. p. 142. Available from: http://www.wcrb2014.org/Portals/0/Conferences/Abstracts/WCRB_2014/WCRB_2014P142.pdf.

[81] Romero S, Smitz J. Exposing cultured mouse ovarian follicles under increased gonadotropin tonus to aromatizable androgens influences the steroid balance and reduces oocyte meiotic capacity. Endocrine. 2010;38:243–253.

[82] Hornick JE, Duncan FE, Shea LD, Woodruff TK. Isolated primate primordial follicles require a rigid physical environment to survive and grow *in vitro*. Human Reproduction. 2012;27:1801–1810.

[83] Abir R, Nitke S, Ben-Haroush A, Fisch B. *In vitro* maturation of human primordial ovarian follicles: clinical significance, progress in mammals, and methods for growth evaluation. Histology and Histopathology. 2006;21:887–898.

[84] Songsasen N, Woodruff TK, Wildt DE. *In vitro* growth and steroidogenesis of dog follicles are influenced by the physical and hormonal microenvironment. Reproduction. 2011;142:113–122.

[85] Smitz J, Dolmans MM, Donnez J, Fortune JE, Hovatta O, Jewgenow K, Picton HM, Plancha C, Shea LD, Stouffer RL, Telfer EE, Woodruff TK, Zelinski MB. Current achievements and future research directions in ovarian tissue culture, *in vitro* follicle development and transplantation: implications for fertility preservation. Human Reproduction Update. 2010;16:395–414.

[86] Kreeger PK, Deck JW, Woodruff TK, Shea LD. The *in vitro* regulation of ovarian follicle development using alginate-extracellular matrix gels. Biomaterials. 2006;27:714–723.

[87] Ting AY, Yeoman RR, Lawson MS, Zelinski MB. *In vitro* development of secondary follicles from cryopreserved rhesus macaque ovarian tissue after slow-rate freeze or vitrification. Human Reproduction. 2011;26:2461–2472.

[88] Tingen CM, Kiesewetter SE, Jozefik J, Thomas C, Tagler D, Shea L, Woodruff TK. A macrophage and theca cell-enriched stromal cell population influences growth and survival of immature murine follicles *in vitro*. Reproduction. 2011;141:809–820.

[89] Tagler D, Tu T, Smith RM, Anderson NR, Tingen CM, Woodruff TK, Shea LD. Embryonic fibroblasts enable the culture of primary ovarian follicles within alginate hydrogels. Tissue Engineering. Part A. 2012;18:1229–1238.

[90] Xu M, Kreeger PK, Shea LD, Woodruff TK. Tissue-engineered follicles produce live, fertile offspring. Tissue Engineering. 2006;12:2739–2746.

[91] Hornick, JFE, Duncan L, Shea LD, Woodruff TK. Multiple follicle culture supports primary follicle growth through paracrine-acting signals. Reproduction. 2013;145:19–32.

[92] Da Silva-Buttkus P, Marcelli G, Franks S, Stark J, Hardy K. Inferring biological mechanisms from spatial analysis: prediction of a local inhibitor in the ovary. Proceedings of the National Academy of the United States of America. 2009;106:456–461.

10

Sperm Motility Regulatory Proteins: A Tool to Enhance Sperm Quality

Sandhya R. Dungdung, Arpita Bhoumik, Sudipta Saha, Prasanta Ghosh, Kaushik Das, Sandipan Mukherjee, Debjani Nath, Jitamanyu Chakrabarty, Chanakyanath Kundu, Bijay Shankar Jaiswal, Mahitosh Mandal, Arunima Maiti, Saswati Banerjee, Madhumita Roychowdhury, Debleena Ray, Debdas Bhattacharyya and Gopal C. Majumder

Abstract

Sperm forward motility is an essential parameter in mammalian fertilization. Studies from our laboratory have identified and characterized a few unique sperm motility regulatory proteins/glycoproteins from the male reproductive fluids and mammalian blood serum. The purified sperm motility-initiating protein (MIP) from caprine epididymal plasma as well as the forward motility-stimulating factor (FMSF) and motility-stimulating protein (MSP) from buffalo and goat serum, respectively, have high efficacy to initiate or increase motility in nonmotile or less motile sperm. Antibody of sperm motility inhibitory factor (MIF-II) has the high potential to enhance sperm vertical velocity and forward motility by increasing intracellular cyclic adenosine monophosphate (cAMP) level. The appearance and disappearance of D-galactose–specific lectin and its receptor along the epididymis has been reported to be involved in motility regulation in spermatozoa. A novel synthetic cryopreservation method and role of lipid to protect membrane damage during cryopreservation have been demonstrated. Motility-promoting proteins may be extremely useful for improving cattle breeding and breeding of endangered species, thereby helping in enhanced production of animal products as well as in the conservation of animals. Isolated proteins and developed cryopreservation technology may also be beneficial in human infertility clinics to increase the chance of fertilization.

Keywords: Spermatozoa, Epididymis, Motility regulatory proteins, Cryopreservation, Reproduction

1. Introduction

Livestock is a very important subsector of Indian agricultural production system. The overall contribution of the livestock is almost 4.11%. India ranks first in milk production in the world (132.4 million tons), which is mostly contributed by cattle and buffalo. The bull concentrates in itself a high economic value and thus need to be maintained on proper nutrition and management to obtain optimum performance in terms of semen production [1]. The demand for the best males has increased considerably due to a shortage in the number of proven bulls having better semen characteristics for sustaining a successful breeding program [2].

Reproductive techniques facilitate the breeding of farm animals, which ultimately improve milk production and growth in dairy industry. Artificial insemination (AI) is a first-generation reproductive biotechnology that profoundly contributed to genetic improvement, particularly in dairy cattle. Such impact would not have been possible without successfully freezing bull semen. Quality control of frozen sperm is of utmost importance for the sperm to be used in AI [3]. Sperm cryopreservation allows prolonged preservation of semen and a wider use of a male gamete [4]. This technique is used for breeding of domestic animals, maintaining the genetic diversity, and establishing gene banks [5]. The success of fertilization with the use of frozen–thawed spermatozoa varies considerably between species and among individuals of the same species [6]. Semen cryopreservation relies on the use of cryoprotectants (such as glycerol and egg yolk), substances to maintain the osmolarity, sources of energy (such as glucose or fructose), and enzymes and antibiotics, which are essential for maintaining the viability of the spermatozoa during cooling, freezing, and thawing [7]. Still, the major disadvantage in the cryopreservation process is its harmfulness to spermatozoa; even using the best preservation techniques, only half of the sperm population survives after freezing and thawing procedures [8].

Mammalian sperm motility is a key factor to determine semen quality and fertilizing capability [9]. Motility of spermatozoa depends on various proteins in blood and male reproductive fluids [10, 11]. For the last few decades, studies in our laboratory focused on the identification and characterization of sperm motility regulatory protein (SMRP) molecules in caprine (*Capra indicus*) sperm [12]. Efforts have been made to raise antibodies against these purified proteins for further investigation of their immunological and functional role expecting to augment motility/fertility of male gametes. Our findings also showed the role of lipid to protect the spermatozoon membrane against damage during cryopreservation. A novel cryopreservation system has also been developed, which may help to improve the existing technology for preservation of spermatozoa across higher mammalian species such as bull and buffalo, as well as in endangered species.

A positive correlation between sperm concentration at semen collection and motility has been reported. In general, fertility rate of frozen thawed semen used for AI is poorer than that obtained with fresh semen, which is partially compensated by using insemination doses with greater numbers of live spermatozoa [13]. Several SMRPs have been isolated, which showed the ability to enhance the motility in cryopreserved semen samples. Based on the hypothesis that these motility enhancer proteins may not be species specific in mammals, it is suggested that these proteins could perform their biological functions in spermatozoa of various species. Therefore, they appear as an alternative to improve sperm fertility after cryopreservation, by incubating sperm with motility enhancing purified proteins.

Sperm motility is a major criterion in semen quality; it is also an important determinant for the success rate of fertilization. Sperm motility is related to the availability of Adenosine triphosphate (ATP): the frequency and amplitude of the tail movement of sperm is closely related to the dephosphorylation of ATP [14].

The epididymal lumen is a complex microenvironment in which synthetically inactive spermatozoa move from the proximal to the distal end and interact with proteins that are synthesized and secreted in a highly regionalized manner in the epididymis get gradually mature and acquire the capacity of progressive motility and fertility [15, 16]. In addition to the integration of epididymal secretory proteins, posttranslational modifications of existing sperm proteins are important for sperm maturation and acquisition of fertilizing potential [17].

2. Sperm Motility Regulatory Proteins (SMRPs)

The epididymis is essential for sperm development and maturation. Sperm taken from the caput show little motility and are infertile, whereas sperm from the cauda are motile and can achieve fertilization. How sperm motility is initiated in the epididymal duct is uncertain, but a sperm motility protein has been demonstrated in both the corpus and the cauda of the epididymis.

Besides having an influence through controlling the extracellular milieu of spermatozoa, proteins secreted by epididymal cells may be incorporated at the sperm surface to exert their regulation. Epididymal secretions suggested to functions as cholesterol carrying proteins that have been identified in the ram [18] and other farm animals [19].

Specific secretory proteins produced in the epididymis seem to remain associated with spermatozoa during the transit through the organ and may play a key role in the mammalian sperm maturation process by conferring to the male gamete the ability to recognize the oocyte [17–20].

2.1. Molecules with sperm motility regulatory properties

Several molecules showing motility regulatory effects on spermatozoa have been identified. Some of them were located in epididymis, but there are several reports also on the occurrence of various types of sperm motility regulatory protein factors in male reproductive fluids and blood serum as well as in follicular fluid [21, 22]. The sperm surface undergoes a sequence of alterations during the epididymal maturation and uterine capacitation process, allowing the male gametes to acquire the potential to fertilize the female egg [23, 24]. Available information on SMRPs has been reviewed below.

Motility of washed rabbit spermatozoa derived from fresh ejaculates was also greatly stabilized by blood serum albumin; rabbit serum albumin was more effective than bovine or human serum albumin [19], suggesting the existence of a species-specific effect.

Of the large number of proteins present in follicular fluid (FF), few have been discovered and most are yet to be nominated. A 52-kDa glycoprotein purified from porcine FF stimulates the motility of boar spermatozoa [25]. We have partially purified a sperm motility enhancer from human FF, which significantly enhances the forward motility and vertical velocity of human and caprine spermatozoa.

Moreover, its importance in the acrosomal reaction and the interaction with specific receptors in oocyte surface at fertilization is also acknowledged [26]. Phosphorylation of proteins catalyzed by protein kinases is recognized as a major regulator of cell functions [27]. It has been demonstrated that it is a cAMP-independent protein kinase (ecto-CIK) and its substrate (MPS) on the external surface of goat epididymal spermatozoa that causes phosphorylation of endogenous membrane-bound phosphoproteins that are externally oriented [28–30]. The presence of lectins (e.g. galactose-specific and N-acetyl-D-galactosamine–specific lectins) on the surface of mature spermatozoa has been reported in few species [31, 32]. We have demonstrated for the first time the epididymal maturational profile of a sperm external surface lectin-like molecule.

2.1.1. Motility initiating protein and motility inhibitory factor

Hoskins and associates [18, 32], in a novel *in vitro* model, triggered forward motility in the immature (immotile) sperm derived from bovine caput epididymis by incubation in presence of epididymal or seminal plasma and theophylline. Subsequently, from caprine epididymal plasma, pursuing to identify the extracellular proteins and their role in biochemical regulation of sperm motility, Jaiswal et al. [33] and Das et al. [34] have purified and characterized a motility initiating protein (MIP) and a motility inhibitory factor (MIF), respectively.

Purification and characterization of the MIP from caprine epididymal plasma was achieved in our laboratory. MIP is a heat-stable, acidic, dimeric protein with a weight of close to 125 kDa, and presenting two subunits: 70 and 54 kDa, with an isoelectric point of 4.75 and maximal activity at pH 8 [33]. It contains mannose, galactose, and N-acetylglucosamine approximately in the ratios of 6:1:6. MIP loses activity by actions of alpha-mannosidase and beta-N-acetylglucosaminidase, thereby showing the sugar side chains responsible for the motility initiating potential. Epididymal plasma is the richest source of MIP. Intrasperm cAMP level was increased by MIP [33].

Immature caput spermatozoa do not show forward motility. Addition of MIP induced forward motility to a significant population of spermatozoa (Figure 1). The number of forward motile cells increased markedly with the increase in MIP concentration. The factor showed maximal activity at concentration as low as 30 μg/ml when it induces forward motility in nearly 22% of the immature spermatozoa.

We have also isolated and purified a different 57-kDa protein from caprine cauda epididymal plasma, which acts as a quiescence factor to immobilize cauda spermatozoa – the sperm motility quiescence factor (QF). This protein also decreases reactive oxygen species (ROS) concentration and thus helps to reduce oxidative stress in cauda spermatozoa, which is prone to damage by ROS due to the presence of high level of polyunsaturated fatty acids (PUFA) in sperm plasma membrane (unpublished data).

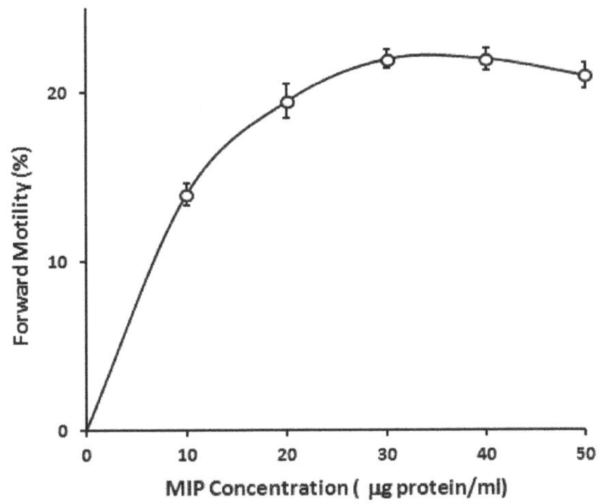

Figure 1. Dose course of MIP for initiation of forward motility in caput-sperm under the standard assay conditions. Reproduced with permission from Jaiswal et. al. (2010).

A rabbit polyclonal antibody raised against purified sperm motility inhibiting factor (MIF-II), at dilution 1:5000, allowed to increase sperm motility by 75% compared to the control within 30 min of incubation (Figure 2a). SPERMA, a sperm motility analyzer, showed a 40% increase in vertical velocity of MIF-II antibody-treated spermatozoa as compared to the control serum (Figure 2b). MIF-II antibody also enhanced the motility of immature caput spermatozoa under *in vitro* initiation media.

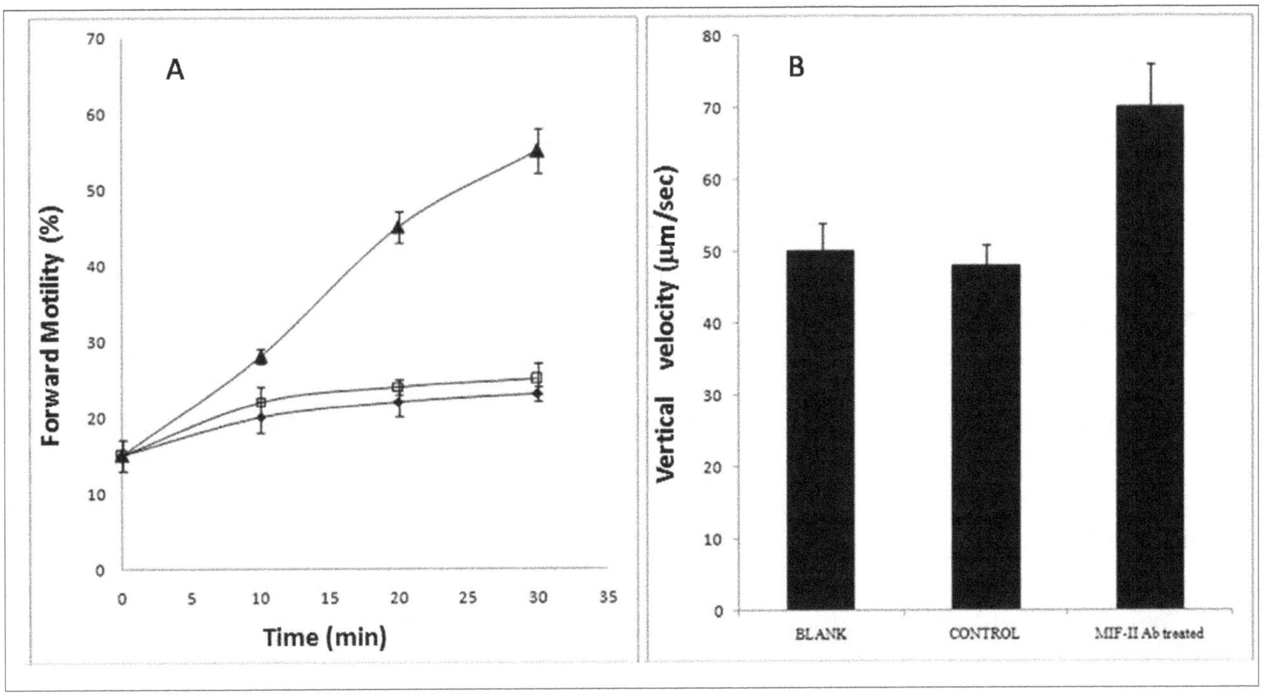

Figure 2. Effect of MIF-II antibody (1:5000) on goat cauda sperm forward motility. a) Microscopic analysis: Blank (untreated) (◆), Control with pre immune sera (□), MIF-II Ab (▲). b) Analysis by SPERMA. Reproduced with permission from Das et. al. (2010).

2.1.2. Forward motility-stimulating factor and motility-stimulating protein

A forward motility-stimulating factor (FMSF) purified to apparent homogeneity from buffalo blood serum, showed high protein specificity and affinity for stimulating forward motility of goat cauda epididymal spermatozoa [35]. A molecule exhibiting similar role was found in goat blood serum and some of its physical, biochemical, physiological, and immunological properties were characterized. This protein was named as sperm forward motility-stimulating protein (MSP) because it stimulated forward motility even in weakly motile spermatozoa [36].

Both FMSF and MSP are 66-kDa monomeric, heat-stable proteins. FMSF is acidic in nature with isoelectric point 3.7. Aspartate, glutamate, and leucine are the amino acids with higher representation in FMSF. FMSF is inhibited when treated with α-mannosidase, which acts on the sugar part of the protein.

In regard to its motility-promoting potential, FMSF is not species specific. Sperm surface has specific receptors of FMSF [37]. FMSF was also immunodetected in uterine fluids of cattle species. A maturation-dependent expression of FMSF receptor and consequential stimulation of forward motility were observed [38]. FMSF binds to the surface of the mature sperm cells to promote forward motility. FMSF initiates a novel signaling cascade to stimulate transmembrane adenylyl cyclase (tmAC) activity that augments intracellular cAMP, which through downstream cross talk of phosphokinases leads to enhanced forward motility in mature spermatozoa [36]. In *in vitro* fertilization, maximum activity of FMSF was observed at 0.5 μM level when nearly 60–70% of spermatozoa showed forward motility [37].

On the other hand, motility-promoting efficacy of MSP is markedly higher than theophylline or bicarbonate or their combination at 0.9 μM and also showed longer motility maintenance. MSP exhibits maximal activation in 1 min or less, whereas theophylline or bicarbonate takes about 3–5 min for maximal activation (Figure 3a). Addition of theophylline and bicarbonate does not activate further MSP-induced motility. Figure 3b shows the maintenance of motility for longer period. Theophylline and bicarbonate have been found to help in the maintenance of motility in a much stabilized manner when added along with MSP. These reagents helped in the storage of sperm cells in motile condition for at least 3 h at room temperature [36].

Studies with CASA or SPERMA evidenced a similar trend of increase in horizontal and vertical velocities, respectively. MSP action was found to be independent of cAMP, in which it differs from FMSF. MSP occurrence was higher in testis, although blood was its richest source.

MSP was localized throughout the cell surface of spermatozoa. *In vitro* fertilization studies were performed with antibody of MSP using mouse oocyte and spermatozoa. The control experiment did not show any inhibition of fertilization, but the MSP antibody treatment inhibited fertilization by 100% at 1:25 dilution [36].

2.1.3. Follicular Fluid Motility Enhancing Protein (FFMP)

FF has a pivotal effect on motility and chemotaxis of spermatozoa for a successful fertilization. FF has been shown to stimulate sperm motility [39, 40], capacitation [41] and acrosome reaction [42], as well as the sperm–oocyte fusion [43]. There are a large number of proteins present in this fluid that are associated with the regulation of sperm function; many of them are yet to be

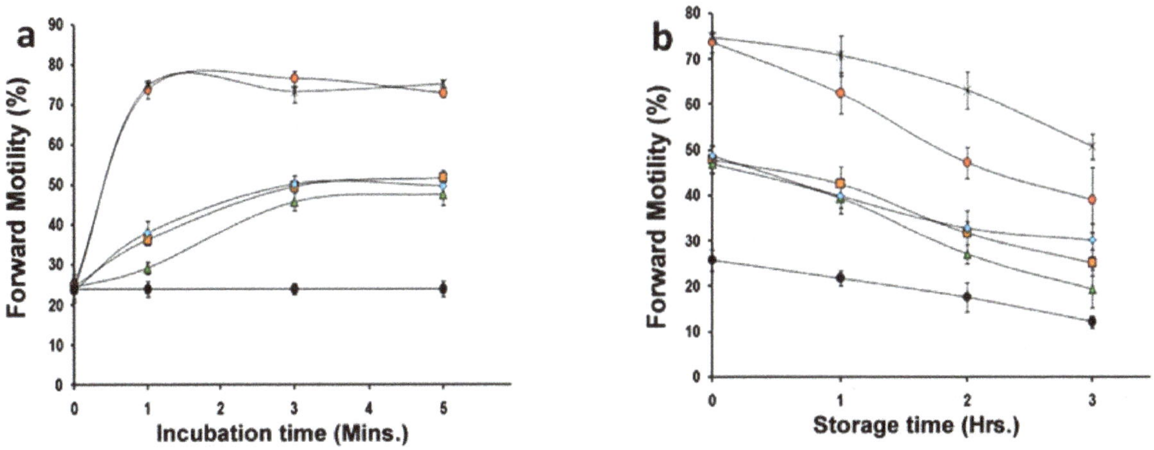

Figure 3. Effect of goat MSP, 0.9 M (-●-) (Red), theophylline 5 mM (-■-) (Orange), bicarbonate 20 mM (-▲-) (Green), theophylline + bicarbonate (-◆-) (Blue), and MSP + theophylline + bicarbonate (-X-) (Black) on sperm motility with respect to control (-●-) (Black). At different times up to 5 minutes MSP-induced motility was found to be more significant with respect to other initiators (panel a). At 30 seconds to the 1 minute time point the statistical significance level is $P<.001$. At storage up to 3 hours, even at the end of 2 hours, there are significant differences in motility in MSP versus theophylline ($P<.05$), MSP versus bicarbonate ($P<.01$), and MSP versus theophylline + bicarbonate ($P<.05$; panel b). Reproduced with permission from Saha et. al. (2013).

nominated or characterized. We have partially purified a sperm motility enhancer molecule from human FF. This protein stimulates sperm motility by more than 50% in a poor motile sample and also increases vertical velocity of human and caprine spermatozoa (unpublished data).

2.1.4. Lectin and lectin receptor

Maturing goat spermatozoa recovered at the distal corpus epididymis showed head-to-head autoagglutination when incubated *in vitro* in a modified Ringer's solution [44]. This is due to a lectin-like molecule located on the sperm surface that specifically interacts with its receptor of neighboring homologous cells. The D-galactose lectin and its receptor were partially purified from the plasma membrane of goat cauda and caput spermatozoa, respectively. The terminal stage of sperm maturation, that is, the induction of flagellar motility, is associated with a sharp disappearance or inactivation of the lectin receptor along with the appearance of the lectin [44], suggesting that the lectin-like molecule acquired by the mature sperm might induce sperm motility, whereas activation of its receptor suppresses the motility-mediating potential of the lectin.

2.2. Applied potential for the different SMRP

Sperm motility regulation is performed by the different SMRP by their respective mode of actions. The motility-promoting effect of all these proteins, including MIF antibody, could support its use to enhance motility in sperm samples with poor motility traits, which could be useful in infertility management. These are physiological activators of sperm motility and thus may be used in biomedical application in human infertility clinics, animal breeding centres, and animal conservation centres.

3. New synthetic model for sperm cryopreservation

Sperm cells are stored for long term by the cryopreservation method. Cattle breeding by AI technique requires ready supply of good quality sperm cells, which are only possible by cryopreservation. Sperm cells are susceptible to less viability and sublethal dysfunction by freezing and thawing events during cryopreservation. Cell deaths during cryopreservation are caused mainly due to membrane damage [45].

Cryopreservation of semen has long been foreseen as a tool to improve breeding of farm animals of economical importance, and it is recognized as a major asset for conservation of endangered species and for solving particular problems of human male infertility. During cryopreservation, a substantial portion of sperm cells undergo damage from thermal, mechanical, chemical, and osmotic stresses [46, 47].

We recently developed a simple sperm cryopreservation method for goat cauda epididymal sperm using a modified Ringer's solution [48, 49]. This cryopreservation model has been developed using a careful manipulation of cooling and freezing rates, with the help of a computer-controlled biofreezer. It is generally accepted that sperm cells were highly sensitive to cooling rates particularly during cooling from room temperature to 5°C and during freezing (5 to −20°C). In this model, the optimized protocol consisted of a drop of 0.25°C/min from ambient temperature to 5°C, a cooling rate of 5°C/min from 5 to −20°C, thereafter a decreased rate of 20°C/min to −100°C, and finally the plunging into liquid nitrogen (Table 1). The cryoprotective ability of several biomolecules, such as amino acids, carbohydrates, and polymers, was assessed. The best cryoprotection was offered by glycerol at the 0.87 M level. Glycerol, dimethyl sulfoxide (DMSO), and amino acids in combination allowed 52% of motility recovery in cryopreserved sperm samples (Figure 4). DMSO and amino acids when used in combination with glycerol increased the cryoprotectant ability of glycerol [50].

PROTOCOL	MOTILITY RECOVERY (%)	
	FORWARD	TOTAL
RT —(5°C/m)→ 5°C —(5°C/m)→ - 20°C —(20°C/m)→ -100°C	0	0
RT —(1°C/m)→ 5°C —(5°C/m)→ - 20°C —(20°C/m)→ -100°C	0	0
RT —(0.5°C/m)→ 5°C —(5°C/m)→ - 20°C —(20°C/m)→ -100°C	0	0
RT —(0.5°C/m)→ 5°C —(5°C/m)→ - 20°C —(20°C/m)→ -100°C	15 ± 1.2	23 ± 2
RT —(0.2°C/m)→ 5°C —(5°C/m)→ - 20°C —(20°C/m)→ -100°C	18 ± 1.5	35 ± 2.5
RT —(0.2°C/m)→ 5°C —(5°C/m)→ - 20°C —(20°C/m)→ -50°C	16 ± 2	34 ± 2
RT —(0.25°C/m)→ 5°C —(5°C/m)→ - 20°C —(20°C/m)→ -100°C	32 ± 3	35 ± 2

(Reproduced with permission from kundu et. al., 2000a)

Table 1. Effect of different cooling rate using 0.87 M glycerol as cryoprotectant

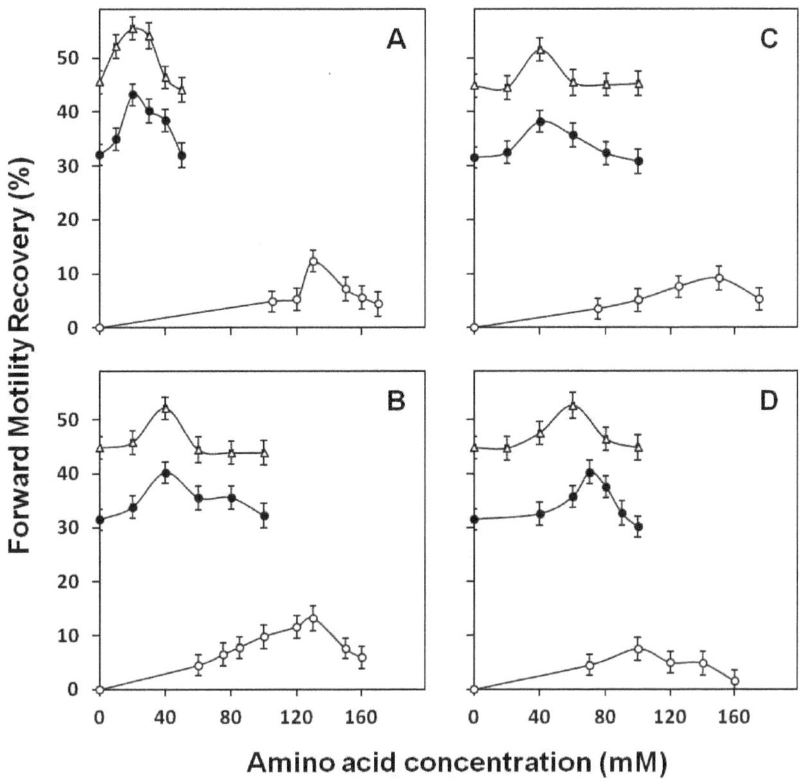

Figure 4. Effect of amino acids on forward motility recovery of spermatozoa after freezing and thawing of goat spermatozoa. (A, B, C and D). Results for L proline, L alanine, glycine and L glutamine, respectively. (○) Effect of amino acids in absence of other cryoprotectants. (●) FM recovery (%) in the presence of glycerol. (Δ) FM recovery (%) in the presence of glycerol + Me_2SO. The values shown are mean ± S.D. of five experiments. Reproduced with permission from Kundu et. al. (2001).

To improve the motility of sperm cells at recovery (after thawing), the combined actions of dextran, glycerol, and DMSO were tested. Addition of different concentrations of 10 kDa dextran (2.0–6.27 mM) to a fixed concentration of glycerol (0.87 M) caused a concentration-dependent increase of the recovery of motility (approximately 22–25%) of sperm cells. The highest recovery of motility (forward motility percent 50 ± 3% and total motility percent 53 ± 3%, respectively) was found at 6.27 mM dextran. Addition of another dextran (40 kDa; at doses between 0.5 and 2.0 mM) to the same concentration of glycerol (0.87 M) also caused a dose-dependent increase of motility recovery [51]. We suggested that glycerol and other penetrating compounds form H-bond with the polar phosphate head group of both sides (exoplasmic and protoplasmic) of the cell membrane forming a protective cushion that protects membrane from ice crystal-mediated damage [52]. The observed additive nature of cryoprotectant efficacies of glycerol and dextran, a polymer of carbohydrate, suggests that these two molecules may exert their cryoprotective effects through different mechanisms. We suggested that dextran cryoprotects the cells by inhibiting ice crystal lattice formation [51].

Because of the high cryoprotectant efficacy, these cryoprotectants may have potential for the cryopreservation of semen of various species, which may finally have a great potential in animal reproduction.

4. Role of lipid to protect membrane damage during cryopreservation

Membrane damage is one of the main reasons for reduced motility and fertility of sperm cells during cryopreservation. Lipid is an important constituent of cell membrane. Dysfunction or malfunction of lipid component may cause lethal damage to different types of cells including spermatozoa. The lipid composition of sperm membrane differs from species to species. Lipid content in the plasma membrane of goat cauda epididymal spermatozoa changes significantly during maturation [53]. Cryopreservation is not a natural phenomenon and sperm cells are not "programmed" for it. So, when sperm cells are subjected to cryopreservation (including both freezing and thawing), they suffer from stress. As a result, a huge population of sperm undergoes irreversible damage during the process. Membrane impairment has been identified as one of the manifestations of such damage.

To study the changes in sperm plasma membrane lipid composition associated to the cryopreservation process, a model system was developed using a synthetic medium, which was devoid of any lipid component [48]. It showed that the total lipid and its components of goat cauda epididymal sperm plasma membrane changed significantly after cryopreservation [54]. The composition in phospholipid (PL) and neutral lipid (NL) was severely altered. The unsaturated fatty acids decreased, whereas the saturated ones augmented; the cholesterol: phospholipid ratio was also increased causing profound enhancement of hydrophobicity of the sperm plasma membrane. Reports from our laboratory also showed that sperm cell membrane selectively sheds off the hydrophilic lipid molecules to remain viable against cryodamage (Table 2) [58]. Therefore, increasing the membrane hydrophobicity of the spermatozoon may enhance its resistance to cryodamage during freezing.

The role of fatty acids in cryopreservation of sperm cells of different species has been discussed elsewhere [55]. Egg yolk is used for a long time as a cryoprotectant [56] although in recent studies it was demonstrated the existence of detrimental effects of egg yolk extender over sperm cells [57, 58]. To circumvent these problems, phospholipids isolated from egg yolk [58] or from other commercial sources [59] were tried instead of the whole egg yolk. Soybean lecithin (phosphatidyl choline) has emerged as a popular choice for replacing egg yolk as cryoprotectant for goat [61, 52].

In a recent study, we have found that a lecithin from marine fish was also effective for cryopreservation of the goat spermatozoa (unpublished data). The loss of plasma membrane lipids was prevented to a great extent by adding this lecithin, prior to cryopreservation, to the sperm suspension in the synthetic media described above. While loss of total, neutral and phospholipids after cryopreservation was 29.5, 32.3, and 33.3%, respectively, addition of marine-origin lecithin reduced it to 7.9 and 19.6% for total lipid and phospholipid, whereas change in neutral lipid was insignificant. According to our observations, the cholesterol did not cryoprotect goat spermatozoa neither did the cholesterol: phospholipid ratio altered appreciably. However, it is possible that integration of exogenous PC into the cell membrane had enhanced its actual amount, suggesting that the integration of exogenous PC might trigger a remodeling of membrane structure to maintain homeostatic condition.

	Before Cryopreservation	After Cryopreservation	Sig. (2- tailed)
Total Lipid[a] (TL) (mg lipid /100 mg protein)	147.92 ± 1.83	115.31 ± 0.7	0.001
% w/w of total lipid			
Neutral Lipid (NL)	50.35 ± 0.57	53.82 ± 0.10	0.008
Glycolipid (GL)	13.28 ± 0.08	15.05 ± 0.16	0.005
Phospholipid (PL)	36.37 ± 0.42	31.13 ± 0.16	0.001
% w/w of total neutral lipid			
Hydrocarbons (HC)	27.94 ± 0.07	41.93 ± 0.45	<0.001
Sterols (ST)	44.55 ± 0.06	41.52 ± 0.16	<0.001
Steryl Esters (SE)	13.61 ± 0.05	6.06 ± 0.21	0.001
Wax Esters (WE)	2.89 ± 0.09	1.92 ± 0.06	<0.001
1-O-alkyl 2,3-diacyl glycerol (ADAG)	6.98 ± 0.08	4.54 ± 0.11	0.005
Triacyl glycerol (TG)	4.03 ± 0.06	4.03 ± 0.04	0.383
% w/w of total phospholipid			
Phosphatidyl choline (PC)	38.91 ± 0.10	36.85 ± 0.20	0.018
Phosphatidyl ethanolamine (PE)	26.95 ± 0.09	24.71 ± 0.03	0.021
Phosphatidyl inositol (PI)	15.81 ± 0.13	16.88 ± 0.05	0.014
Sphingomyelin (Sph)	18.33 ± 0.09	21.56 ± 0.05	0.004

(Reproduced with permission from Chakrabarty et. al., 2007)

Table 2. Lipid composition of sperm plasma membrane

5. Conclusion

Animal products have an important role in the global economy and any improvement in animal breeding will lead to production of more and better-quality animal products such as milk, butter, meat, wool, leather, etc. Motility promoting proteins such as MSP, FMSF, MIP, and FFMP have a direct role in motility enhancement of weak sperm cells. MIF antibody also enhances motility by neutralizing the activity of MIF, which is a potent motility inhibitor. Membrane proteins such as ecto-CIK and its substrate MPS regulate the sperm motility as well as increase the acrosomal reaction, thereby helping in fertilization. All these proteins have the potential for improving the breeding of animals in animal husbandry or poultry for better yield or for conservation purpose of almost extinct species in the world.

Motility regulatory proteins may be useful in human infertility clinics to solve some of the problems of human infertility. The purified motility regulatory proteins from our laboratory are expected to have potentially important applications because it will improve the quality of mammalian semen by activating sperm motility essential for fertility of the male gametes.

Cryopreservation has been widely used in the modern cattle industry for animal production and AI has been the most widely applied tool in facilitating the extensive utilization of frozen semen. The sperm cells of different species have different inherent resistance toward cryopreservation like high sensitivity to cooling rates particularly before and during freezing. The synthetic model developed by our laboratory has a great promise for better understanding the mechanism of cryoprotection. The development of an optimum cryoprotectant formulation of the combined action of glycerol, DMSO, Ficoll, and amino acids provided the high motility recovery.

Acknowledgements

We take this opportunity to thank the Department of Science and Technology, Indian Council of Medical Research, New Delhi, and the Council of Scientific and Industrial Research (CSIR), Government of India, New Delhi, for the financial support by Network Project (BSC0101) and fellowships provided to do these works. Emeritus fellowship offered to Gopal C. Majumder by the All India Council for Technical Education (AICTE), New Delhi, and the Support of Centre for Rural and Cryogenic Technologies (CRCT), Jadavpur University, Kolkata, are gratefully acknowledged. The authors are also grateful to the Director of Indian Institute of Chemical Biology (IICB), Kolkata, for his support and interest in our research work.

Author details

Sandhya R. Dungdung[1*], Arpita Bhoumik[1], Sudipta Saha[1], Prasanta Ghosh[1], Kaushik Das[1], Sandipan Mukherjee[1], Debjani Nath[2], Jitamanyu Chakrabarty[3], Chanakyanath Kundu[4], Bijay Shankar Jaiswal[5], Mahitosh Mandal[6], Arunima Maiti[1], Saswati Banerjee[1], Madhumita Roychowdhury[7], Debleena Ray[7], Debdas Bhattacharyya[8] and Gopal C. Majumder[8]

*Address all correspondence to: srdungdung@iicb.res.in; sandhya2308@hotmail.com

1 Cell biology and Physiology Division, CSIR-Indian Institute of Chemical Biology, Kolkata, India

2 Department of Zoology, University of Kalyani, Kalyani, West Bengal, India

3 National Institute of Technology, Durgapur, West Bengal, India

4 School of Biotechnology, KIIT University, Odissa, India

5 Department of Molecular Biology, Genentech Inc., CA, USA

6 Indian Institute of Technology, Kharagpur, India

7 A. H. IVF & Infertility Research Centre, Kolkata, India

8 Centre for Rural and Cryogenic Technologies, Jadavpur University, Kolkata, India

References

[1] Naha BC, Chakravarty AK, Mir MA, Jamuna V, Singh AP, Maher D. Identifying factors affecting age at first semen freezing and age at first semen use in Sahiwal bulls. Veterinary World. 2015; 87: 928–931.

[2] Rodríguez-Martínez H. Semen evaluation techniques and their relationship with fertility. Anim Reprod. 2013; 10: 148–159.

[3] Kumaresan A. Advancements in quality assessment of frozen semen and correlations between lab tests and field fertility data. In: Proceedings of International workshop on Bovine Semen production and processing – Compendium on 16-17 Oct, 2014, at National Dairy Development Board, Anand, Gujarat, India.

[4] Haugan T, Grohn YT, Kommisrud E, Ropstad E, Reksen O. Effects of sperm concentration at semen collection and storage period of frozen semen on dairy cow conception. Anim Reprod Sci. 2007; 97: 1–11.

[5] Shivaji S, Kholkute SD, Verma SK, Gaur A, Umapathy G, Singh A, Sontakke S, et al. Conservation of wild animals by assisted reproduction and molecular marker technology. Indian J Exp Biol. 2003; 41(7): 710–723.

[6] Thurston LM, Siggins K, Mileham AJ, Watson PF, Holt WV. Identification of amplified restriction fragment length polymorphism markers linked to genes controlling boar sperm viability following cryopreservation. Biol Reprod. 2002; 66: 545–554.

[7] Vishwanath R, Shannon P. Storage of bovine semen in liquid and frozen state, Anim Reprod Sci. 2000; 62(1–3): 23–53.

[8] Mathevon M, Buhr MM, Dekkers JCM. Environmental, management and genetic factors affecting semen production in Holstein bulls. J Dairy Sci. 1998; 81: 3321–3330.

[9] Petrunkina AM, Waberski D, Günzel-Apel AR, Töpfer-Petersen E. Determinants of sperm quality and fertility in domestic species. Reproduction. 2007; 134: 3–17.

[10] de Souza FF, Barreto CS, Lopes MD. Characteristics of seminal plasma proteins and their correlation with canine semen analysis. Theriogenology. 2007; 68: 100–106.

[11] Tso WW, Lee WM. Seminal plasma and progressive motility of boar spermatozoa. Int J Androl. 2008; 3: 243–250.

[12] Majumder GC, Das K, Saha S, Nath D, Maiti A, Das S, Dey N, et al. Purification and characterization of novel sperm motility-related proteins. In: Benitez M, Aguiree V, editors. Protein Purification. Nova Science Publishers: USA. 2012. p. 1–90.

[13] Haugan T, Grohn YT, Kommisrud E, Ropstad E, Reksen O. Effects of sperm concentration at semen collection and storage period of frozen semen on dairy cow conception. Anim Reprod Sci. 2007; 97: 1–11.

[14] Gibbons BH, Gibbons IR. Flagellar movement and adenosine triphosphatase activity in sea urchin sperm extracted with Triton X-100. From the Pacific Biomedical Research Center, University of Hawaii, Honolulu, Hawaii 96822.

[15] Orgebin-Crist MC, Fournier-Delpech S. Sperm-egg interaction. Evidence for maturational changes during epididymal transit. J Androl. 1982; 3: 429–433.

[16] Cuasnicú PS, González Echeverría F, Piazza A, Piñeiro L, Blaquier JA. Epididymal proteins mimic the androgenic effect on zona pellucida recognition by immature hamster spermatozoa. J Reprod Fertil. 1984b; 71(2): 427–431.

[17] González-Echeverría F, Cuasnicú PS, Piazza A, Pineiro L, Blaquier JA. Addition of an androgen-free epididymal protein extract increases the ability of immature hamster spermatozoa to fertilize in vivo and in vitro. J Reprod Fertil. 1984; 71: 433–437.

[18] Hoskins DD, Brandt H, Acott TS. Initiation of sperm motility in the mammalian epididymis. Fed Proc. 1978; 37: 2534–2542.

[19] Muller B, Kirchner C. Influence of seminal plasma proteins on motility of rabbit spermatozoa. J Reprod Fertil. 1978; 54(1): 167–172.

[20] Rochwerger L, Cuasnicu P, Conesa D. Participation of a rat sperm epididymal glycoprotein in the acquisition of sperm-egg fusion ability during maturation. In: Alexander NJ, Griffin D, Spieler JM, Waites GMH, editors. Gamete interaction. Prospects for immunocontraception. John Wiley Liss & Sons, Inc: New York. 1990. p. 630–631.

[21] Mandal M, Banerjee S, Majumder GC. Stimulation of forward motility of goat cauda epididymal spermatozoa by a serum glycoprotein factor. Biol Reprod 1989; 41:983-9.

[22] Majumder GC, Mandal M, Banerjee S. A novel sperm motility-promoting protein from blood serum and its applied potential. In: Ghosal SK, Ray D, editors. Recent Advances in Animal Science Research. Orion Press International: Chandernagore, West Bengal. 2001. p. 61–67.

[23] Hammerstedt RH, Parks JE. Changes in sperm surfaces associated with epididymal transit. J Reprod Fertil Suppl. 1987; 34: 133–149.

[24] Majumder GC, Dey CS, Haldar S, Barua M. Biochemical parameters of initiation and regulation of sperm motility. Arch Androl. 1990; 24(3): 287–303.

[25] Lee SL, Kuo YM, Kao CC, Hong CY, Wei YH. Purification of a sperm motility stimulator from porcine follicular fluid. Comp Biochem Physiol B. 1992; 101(4): 591–594.

[26] Walsh DA, Cooper RH. The physiological regulation and function of cAMP dependent protein kinase. In: Litwack G, editor. Biochemical Actions of Hormones. Academic Press: New York. 1979; vol. 6: p. 2–75.

[27] Halder S, Majumder GC. Phosphorylation of external cell surface proteins by an endogenous ecto-protein kinase of goat epididymal intact spermatozoa. Biochem Biophys Acta. 1986; 887: 291–303.

[28] Halder S, Dey CS, Majumder GC. An ecto-cyclic AMP-independent protein kinase in goat spermatozoa and its change of activity during forward motility. Biochem Int. 1986; 13(5): 809–817.

[29] Mitra S, Nath D, Majumder GC. Purification and characterization of a protein kinase from goat sperm plasma membrane. Biochem Cell Biol. 1994; 72(5–6): 218–226.

[30] Goluboff ET, Mertz JR, Tres LL, Kierszenbaum AL. Galactosyl receptor in human testis and sperm is antigenically related to the minor C-type (Ca2+-dependent) lectin variant of human and rat liver. Mol Reprod Dev. 1995; 40: 460–466.

[31] Yuuji S, Harumi S, Takashi S. Purification of a sperm lectin extracted from spermatozoa of the sea urchin *Hemicentrotus pulcherrimus*. Dev Growth Differ. 1992; 34: 285–291.

[32] Acott TS, Hoskins DD. Bovine sperm forward motility protein: partial purification and characterization. J Biol chem. 1978; 253: 6744–6750.

[33] Jaiswal BS, Das K, Saha S, Dungdung SR, Majumder GC. Purification and characterization of a motility initiating protein from caprine epididymal plasma. J Cell Physiol. 2010; 222: 254–263.

[34] Das S, Saha S, Majumder GC, Dungdung SR. Purification and characterization of a sperm motility inhibiting factor from caprine epididymal plasma. PLoS One. 2010; 5: 1–12.

[35] Mandal M, Saha S, Ghosh AK, Majumder GC. Identification and characterization of a sperm motility promoting glycoprotein from buffalo blood serum. J Cell Physiol. 2006; 209: 353–362.

[36] Saha S, Das S, Bhoumik A, Ghosh P, Majumder GC, Dungdung SR. Identification of a novel sperm motility-stimulating protein from caprine serum: its characterization and functional significance. Fertil Steril. 2013; 100: 269–279.

[37] Dey S, Roy D, Majumder GC, Bhattacharyya D. Receptor expression is essential for forward motility in the course of sperm cell maturation. Int J Biochem Cell Biol. 2014a; 92: 43–52.

[38] Dey S, Roy D, Majumder GC, Bhattacharyya D. Extracellular regulation of sperm transmembrane adenylyl cyclase by a forward motility stimulating protein. PLOS One 2014b; 9(10): e110669.

[39] Chao HT, Ng HT, Kao SH, Wei YH, Hong CY. Human follicular fluid stimulates the motility of washed human sperm. Andrologia. 1991; 26: 61–65.

[40] Jeon BG, Moon JS, Kim KC, Lee HJ, Choe SY, Rho GJ. Follicular fluid enhances sperm attraction and its motility in human. J Assist Reprod Genet. 2001; 18: 407–412.

[41] Ravnik SE, Zarutskie PW, Muller CH. Lipid transfer activity in human follicular fluid lipid transfer protein that stimulates human sperm capacitation. J Androl. 1990; 11: 216–226.

[42] Suarez SS, Wolf DP, Meizel S. Induction of the acrosome reaction in human spermatozoa by a fraction of human follicular fluid. Gamete Res. 1986; 14: 107–121.

[43] Lambert H, Steinleitner A, Eiserman J, Serpa N, Cantor B. Enhanced gamete interaction in the sperm penetration assay after co-incubation with pentoxifylline and human follicular fluid. Fertil Steril. 1992; 58: 1205–1208.

[44] Banerjee S, Dungdung SR, Das K, Majumder GC. Synchronous modulation of cell-surface lectin and its receptor in a homologous cell population: a novel mechanism of cellular regulation. Exp Cell Res. 2006; 312: 2299–2308.

[45] Radojcic L, Vukotic-Maletic V, Balint B. Current knowledge on cryopreservation of spermatozoa, ovum cells and zygotes. Med Pregl. 1998; 51: 29–36.

[46] Watson PF. Recent developments and concepts in the cryopreser-vation of spermatozoa: the assessment of their post-thawing function. Reprod Fertil Dev. 1995; 7: 871–891.

[47] Kundu CN, Chakrabarty J, Dutta P, Bhattacharyya D, Ghosh A, Majumder GC. Development of a simple sperm cryopreservation model using a chemically defined medium and goat cauda epididymal spermatozoa. Cryobiology. 2000a; 40: 117–125.

[48] Kundu CN, Chakrabarty J, Dutta P, Bhattacharyya D, Ghosh A, Majumder GC. A novel epididymal sperm model for analyzing the cryoprotecting potential of various reagents. In: Proceedings of the 18th International Cryogenic Engineering Conference. Mumbai, India. 2000b; p. 695–699.

[49] Kundu CN, Das K, Majumder GC. Effect of amino acids on goat cauda-epididymal sperm cryopreservation using a chemically defined model system. Cryobiology. 2001; 1: 21–27.

[50] Kundu CN, Chakrabarty J, Dutta P, Bhattacharyya D, Ghosh A, Majumder GC. Effect of dextrans on cryopreservation of goat cauda epididymal spermatozoa using a chemically defined medium. Reproduction. 2002; 123: 907–913.

[51] Rana APS, Misra S, Majcunder GC, Ghosh A. Phospholipid asymmetry of goat sperm plasma membrane during epididymal maturation. Biochem Biophys Acta. 1993; 1210: 1–7.

[52] Kmenta I, Strohmayer C, Muller-Schlosser F, Schafer-Somi S. Effects of a lecithin and catalase containing semen extender and a second dilution with different enhancing

buffers on the quality of cold-stored canine spermatozoa. Theriogenology. 2011; 75: 1095–1103.

[53] Chakrabarty J, Banerjee D, Pal D, De J, Ghosh A, Majumder GC. Shedding off specific lipid constituents from sperm cell membrane during cryopreservation. Cryobiology. 2007; 54: 27–35.

[54] Mandal R, Badyakar D, Chakrabarty J. Role of membrane lipid fatty acids in sperm cryopreservation. Adv Androl. 2014; Article ID 190542, 9 pages.

[55] Phillips PH, Lardy HA. A yolk-buffer pabulum for the preservation of bull semen. J Dairy Sci. 1940; 23: 399–404.

[56] Benchmark D, Amirat L, Pascal O, Anton M, Schmitt E, Desherces S, Delhomme G, et al. The advantages of LDL (low density lipoproteins) in the cryopreservation of canine semen. Theriogenology. 2008; 70: 1478–1488.

[57] Forouzanfar MM, Sharafi SM, Hosseini S, Ostadhosseini M, Hajian L, Hosseini P, Abedi N, et al. In vitro comparison of egg yolk-based and soybean lecithin-based extenders for cryopreservation of ram semen. Theriogenology. 2010; 73: 480–487.

[58] Pillet E, Labbe C, Batellier F, Duchamp G, Beaumal V, Anton M, Desherces S, et al. Liposomes as an alternative to egg yolk in stallion freezing extender. Theriogenology. 2012; 77(2): 268–279.

[59] Röpke T, Oldenhof H, Leiding C, Sieme H, Bollwein H, Wolkers WF. Liposomes for cryopreservation of bovine sperm. Theriogenology. 2011; 76(8): 1465–1472.

[60] Vidal AH, Batista AM, da Silva ECB, Gomes WA, Pelinca MA, Silva SV, Guerra MMP. Soybean lecithin-based extender as an alternative for goat sperm cryopreservation. Small Ruminant Res. 2013; 109: 47–51.

[61] Salmani H, Towhidi A, Zhandi M, Bahreini M, Sharafi M. In vitro assessment of soybean lecithin and egg yolk based diluents for cryopreservation of goat semen. Cryobiology. 2014; 68(2): 276–280.

Permissions

All chapters in this book were first published in IAR, by InTech Open; hereby published with permission under the Creative Commons Attribution License or equivalent. Every chapter published in this book has been scrutinized by our experts. Their significance has been extensively debated. The topics covered herein carry significant findings which will fuel the growth of the discipline. They may even be implemented as practical applications or may be referred to as a beginning point for another development.

The contributors of this book come from diverse backgrounds, making this book a truly international effort. This book will bring forth new frontiers with its revolutionizing research information and detailed analysis of the nascent developments around the world.

We would like to thank all the contributing authors for lending their expertise to make the book truly unique. They have played a crucial role in the development of this book. Without their invaluable contributions this book wouldn't have been possible. They have made vital efforts to compile up to date information on the varied aspects of this subject to make this book a valuable addition to the collection of many professionals and students.

This book was conceptualized with the vision of imparting up-to-date information and advanced data in this field. To ensure the same, a matchless editorial board was set up. Every individual on the board went through rigorous rounds of assessment to prove their worth. After which they invested a large part of their time researching and compiling the most relevant data for our readers.

The editorial board has been involved in producing this book since its inception. They have spent rigorous hours researching and exploring the diverse topics which have resulted in the successful publishing of this book. They have passed on their knowledge of decades through this book. To expedite this challenging task, the publisher supported the team at every step. A small team of assistant editors was also appointed to further simplify the editing procedure and attain best results for the readers.

Apart from the editorial board, the designing team has also invested a significant amount of their time in understanding the subject and creating the most relevant covers. They scrutinized every image to scout for the most suitable representation of the subject and create an appropriate cover for the book.

The publishing team has been an ardent support to the editorial, designing and production team. Their endless efforts to recruit the best for this project, has resulted in the accomplishment of this book. They are a veteran in the field of academics and their pool of knowledge is as vast as their experience in printing. Their expertise and guidance has proved useful at every step. Their uncompromising quality standards have made this book an exceptional effort. Their encouragement from time to time has been an inspiration for everyone.

The publisher and the editorial board hope that this book will prove to be a valuable piece of knowledge for researchers, students, practitioners and scholars across the globe.

List of Contributors

Celeste Eno and Francisco Pelegri
Laboratory of Genetics, University of Wisconsin - Madison, Madison, WI, USA

Sang-Kon Lee
Department of Urology, Sacred Heart Hospital, School of Medicine, Hallym University, Gangwon, Republic of Korea

Sungman Park
Institute of Medical Science, School of Medicine, Hallym University, Chuncheon, Gangwon, Republic of Korea

Yoon-Won Kim
Institute of Medical Science, School of Medicine, Hallym University, Chuncheon, Gangwon, Republic of Korea
Department of Microbiology, School of Medicine, Hallym University, Chuncheon, Gangwon, Republic of Korea

Izabela Szczerbal and Marek Switonski
Department of Genetics and Animal Breeding, Poznan University of Life Sciences, Poznan, Poland

Alexandre R. Silva, Alexsandra F. Pereira, Gabriela L. Lima, Gislayne C. X. Peixoto and Ana Liza P. Souza
Laboratory of Animal Germplasm Conservation - UFERSA, Mossoró, RN, Brazil

Ricardo Romão and Elisa Bettencourt
Escola de Ciências e Tecnologia, Instituto de Ciências Agrárias e Ambientais Mediterrânicas (ICAAM), Universidade de Évora, Polo da Mitra, Évora, Portugal

Carla C. Marques
Unidade de Biotecnologia e Recursos Genéticos, INIAV-Santarém, Quinta da Fonte Boa, Portugal

Rosa M.L.N. Pereira
Unidade de Biotecnologia e Recursos Genéticos, INIAV-Santarém, Quinta da Fonte Boa, Portugal
CIISA, Faculdade de Medicina Veterinária da Universidade de Lisboa, Avenida da Universidade Técnica, Lisboa, Portugal

Vicente J. F. Freitas, Luciana M. Melo, Dárcio I.A. Teixeira and Maajid H. Bhat
Laboratory of Physiology and Control of Reproduction, Faculty of Veterinary, State University of Ceará, Fortaleza, Brazil

Irina A. Serova and Oleg L. Serov
Institute of Cytology and Genetics, Russian Academy of Sciences, Novosibirsk, Russia

Lyudmila E. Andreeva
Institute of Molecular Genetics, Russian Academy of Sciences, Moscow, Russia

María-José Argente
Departamento de Tecnología Agroalimentaria, Universidad Miguel Hernández de Elche, Orihuela, Alicante, Spain

Maria dos Anjos Pires and Fernanda Seixas
CECAV, Animal and Veterinary Research Centre, Universidade de Trás-os-Montes e Alto Douro, Vila Real, Portugal
Veterinary Sciences Department, Universidade de Trás-os-Montes e Alto Douro, Portugal

Hugo Vilhena and Sónia Miranda
CECAV, Animal and Veterinary Research Centre, Universidade de Trás-os-Montes e Alto Douro, Vila Real, Portugal
Department of Veterinary Medicine, Escola Universitária Vasco da Gama, Coimbra, Portugal
Baixo Vouga Veterinary Hospital, Águeda, Portugal

Miguel Tavares Pereira
Vet4, Estarreja, Portugal

Ana Laura Saraiva
Veterinary Sciences Department, Universidade de Trás-os-Montes e Alto Douro, Portugal
Department of Veterinary Medicine, Escola Universitária Vasco da Gama, Coimbra, Portugal

Duda Malgorzata and Knet-Seweryn Malgorzata
Department of Endocrinology, Institute of Zoology, Jagiellonian University, Kraków, Poland

Grzesiak Malgorzata
Department of Animal Physiology and Endocrinology, University of Agriculture, Kraków, Poland

Zbigniew Tabarowski
Department of Experimental Hematology, Institute of Zoology, Jagiellonian University, Kraków, Poland

Sandhya R. Dungdung, Arpita Bhoumik, Sudipta Saha, Prasanta Ghosh, Kaushik Das, Sandipan Mukherjee, Arunima Maiti and Saswati Banerjee
Cell biology and Physiology Division, CSIR-Indian Institute of Chemical Biology, Kolkata, India

Debjani Nath
Department of Zoology, University of Kalyani, Kalyani, West Bengal, India

Jitamanyu Chakrabarty
National Institute of Technology, Durgapur, West Bengal, India

Chanakyanath Kundu
School of Biotechnology, KIIT University, Odissa, India

Bijay Shankar Jaiswal
Department of Molecular Biology, Genentech Inc., CA, USA

Mahitosh Mandal
Indian Institute of Technology, Kharagpur, India

Madhumita Roychowdhury and Debleena Ray
A. H. IVF & Infertility Research Centre, Kolkata, India

Bhattacharyya and Gopal C. Majumder
Centre for Rural and Cryogenic Technologies, Jadavpur University, Kolkata, India

Index